Geochemistry of Travertines and Calcareous Tufas

Geochemistry of Travertines and Calcareous Tufas

Editors

Francesca Giustini
Mauro Brilli

MDPI

Basel • Beijing • Wuhan • Barcelona • Belgrade • Novi Sad • Cluj • Manchester

Editors

Francesca Giustini
Istituto di Geologia Ambientale
e Geoingegneria (IGAG)
Consiglio Nazionale
delle Ricerche (CNR)
Rome
Italy

Mauro Brilli
Istituto di Geologia Ambientale
e Geoingegneria (IGAG)
Consiglio Nazionale
delle Ricerche (CNR)
Rome
Italy

Editorial Office
MDPI AG
Grosspeteranlage 5
4052 Basel, Switzerland

This is a reprint of articles from the Special Issue published online in the open access journal *Minerals* (ISSN 2075-163X) (available at: www.mdpi.com/journal/minerals/special_issues/GTCT).

For citation purposes, cite each article independently as indicated on the article page online and as indicated below:

Lastname, A.A.; Lastname, B.B. Article Title. *Journal Name* **Year**, *Volume Number*, Page Range.

ISBN 978-3-7258-2100-6 (Hbk)
ISBN 978-3-7258-2099-3 (PDF)
doi.org/10.3390/books978-3-7258-2099-3

Cover image courtesy of Francesca Giustini
Travertine terraces, Pamukkale, Turkey.

Contents

About the Editors

Francesca Giustini

Francesca Giustini, PhD, is a geochemist and an isotope geochemist; she has been working as researcher in Rome at the Istituto di Geologia Ambientale e Geoingegneria of the CNR (National Research Council) since 2018. Her research activity focuses on environmental radioactivity and stable isotopes, as well as their applications to cultural heritage and global changes.

Mauro Brilli

Mauro Brilli, PhD, is an isotope geochemist; he works in Rome at the Istituto di Geologia Ambientale e Geoingegneria of the CNR (National Research Council). In charge of the Laboratory of Mass Spectrometry, his research activity focuses on the application of light stable isotopes to earth sciences and cultural heritage.

minerals

MDPI

Editorial

Editorial for Special Issue "Geochemistry of Travertines and Calcareous Tufas"

Francesca Giustini * and Mauro Brilli *

Istituto di Geologia Ambientale e Geoingegneria (IGAG), Consiglio Nazionale delle Ricerche (CNR), Area della Ricerca di Roma 1, Via Salaria km 29,300, Monterotondo Street, 00015 Rome, Italy
* Correspondence: francesca.giustini@cnr.it (F.G.); mauro.brilli@cnr.it (M.B.)

Travertine and calcareous tufa are the lithological terms generally used to describe continental carbonates. They form in subaerial environments, located around groundwater seepages, springs, and along streams and rivers, via the precipitation of calcite/aragonite from waters that range in temperature from ambient to boiling [1]. These deposits exist across all continents, some of them are confined to small, local discharge aprons below spring emergences, whereas others cover extensive areas and form impressive structures (e.g., Tivoli in Italy, Pamukkale in Turkey, Mammoth Hot Springs in USA, Plitvice in Croatia). Travertine and calcareous tufa are complex systems, which are generally investigated using a multidisciplinary approach in order to elucidate the formation processes, environments, depositional models, and diagenesis. The multidisciplinary approach involves geochemistry (elemental and isotope compositions, dating), meso- and macro-scale morphological observations, mineralogy, petrography and sedimentology, and the analysis of biotic content. The active carbonate-depositing systems are natural laboratories, enabling the study of the geochemical characteristics of forming-travertine/tufa and their parental waters. Fossil deposits are important terrestrial archives that are capable of recording past climates, environmental changes, neotectonic activity, and hydrological–hydrothermal circulations. They indicate the possible locations of geothermal resources or contribute to the reconstruction of the carbon cycle.

This Special Issue, entitled "Geochemistry of travertines and calcareous tufas", includes eleven contributions about travertine and calcareous tufa. The topics discussed included fossil and/or active deposits of natural or anthropogenic origin from three continents (Europe, Asia and America). These carbonate deposits were investigated using several geochemical methods, such as the determination of major and trace element composition, mineralogy, stable (oxygen, carbon, strontium) and clumped isotopes, and radioisotopes (radiocarbon, uranium series). In many cases, the geochemistry of the parental waters of the precipitating carbonates was also considered. Combining different geochemical methods with results from field survey and facies analysis, several contributions aim to determine the genesis of the deposits, the characteristics of the precipitating waters, and the relationship between carbonates and the geological/tectonic settings (Contributions 1, 5, 6, 7, 8, 9, 11). Some studies focus on the use of travertine as a paleoclimate record (Contributions 2, 6, 10), or as an indicator of geothermometric sequence, via the analysis of the Mg/Ca ratio (Contribution 4). With a different purpose, a study on clumped isotopes focuses on a methodological aspect regarding the use of this new technique on carbonate systems of anthropogenic origin (Contribution 3). The main features of these articles are reported in the summary below.

The paper of Porras-Toribio et al. (Contribution 1) investigates the laminated aragonite travertines currently precipitating from a high-temperature hot spring (70 °C) with active degassing on the coast of Tamiahua Lagoon (Mexico). The authors assess mineralogy and petrography, trace elements, stable isotopes (C, O, S) on carbonates, elemental sulfur, and sulphates, as well as the chemical (major and trace elements) and isotopic characterization

Citation: Giustini, F.; Brilli, M. Editorial for Special Issue "Geochemistry of Travertines and Calcareous Tufas". *Minerals* **2024**, *14*, 442. https://doi.org/10.3390/min14050442

Received: 18 April 2024
Accepted: 22 April 2024
Published: 24 April 2024

of freshwater and hydrothermal fluids. The authors conclude that aragonitic travertines are generated by CO_2 and H_2S emanations of deep origin and by oxidation–reduction reactions that are linked to surficial bacterial activity.

The paper of Li et al. (Contribution 2) presents detailed petrographic and mineralogical observations, stable C and O isotopes, and U/Th dating of a travertine dam from Zabuye Salt Lake, which is a semi-dry salt lake located in the southwest of the Tibetan Plateau (China). This study tries to explain the precipitation sequence of Zabuye Salt Lake over the last 800 years (1191–1982 AD), suggesting that travertine might record the evolution of meteoric paleo-precipitation (paleo-monsoon) on a decadal–centennial scale at least.

The paper of Holdsworth et al. (Contribution 3) investigates the clumped isotope values recorded in anthropogenic tufa from high-pH (>10), hydroxylation-related carbonates sampled in northeast England. The authors present the test results of these human-derived carbonates as an archive for evaluating the impact of kinetic fractionations on $\Delta47$.

The paper of Auqué et al. (Contribution 4) focused on the utility of the Mg/Ca elemental ratio of calcite as a temperature indicator in continental carbonates. Recent tufa stromatolite records from four rivers on the Iberian Peninsula are selected for trace element analysis based on six-monthly sampling. The authors report that the seasonal variations in mineralogy, stable isotopes (C and O) of carbonate and water, and hydrochemistry offered the opportunity to decipher the behavior of Mg/Ca ratio in the continental carbonates. The authors concluded that the wide variety of parameters that are involved in the $(Mg/Ca)_{calcite}$ limit it as a geochemical thermometer in continental sedimentary environments.

The paper of You et al. (Contribution 5) investigates the lithofacies and stable carbon and oxygen isotopic compositions of banded travertines sampled from the fissure ridge of Xiagei (southwestern China) in order to examine their formation mechanisms. The parental CO_2 and the paleotemperature are determined, and two lithofacies are recognized. The latter show distinct calculated paleotemperatures, suggesting that temperature is a non-negligible control factor in the deposition of the two types of travertines, with important implication for paleo-environment studies.

The paper of Lv et al. (Contribution 6) investigates the Jiuzhaigou calcareous tufa deposit, located at a high altitude (about 2200 m a.s.l.) in the eastern margin of the Tibetan Plateau (China); the study is focused on a profile recently that was exposed by an Mw 7.0 earthquake that occurred in the region on 8 August 2018. Mineralogy, petrography, major/trace element content, stable isotope composition, and radiocarbon dating allow the authors to investigate the origin of this deposit, its paleoclimate significance, and its relationship with the seismic activity of the region.

The paper of Lu et al. (Contribution 7) studies a travertine mound from Sobcha, a high-altitude deposit (about 4700 m a.s.l.) in central Tibet (China). The Sobcha travertines are fossil deposits, but modern hot springs, with alkaline pH and deficient in Ca^{2+}, are still present near the studied deposit. Using stable isotopes (C, O, Sr), trace elements including the rare-earth elements (REE), U/Th dating, mineralogy, and petrography, the authors reconstruct the genesis, the paleofluid source(s), and fluid evolution of this deposit.

The paper of Deev et al. (Contribution 8) studies the travertine cements that occur within the zones of active faults bordering two intramontane basin in the Gorny Altay area (Russia). The authors present results for petrography and mineralogy, major and trace elements including REE, stable C and O isotopes. They also perform U/Th dating on carbonates and analyze major and trace elements and stable O and H of present-day waters. Combining the results of their analyses, the authors discuss the origins of paleofluid, the factors influencing the precipitation of calcite and aragonite, the relationship of travertine formation with paleoearthquake and paleoclimate, the possibility of use travertine cement as important tool for paleoseismological research in seismically active regions.

The paper of Giustini and Brilli (Contribution 9) reviews the literature data available on the uranium content and $^{234}U/^{238}U$ initial activity ratio of 15 travertine and calcareous tufa deposits found in Italy. Using a graphical method, data are interpreted considering the U geochemistry in natural environments as well as the geological, hydrogeological, and hydrogeochemical features of each site. The paper discusses the factors that appears to affect the U content and $^{234}U/^{238}U$ initial activity ratio in travertine and tufa, such as the availability of U in the aquifer rocks, the redox state of the waters, and the alpha-active radionuclide recoil phenomenon. The results suggest an interpretative model that might contribute to the paleo-environmental reconstruction of fossil travertine and calcareous tufa-depositing systems.

The paper of Brilli and Giustini (Contribution 10) investigates the stratigraphy of a buried travertine deposit, identified in a well at Prima Porta (north of Rome, Italy), using stable C and O isotopes, trace elements, and U/Th dating. The resulting isotope and chemical stratigraphy describe the temporal evolution of events that are correlated with global climatic variability. The authors reconstruct how paleoclimatic changes influenced the hydrological regime and indirectly altered tectonic activity by modulating the emission of deep CO_2, the chemistry of the groundwater, and ultimately the precipitation of travertine.

Using stable and clumped isotopes, U/Th dating of carbonates, and the isotopic composition of water sampled at active travertine spring mounds, the paper of Vieira et al. (Contribution 11) proposes a geochemical study of active and fossil travertine mounds found in several localities of Slovakia. The authors outline the age, origin, precipitation conditions, and temperature of these travertines, reconstructing the paleoclimate and paleo-environment during their deposition.

Overall, the contributions published in the Special Issue of *Minerals*, entitled "Geochemistry of travertines and calcareous tufas", have confirm the ability of geochemical tracers to integrate and deepen the knowledge obtained from the geomorphological and sedimentological studies of travertine and calcareous tufa deposits. We are particularly pleased to present the contributions relating to the new clumped isotope technique and to the use of trace elements which, due to their behavioral complexity, have been outpaced over the years by the more easily interpretable conventional isotopes, with the idea being that "an intelligent application of complementary techniques is a powerful means of constraining the validity of competing theoretical options" [2].

As the Guest Editors of this Special Issue, we hope you enjoy reading these varied and excellent contributions. Our sincere thanks go to the *Minerals* staff for their help and guidance, and to all reviewers for their hard work.

Conflicts of Interest: The authors declare no conflicts of interest.

List of Contributions

1. Porras-Toribio, I.; Pi-Puig, T.; Villanueva-Estrada, R.E.; Rubio-Ramos, M.A.; Solé, J. Mineralogy, Geochemistry, and Stable Isotopes (C, O, S) of Hot Spring Waters and Associated Travertines near Tamiahua Lagoon, Veracruz, Gulf of Mexico (Mexico). *Minerals* **2022**, *12*, 822; https://doi.org/10.3390/min12070822.
2. Li, M.; Zheng, M.; Ye, C.; Wang, C.; Zhang, X.; Wang, X.; Zhao, Y.; Zhang, Y. Response of Travertine Dam to Precipitation over the Past 800 Years in Zabuye Salt Lake, Southwestern Tibetan Plateau. *Minerals* **2022**, *12*, 916; https://doi.org/10.3390/min12070916.
3. Holdsworth, C.; MacDonald, J.; John, C. Non-Linear Clumped Isotopes from DIC Endmember Mixing and Kinetic Isotope Fractionation in High pH Anthropogenic Tufa. *Minerals* **2022**, *12*, 1611; https://doi.org/10.3390/min12121611.
4. Auqué, L.F.; Osácar, M.C.; Arenas, C.; Cukrov, N.; Lojen, S.; Sancho, C. Controls on Mg/Ca Ratios in Recent Stromatolites: Insights from Fluvial Systems in the Iberian Range (Spain). *Minerals* **2023**, *13*, 57; https://doi.org/10.3390/min13010057.
5. You, Y.; Wen, H.; Luo, L.; Lu, Z.; Li, L. Stable Carbon and Oxygen Isotopic Features of Banded Travertines from the Xiagei Fissure Ridge System (Shangri-La, China). *Minerals* **2023**, *13*, 76; https://doi.org/10.3390/min13010076.
6. Lv, C.; Zhao, X.; Jiang, Y.; Zhu, H.; Zhang, H.; Wang, F.; Li, Q.; Hou, K. Insights into Alpine-Karst-Type Tufa Deposits in Geological Environmental Records: A Case Study of the Calcareous Tufa Profile of the Jiuzhaigou Natural Reserve on the Eastern Margin of the Tibetan Plateau. *Minerals* **2023**, *13*, 120; https://doi.org/10.3390/min13010120.
7. Lu, Z.; Wen, H.; Luo, L.; Li, L.; Nie, Y. Geochemical Characterization of Laminated Crystalline Crust Travertines Formed by Ca2+-Deficient Hot Springs at Sobcha (China). *Minerals* **2023**, *13*, 220; https://doi.org/10.3390/min13020220.
8. Deev, E.V.; Kokh, S.N.; Dublyansky, Y.; Sokol, E.V.; Scholz, D.; Rusanov, G.G.; Reutsky, V.N. Travertines of the South-Eastern Gorny Altai (Russia): Implications for Paleoseismology and Paleoenvironmental Conditions. *Minerals* **2023**, *13*, 259; https://doi.org/10.3390/min13020259.

9. Giustini, F.; Brilli, M. Uranium Geochemistry of Italian Travertines and Calcareous Tufas: Exploring the Relationship between Carbonate Deposition, Groundwater Circulation and Subsurface Geology. *Minerals* **2023**, *13*, 782; https://doi.org/10.3390/min13060782.
10. Brilli, M.; Giustini, F. Geochemical Stratigraphy of the Prima Porta Travertine Deposit (Roma, Italy). *Minerals* **2023**, *13*, 789; https://doi.org/10.3390/min13060789.
11. Vieira, D.S.C.; Pivko, D.; Rinyu, L.; Palcsu, L.; Kiss, G.I.; Hu, H.-M.; Shen, C.-C.; Kele, S. Age and Depositional Temperature of Quaternary Travertine Spring Mounds from Slovakia. *Minerals* **2023**, *13*, 794; https://doi.org/10.3390/min13060794.

References

1. Pentecost, A. *Travertine*; Springer: Berlin, Germany, 2005; p. 446.
2. Veizer, J. Trace elements and isotopes in sedimentary carbonates. *Rev. Mineral. Geochem.* **1983**, *11*, 265–299.

minerals

MDPI

Article

Mineralogy, Geochemistry, and Stable Isotopes (C, O, S) of Hot Spring Waters and Associated Travertines near Tamiahua Lagoon, Veracruz, Gulf of Mexico (Mexico)

Israel Porras-Toribio [1], Teresa Pi-Puig [2,*], Ruth Esther Villanueva-Estrada [3], Marco Antonio Rubio-Ramos [4] and Jesús Solé [2]

1 Posgrado en Ciencias de la Tierra, Instituto de Geología, Universidad Nacional Autónoma de México, Ciudad Universitaria, Coyoacán, Ciudad de México 04510, Mexico; israel.porras.t@gmail.com
2 Laboratorio Nacional de Geoquímica y Mineralogía, Instituto de Geología, Universidad Nacional Autónoma de México, Ciudad Universitaria, Coyoacán, Ciudad de México 04510, Mexico; jsole@geologia.unam.mx
3 Instituto de Geofísica Unidad Michoacán UNAM, Km 8 Antigua Carretera a Pátzcuaro 8701, Col. Ex Hda San José de la Huerta, Morelia 58190, Michoacán, Mexico; ruth@igeofisica.unam.mx
4 Facultad de Ingeniería, Universidad Nacional Autónoma de México, Ciudad Universitaria, Coyoacán, Ciudad de México 04510, Mexico; tlatoani.mr@gmail.com
* Correspondence: tpuig@geologia.unam.mx

Abstract: Laminated travertine forms in and around an active hot spring on the west coast of Tamiahua Lagoon, north of the state of Veracruz, Mexico. Fluid chemistry is characterized by discharging slightly acidic pH hot water and gas at a constant flow rate. Moreover, finely interbedded mineralogical products from discharging waters at 70 °C host scattered hydrocarbons. The mineralogy and geochemistry of the travertine formations were characterized to determine their origin. Rock samples were collected and further studied by transmitted light petrography, X-ray diffraction, and EDS-coupled scanning electron microprobe. Identified mineralogy from outcrop samples includes aragonite, gypsum, anhydrite, and elemental sulfur as essential minerals, with calcite, celestine, barite, jarosite, opal, and fluorite as accessory minerals. Isotopic analyses for C and O were determined in carbonates, S isotope ratios on both elemental sulfur and sulfates, whereas measurements for trace elements and lanthanides were performed on carbonates. A suit of brines and condensates from gas samples was collected for H and O isotopic analyses and concentration determinations of the main ions and major and trace elements. Isotopic values of $\delta^{13}C$ and $\delta^{18}O$ of aragonite are in the range of +1.75‰ to +2.37‰ and −1.70‰ to −0.78‰, respectively. The $\delta^{34}S$ isotopic values of native sulfur and sulfates ranged from −4.0‰ to +1.2‰. The isotopic values of $\delta^{2}H$ (−5.50‰) and $\delta^{18}O$ (+7.77‰) of hot water samples collected in terraces where aragonite precipitates suggest a mixture between meteoric water and the Gulf of Mexico oil-field related waters. It was concluded that the aragonitic formations near Tamiahua Lagoon are hypogenic and were generated by CO_2 and H_2S emanations of deep origin and by oxidation-reduction reactions that can be linked to surficial bacterial activity.

Keywords: travertine; meteogene carbonates; thermogene carbonates; hydrocarbon; native sulfur; dendritic aragonite; oil-field waters; sulfates

Citation: Porras-Toribio, I.; Pi-Puig, T.; Villanueva-Estrada, R.E.; Rubio-Ramos, M.A.; Solé, J. Mineralogy, Geochemistry, and Stable Isotopes (C, O, S) of Hot Spring Waters and Associated Travertines near Tamiahua Lagoon, Veracruz, Gulf of Mexico (Mexico). *Minerals* **2022**, *12*, 822. https://doi.org/10.3390/min12070822

Academic Editors: Francesca Giustini and Mauro Brilli

Received: 19 May 2022
Accepted: 24 June 2022
Published: 28 June 2022

1. Introduction

Travertines are very active sedimentary continental systems around the world, where above ambient temperature carbonates (calcite and aragonite) chemically precipitate by rapid CO_2 degassing from highly supersaturated calcium carbonate brines. Travertine formation results from a complex interaction between geochemical, physical, and microbial processes [1–5]. Carbonates precipitated from hot spring water, seeps, and fumaroles are classified as thermogenic travertines [6–8], while those derived from ambient-temperature water are known as tufas. According to Pentecost (2005) [6] and Pentecost and Viles

(1994) [9], those carbonates formed from groundwater loaded with some meteoric component and where atmospheric carbon dioxide is fixed are called meteogene. The carbonic acid formed in this process dissolves the regional deep limestone layers and enriches the groundwater with CO_2 [6]. On the other hand, thermogene travertines are formed by CO_2 [8,10] released by thermal processes within the earth's crust. When CO_2 is dissolved in groundwater under higher-pressure conditions, it dissolves rocks rich in calcium and rises to the surface as hot springs, often in zones of recent volcanic or hydrothermal activity [6,11,12]. It is worth mentioning that different authors [13] choose to use the acronym CATT (Calcite and Aragonite Travertine and Tufa) to designate these deposits jointly.

The detailed study of travertine formations requires the combined use of different disciplines and analytical methods such as mineralogy, sedimentology, isotope geochemistry, carbonate chemistry kinetics, paleoclimatology, and geomicrobiology ([8,14] and references therein). The most frequent low-temperature travertine morphologies are mounds, pendants, and terraces associated with dams. The laminations are the most common textural lithofacies observed in many of these formations. Their cyclicity, origin, and relationship with biogenic activity are still discussed [8,15–26].

Travertines have historically been recognized worldwide as construction material [27,28]; they are also located in sightseeing areas, and their scientific importance is even more relevant. Travertine formation results from complex interactions between the upper crust lithologies, aqueous chemical processes in the sub-surface environment, and microorganism metabolic processes. According to some authors, current continental deposits represent a recent analog of the non-marine stromatolites from the past [29]. In addition, travertine characterization is important to understand the effects of past climate change based on its paleoformation, the survival mechanisms of extremophile microorganisms at high temperatures, and the natural mechanism of CO_2 sequestration [6].

The Gulf of Mexico is a well-known zone characterized for hosting a series of hydrocarbon oil fields at depth. On this area, on the west coast of Tamiahua Lagoon—north of the state of Veracruz and just on the shoreline—laminated travertines currently precipitate near a high-temperature hot spring with active degassing. Although it was previously reported solely as a hydrothermal feature, the present work is the first study to characterize the hydrothermal precipitates in the margin of Tamiahua Lagoon. The present work aims to: (a) determine the mineralogy, fluid chemistry, and classification of the travertine formation; (b) characterize the compositional and isotopic signatures of the different types of water (brines) related to the travertine precipitation; and (c) determine travertine genesis, establishing its relationship with the regional fault system and with the hydrocarbon emanations of the area.

2. Geological Context

The studied area is located in the northern area of the state of Veracruz, east of the Campo del Mamey oil field, on the eastern side of Tamiahua Lagoon, and between the coordinates 21°31′27″ N and 97°36′26″ W (Figure 1).

Non-consolidated detrital sedimentary units and carbonates from the Eocene to the Pleistocene ages outcrop in the study area. Recent lacustrine-type material is recorded, given the constant influence of the Tamiahua Lagoon and the rivers that flow into it, whereas basement units are not superficially exposed (Figure 2).

At the regional scale, Tamiahua Lagoon is located on the Gulf of México coast and belongs to the "Eastern Intraplate Volcanism" sector of the Trans-Mexican Volcanic Belt (TMVB), being the area therefore dominated by a convective-type heat flow [30]. The center and the southeast area of this province are characterized by recent volcanic activity of the Miocene to the Quaternary age. The most important volcanic fields in the zone are San Carlos, Sierra de Tamaulipas, Tlanchinol-Tantima-Alamo, Chiconquiaco-Palma Sola, Anegada High, and Los Tuxtlas Volcanic Field, all of which are associated with an extensional tectonic setting [31], which in turn determines the existence of the main scale tectonic lineaments with regional NW-SE orientation. The second type of lineament is associated with the Otontepec or Tántima mountains, which display 40° to 60° NE-SW direction. Finally, a third younger lineation from NE 80° SW to E-W [32] is recorded, tectonically unrelated to the previous ones.

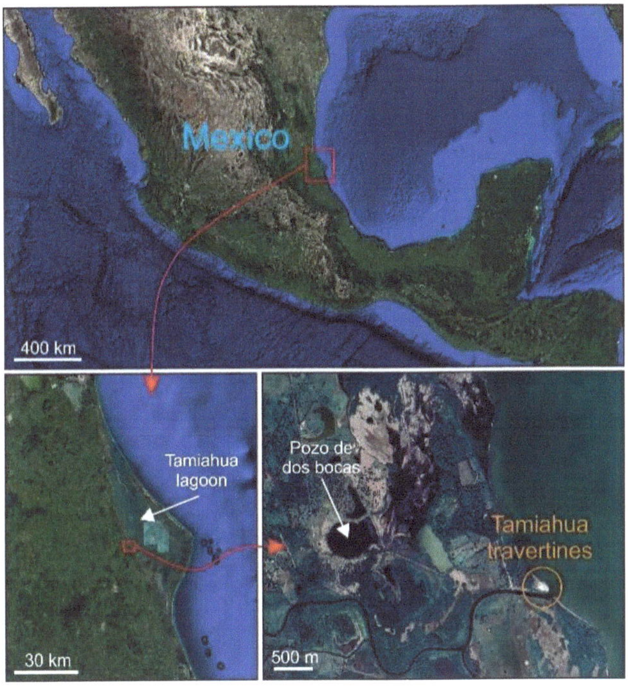

Figure 1. Location of the studied area: Tamiahua Lagoon, Veracruz, Mexico.

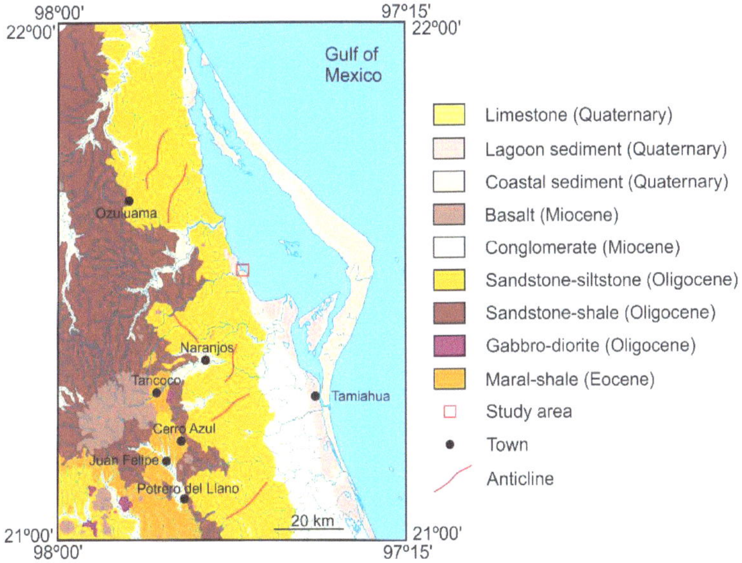

Figure 2. Geologic map of the area. Modified from Maldonado Lee et al., (2004) [32].

The travertine formations of Tamiahua Lagoon could indirectly be associated with an anthropogenic event that occurred in July 1908, during the drilling of an oil well called San Diego de la Mar Number 3, which was conducted by the "Pennsylvania Oil Company of Mexico". When the drilling reached a depth of approximately 556 m, it suddenly went out of control, causing the release of the oil and setting fire to the areas where the hydrocarbon spilled out. Furthermore, due to the reservoir's internal pressure, a second outlet was generated, causing the collapse of the surface and the formation of a crater denominated "Pozo de dos Bocas", situated at 2.65 km of the studied area (Figure 1).

The spilled oil flowed to the east into Tamiahua Lagoon, covering an area of 30 km^2. After the explosion, it left an artificial lagoon with a diameter of ~500 m, where the fire lasted for approximately three years until the oil field was exhausted and the fire extinguished. Currently, this artificial lake is a remnant of the well San Diego de la Mar Number 3, which intermittently emanates crude oil (chapopotera) and sulfurous saltwater [33].

Outcrop Description

The studied travertine formation is on the eastern shoreline of Tamiahua Lagoon, and—according to Pentecost's [6] classification—it can be described as a small composite autochthonous flat mound with subcircular morphology and an irregular sub-vertical cone (Figure 3).

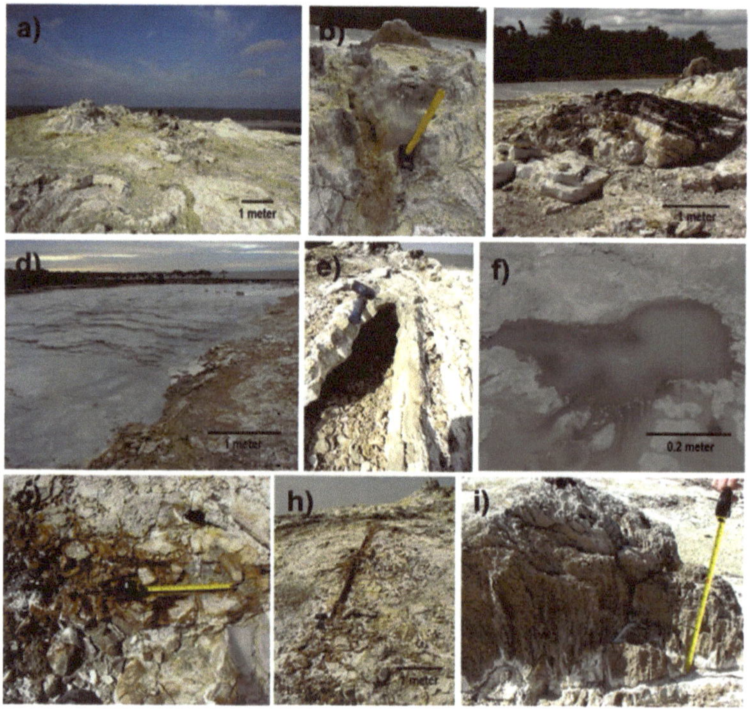

Figure 3. General and detailed views of the travertine formation at Tamiahua. (**a**) General view of the travertine formation; (**b**) main fumarole where the steam outlet can be seen (E-W direction); (**c**) sub-horizontal carbonate and black hydrocarbon layers near the vent; (**d**) general view of aragonite terraces; (**e**) cavity in the superficial part of the principal mound; (**f**) general view of the spring water; (**g**) iron mineral deposit associated with fractures near the vent; (**h**) fault with NW-SE orientation; and; (**i**) vertical layers of aragonite (banded palisade carbonate) travertine near the fumarole cone (E-W direction).

In 2017, the original cone elevation was less than a meter high and was almost inactive (probably by sealing at depth), but later raised, and lateral emanations on lower cones have been developed. Morphologically, it is characterized by a slight elevation (approx. 2.5 m above sea level) towards the center, where small fumaroles that exhale vapors stand out (Figure 3b). The surficial outcrops of older precipitates are laterally exposed as thin terraces concentrical to the main fumarole and have a diameter of nearly 50 m. The present surface is covered by elemental sulfur, giving a characteristic yellow–green color (Figure 3a,b,e). Small fissure walls were observed surrounding the original cone, and the whole system is crossed by fractures enriched in iron oxides (Figure 3g,h). The carbonate layers are intercalated with very dark ones and related to hydrocarbon emanations (Figure 3c).

The current active spring is situated at sea level to the west of the fossil formation and covers an area of approximately 12,000 m^2. Additionally, the area is characterized by the presence of mud terraces (0.5 to 1.5 m in height), shallow pools (~10–30 cm), and low-temperature hot waters flow, from which the hydrothermal fluid emanates (Figure 3d,f). The measured surface temperature is ~70 °C. The deposition rate in the terraces is very high and the carbonate accretion remains ongoing.

3. Materials and Methods

Aiming to reveal the origin of the hot spring precipitates, investigations were carried out during two field trip periods corresponding to November 2017 and September 2018.

3.1. Rock System

Forty-seven rock samples from the travertine formation were collected from several sites (Figure 4), cut, and polished to characterize mesofabric textures, after which they were selected as dedicated samples for the transmitted light petrographic analysis (Figure 4). Polished slabs and thin sections were observed with a Zeiss Discovery V8 and a Zeiss Axio Imager A2m, respectively; both are equipped with an AxioCam camera (Carl-Zeiss Micrscopy, Oberkochen, Germany). Microscopic features were described using petrographic and scanning electronic microscope observations (SEM). The SEM operated at low vacuum (as environmental electronic microscope) and this allowed the samples to be visualized without coating them. Nitrogen gas and small pressure variations were used to increase conductivity. Samples were studied using both backscattered electrons (BSE) and semi-quantitative analysis by energy-dispersive X-rays (EDS). The equipment used was a Zeiss model EVO MA10 (Carl-Zeiss Micrscopy, Oberkochen, Germany) from the Laboratory of Electron Microscopy and Nacional de Geoquímica y Microanalysis of LANGEM (Laboratorio Mineralogía, UNAM, Mexico City, Mexico).

For X-Ray diffraction (XRD) analysis, 36 selected samples were finely grounded using an agate mortar and pestle, sieved (<45 μm), and mounted as random fractions using a double-sided aluminum holder to determine their mineralogy. The fine material with banded texture was sampled using a micro-drill with a DREMEL 4000 tungsten carbide bit. XRD measurements were made using a Malvern Panalytical Empyrean diffractometer (Malvern Panalytical, Malvern, UK) at the LANGEM, operating with an accelerating voltage of 45 kV and a filament current of 40 mA, using Co$K\alpha$ radiation and iron filter (samples rich in iron) and Cu$K\alpha$ radiation and nickel filter (all the samples). All samples were measured using high-resolution routines over a 2θ angular range of 4–80° at a step scan of 0.003° and an integration time of 40 s by step. Phase identification and Rietveld quantification were made with a PDF-2 database using Highscore v.4.5 software. Relevant data from the diffractometer used in the refinement are shown in Table 1. The refined specimen-dependent parameters were the zero error, displacement error, polynomial fitting for the background with six coefficients, cell parameters, crystallite size, atomic coordinates, and isotropic temperature factors. The GOF values (Goodness of Fitting) show the fit between Rietveld refinement and the experimental profile.

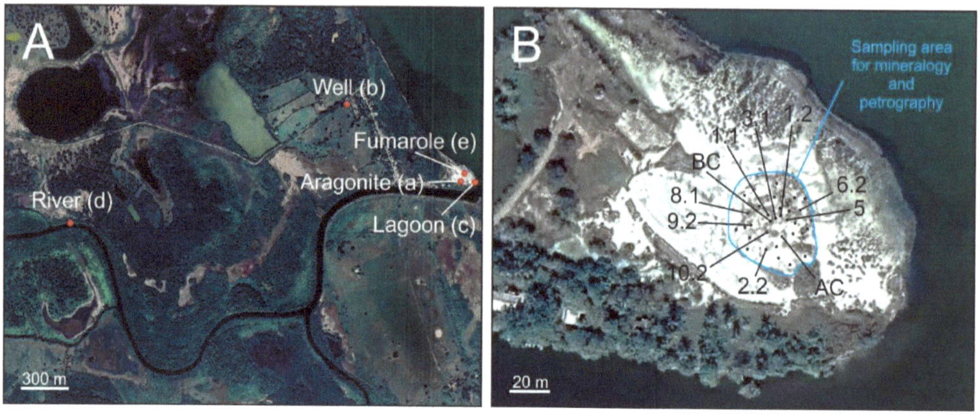

Figure 4. Location of sampled waters (**A**) and rocks (**B**) of the studied site at Tamiahua Lagoon. In the left image (**A**): (a) thermal water that emanates from the aragonite terraces; (b) freshwater from a well close to the area; (c) seawater from the coast of Tamiahua Lagoon; (d) water from a small river that flows next to the carbonate formation and flows into the lagoon; and (e) condensate from the fumarole obtained using a titanium bell. In the right image (**B**): to avoid overlapping samples in the figure, only the rock samples that have been cited in the body of the manuscript have been labeled.

Table 1. Relevant data from X-ray diffractometer EMPYREAN XRD used for the Rietveld refinement.

Geometry	Bragg–Brentano
Goniometer radius	240 mm
Radiation source	CoKα & CuKα
Generator	45 kV, 40 mA
Tube	Fine Focus
Divergence Slit	1/2° (fixed)
Soller Slits	0.04 rad (incident and diffracted beam)
Incident beam optics	Parallel mirror
Filter	Iron filter & Nickel filter
Detector	PIXcel3D
Step Size	0.002°
Integration Time	40 s

Trace element analyses were obtained from ACTLABS (Activation Laboratories, Ontario, CA, Canada) in four aragonite samples (Lithogeochemistry package 4B2-Res). The samples were melted with a flow of lithium metaborate and tetraborate using an induction furnace. Once this process was completed, digested samples were continuously mixed with a 5% nitric acid solution containing an internal standard until the sample was completely dissolved. Dissolved samples were analyzed with a Perkin Elmer Sciex ELAN 6000, 6100, or 9000 ICP–MS (PerkinElmer, Waltham, MA, USA). Two samples were analyzed with blank and three controls (two before the sample group and one after).

The carbon and oxygen isotopic measurements were determined in six calcium carbonate powders obtained with a DREMEL 4000 tungsten carbide bit. First, each sample was cleaned with acetone to remove surface organic residues. Then, the analyses were carried out (by triplicate) using the procedure described by Révész et al., (2001, 2002) [34,35] using a Gas Bench II coupled to a Thermo Finnigan MAT 253 (Thermo Electron, Bremen, Germany) stable isotope mass spectrometer at LANGEM Stable Isotope Laboratory. An Oztech CO_2 tank of certified isotopic composition ($\delta^{18}O_{VPDB} = -9.78$ and $\delta^{13}C_{VPDB} = -10.99$) was used as a working standard. Two types of calcium carbonate internal standards (Sigma and Merck) were included, each with a different isotopic fingerprint. $\delta^{18}O_{VPDB}$ and $\delta^{13}C_{VPDB}$ results of carbonates were normalized using NIST reference materials NBS-19, NBS-18, and LSVEC to the VPDB scale according to the corrections described by Coplen (1988) [36] for $\delta^{18}O_{VPDB}$ and scale change to $\delta^{18}O_{VSMOW}$ (Vienna Standard Mean Ocean Water) and Coplen et al., (2006) [37] for $\delta^{13}C_{VPDB}$. A standard deviation of 0.2% was obtained for oxygen and carbon with this technique.

The determination of $\delta^{13}C$ and $\delta^{34}S$ in solid hydrocarbons (chapopote) was carried out in two samples of approximately 2 g each. The analyses were performed at the GeoMark Research laboratory in Houston, Texas, USA, using a Vario ISOTOPE Cube elemental analyzer and a VISION isotope ratio mass spectrometer. Carbon isotopic analyses were run twice each and calibrated using international standards (USGS24 and NBS22) and three internal standards, with an analytical error of ±0.1‰ on the VPDB scale. Sulfur isotopes were measured in triplicate and were calibrated using international standards (IAEA-S-1, IAEA-S-2, and NBS127) with an analytical error of ±0.3‰ and reported on the VCDT scale.

Seven samples of native sulfur and three samples of sulfates (gypsum and anhydrite) were selected to obtain their $\delta^{34}S$ isotopic fingerprints; all these samples were previously sieved to <45 microns. The method used was described by Grinenko (1962) [38], using cuprous oxide (Cu_2O) as an oxidizing agent. For the samples to enter the sulfur line, each one was mixed with copper oxide and quartz in an agate mortar, and later the material was introduced into glass tubes with quartz wool at the ends to allow the gas to flow. In the sulfate samples (30 mg), 200 mg of cuprous oxide and 600 mg of quartz were used, whereas only 200 mg of cuprous oxide was added for the elemental sulfur (10 mg). The sulfur and sulfate in the samples were transformed to SO_2 through a furnace, employing oxide-reduction processes at temperatures above 1000 °C (1080 °C for native sulfur and 1150 °C for sulfates). The SO_3 was reduced to SO_2 in a second copper furnace at 750 °C to avoid fractionation, and the water was trapped in an acetone trap cooled with solid CO_2 while the vacuum pumps discarded the gaseous CO_2 through a pentane trap. Final SO_2 removal was performed by passing gas trapped on a cold finger (liquid nitrogen) into a bottle. Subsequently, SO_2 was isotopically analyzed in a SIRA II mass spectrometer at the University of Salamanca, Spain.

3.2. Hydrothermal System near Tamiahua Lagoon

Fluid sampling consisted of acquiring five samples from hydrothermal brines (Figure 4A): (a) thermal water that emanates at constant flow from the aragonite terraces; (b) freshwater from a well close to the studied area; (c) seawater from the coast of Tamiahua Lagoon; (d) water from a small river that flows next to the carbonate formation and into the lagoon; and (e) condensate from the fumarole obtained using a titanium bell. According to the pursued chemical analysis, samples were placed on polyethylene containers that were previously washed. Four bottles of 120 mL (analysis of stable isotopes, trace elements, anions, and cations) and one of 250 mL (analysis of bicarbonates) were filled with brine for each sample. It was difficult to accumulate enough volume of condensate from the fumarole, hence only stable isotopes (H and O) and major ions could be measured in this sample. Type 1 water was used as a blank.

Additionally, a HANNA Instruments Waterproof Tester was used to obtain in situ physicochemical measurements (pH, temperature, conductivity, salinity, and total dissolved solids) of the different waters.

Samples for trace elements and isotopic analysis were filtered using a 0.45 µm cellulose acetate membrane. Major ions, trace, and isotopic analysis samples were filtrated using cellulose acetate membrane. After filtration, the water samples for cation and trace elements determinations were acidified (pH = 2) with ultrapure HNO_3 (70% w/w%). The bicarbonate/carbonate concentration was determined in the field by volumetric acid-base titration with pre-normalized HCl and visual indicators to identify the endpoint.

Using a Thermo brand Dionex ICS-5000 ion chromatography instrument, anions and cations were determined. The eluent used for the anions was sodium hydroxide (NaOH) and methanesulfonic acid for the cations. The calibrating solutions used were High Purity

Standard with a stock concentration of 1000 mg/L. As a quality criterion of the standards and sample repetitions, the variation coefficient was less than or equal to 2%.

For the $\delta^{18}O$ and δ^2H analyses, the containers were filled, avoiding the formation of bubbles. Furthermore, each flask was sealed and stored in a cooler to refrigerate the samples at 5 °C, and later taken to the laboratory for analysis. Stable isotope analyses of $\delta^{18}O$ and δ^2H in water samples were carried out using an intracavity laser absorption spectrometer (CRDS) Picarro L2130i with a high precision vaporizer from the Geothermal Fluid Geochemistry Laboratory of the Instituto de Geofísica, UNAM, using international standards (VSMOW2 and GRESP). The analyses performed on the water samples for isotopes were done in triplicate, with reproducibility of ±0.1‰ and ±1.0‰ for $\delta^{18}O$ and δ^2H, respectively.

4. Results

4.1. Rock System

4.1.1. Mesofabric

The Tamiahua Lagoon travertines are constituted mainly by abiogenic cyclic crystalline multi-layered crusts. The principal mound is formed by well-bedded and finely laminated compact carbonates, intercalated with very thin layers of native sulfur (yellow–orange color) and sulfates (mainly white). In addition, layers of variable thickness and very dark color are scarce and related to the emanation and deposition of asphaltene-like hydrocarbons (cookie oils). Microbialitic bio-formations were not observed on the fumarole site, although active biogenic processes cannot be ruled out due to small concentrations of algal mats being found closer to the main discharging high-temperature hot spring. Lime-mudstone layers dominate the facies and, bordering the fumaroles, cross-cutting bands (banded palisade carbonate, [39–41]) were observed (Figure 3i).

4.1.2. Microfabric

The lamination at different scales (microns to centimeters) is the most noticeable textural feature of the deposits and it can be associated with a periodic change in precipitation rates (Figure 5a–f) or to changes in the dynamics of the hydrological cycle, associated with processes of natural or anthropogenic origin. The layers are plane, undulated, or wavy and, in general, show alternate (two fabrics and colors) or cyclic (more than two fabrics) repetitions [28]. The process of formation of wavy and porous lamination can be observed in the terraces zone, carbonate layers being white or with different hues of blue. The sequences with layers of different thicknesses (heteropachous) are predominant but not exclusive on the site (Figure 5a,b).

The size and distribution of porosity are very heterogeneous. The cavities have greater length than width, are parallel to the bedding, and seem related mainly to the inclusion and emission of gas, causing fenestral type porosity (Figure 5c). Characteristic soft-sediment deformation textures associated with the recurrence of gas bubbles (e.g., [12]) were also frequently observed (Figure 5b,e).

In the petrographic observation, it was possible to verify that the aragonite is the predominant mineral (Figure 6g). Crystalline crusts can be defined as bundles of acicular to dendritic crystals of aragonite (e.g., [18,42–46] and references therein). This mineral occurs as euhedral crystals with a fibrous habit (length to width > 6:1) (Figure 6a,b). These fine needle-shaped crystals of tens of microns in length are rarely found as single crystals and form regular arrays as plume-like fans, and spherulites made up of hundreds of fibrous crystals (Figure 6d,h). Sometimes, in the center of these structures, small particles of other mineral phases are observed; in some cases, the structures are a little more complex and of a branched or dendritic type (Figure 6c). Acicular aragonite crystal growth was observed directly in the water from terraces. The aragonite crystals are oriented perpendicularly to the surface and can be associated exclusively with a rapid and inorganic crystallization process [20,44,47–49]. In the water–air interface at the terraces zone and where the hot spring is active, the rapid formation of aragonite paper-thin raft fabrics (also described as 'hot water ice') and its continuous accumulation at the bottom was identified [6,27,50,51]. These textures were also recorded in the travertine's fossil outer part (Figure 6e,f). This texture is not very compact; therefore, sulfate crystals (gypsum, anhydrite, and, more specifically, celestine) have been most frequently identified in the cavities between the aragonite crystals (Figure 7a–c).

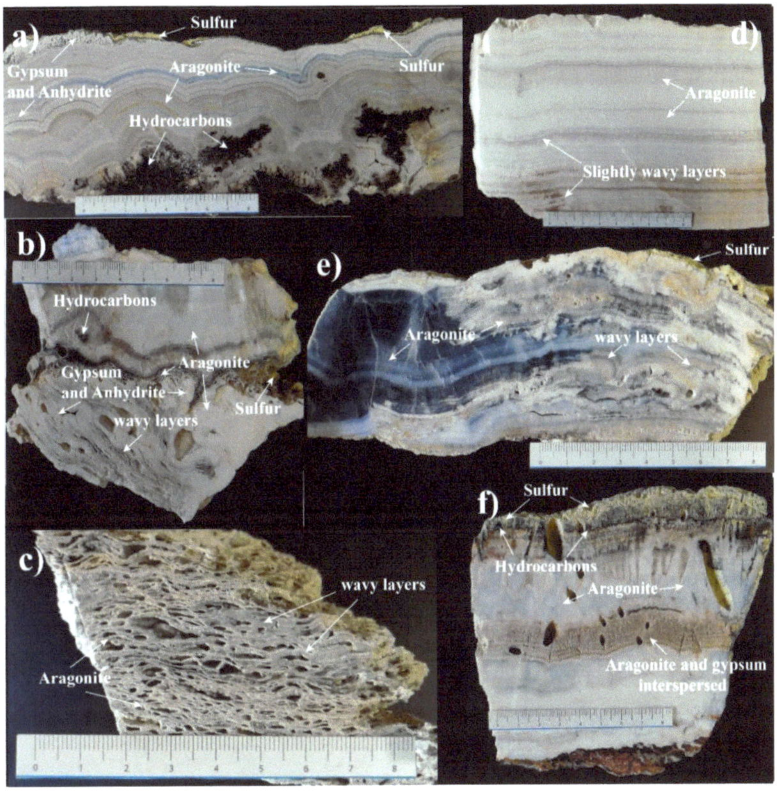

Figure 5. Main textural varieties observed in travertines. (**a**) Wavy homopachous aragonite layers of different colorations and some very thin interspersed layers of sulfates (gypsum and anhydrite). Native yellow sulfur crystals can be observed on the surface of the sample. The crystals of aragonite have different sizes and are mainly arranged in radial aggregates of fibrous crystals. The radial fans have an approximate diameter of 1 cm and their core is formed by solid hydrocarbon fragments (black coloration) (Tami. 3.1). (**b**) Cross-section of the sample Tami 1.2, formed by slightly wavy homogeneous and compact layers of aragonite in the upper section and more porous sheets with evidence of soft-sediment deformation in the lower section. (**c**) Cross-section of sample Tami 1.1, formed by porous wavy layers of aragonitic composition. The porosity is of fenestral type. This sample was taken in the highest part of the formation near the gas emanation. (**d**) Plane heteropachous aragonite layers with different thicknesses and some interspersed thin layers of sulfates (gypsum and anhydrite) and low native sulfur. A sample was taken from the thicker white central band for lanthanide analysis (Tami BC). (**e**) Blue wavy layers of aragonite with evidence of soft-sediment deformation on the left section. A sample was taken from the thicker blue central band for lanthanide analysis (Tami AC). (**f**) Little porous and almost horizontal levels of aragonite (mainly blue) interspersed with sulfates (pure white) in the upper section; native sulfur (dirty yellow) and intercalated hydrocarbons (black) can be observed (Tami 6.2). Table S1 (Supplementary Material) shows the mineralogic results obtained by XRD, with the exception of the Tami BC and Tami AC samples, which were used to measure lanthanides.

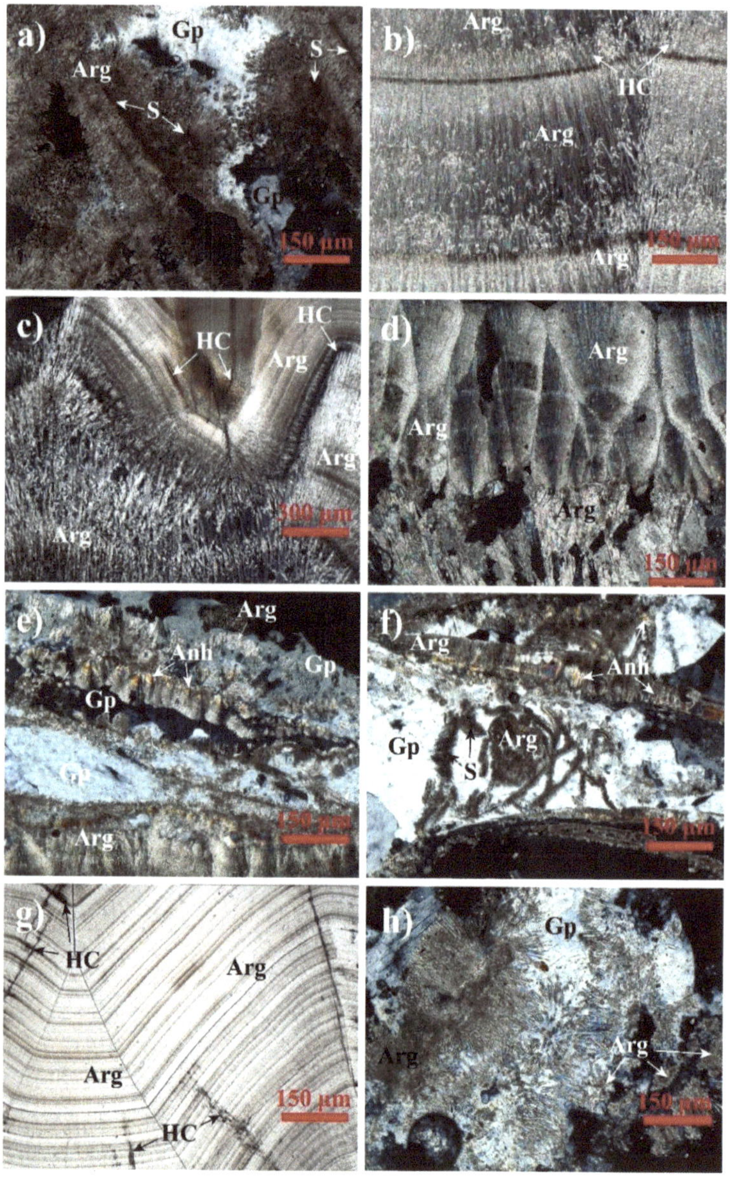

Figure 6. Aragonite crystals with the acicular (**a**) (Tami 2.2); fibrous (**b**) (Tami 6.2); or dendritic (**c**) (Tami 9.2) habit and forming aggregates with fan morphology (**d**) (Tami 1.2). Paper-thin raft fabrics (**e**,**f**) of aragonite with gypsum filling the cavities (Tami 1.1). Regular, cyclic, and non-porous aragonite layers of different colorations (**g**) (Tami 9.2). Pseudo spherulitic aggregates of aragonite associated with gypsum (Tami 10.2) (**h**). Arg—aragonite; Anh—anhydrite; Gp—gypsum; HC—hydrocarbons; S—native sulfur. Table S1 (Supplementary Material) shows the mineralogic results obtained by XRD, with the exception of the Tami 9.2 and Tami 10.2 that were not analyzed.

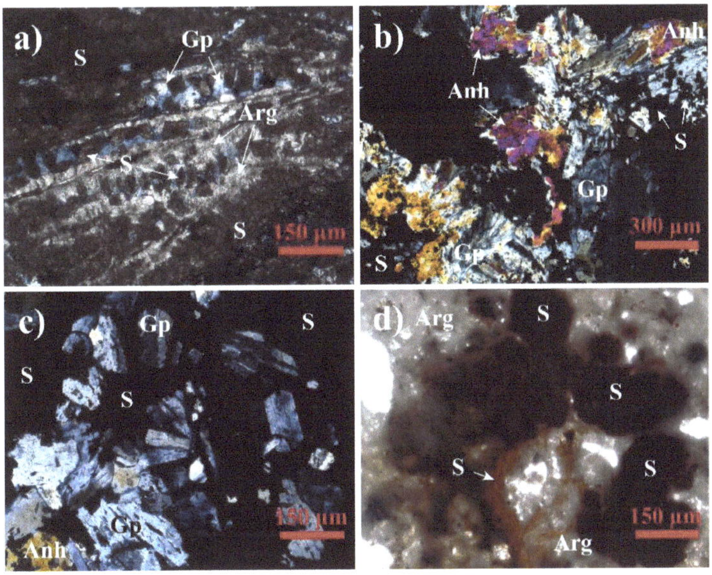

Figure 7. Small layers of native sulfur between the different aragonite layers (**a**) (Tami. 8.1). Gypsum, anhydrite, and native sulfur filling cavities (**b**,**c**) (Tami #5). Native sulfur appears mainly as irregular or botryoidal aggregates of small crystals with low crystallinity (**d**) (Tami #5). Arg—aragonite; Anh—anhydrite; Gp—gypsum; S—native sulfur. Table S1 (Supplementary Material) shows the mineralogic results obtained by XRD, except for the Tami 8.1 that was not analyzed.

In the petrographic study, it was also possible to verify that calcite is a very minor phase (less than 10%); it appears as aggregates of very small anhedral crystals (a few microns) of the micrite type. X-ray diffraction shows that such carbonate has low crystallinity (very wide peaks in the diffractograms), and thus it is not ruled out that its precipitation may be related to some process associated with biological activity. However, no petrographic evidence of this hypothesis could be found.

Sulfur is abundant, unevenly distributed, and occurs as botryoidal aggregates of anhedral crystals, often associated with iron oxides, making it difficult to observe in a transmitted light microscope (Figure 7d). Small layers of native sulfur and hydrocarbons (Figure 6g) are also found between the different aragonite layers. Associated with the presence of sulfur, it is also common to find aggregates of euhedral to subhedral crystals of gypsum and anhydrite (Figure 7b,c). These minerals are mainly found filling the cavities that exist between the aragonite crystals, but it is also common to find them forming thin layers interspersed between the carbonate layers

4.1.3. Mineralogy

The predominant minerals identified by XRD analysis were aragonite, gypsum, anhydrite, and elemental sulfur. Aragonite is mainly found forming regular aggregates of acicular crystals. Gypsum and anhydrite occur mainly as subhedral crystals forming irregular aggregates and filling cavities. Finally, elemental sulfur appears as microcrystalline and botryoidal aggregates associated with iron oxides and sulfates. The accessory minerals, generally filling cavities or in the interior of the main phases, were calcite, celestine, barite, quartz, and opal (Figures 7 and 8 and Table S1 of Supplementary Material). In addition, hematite, goethite, and local hydronium jarosite-group minerals have been identified in the small fractures present in the area.

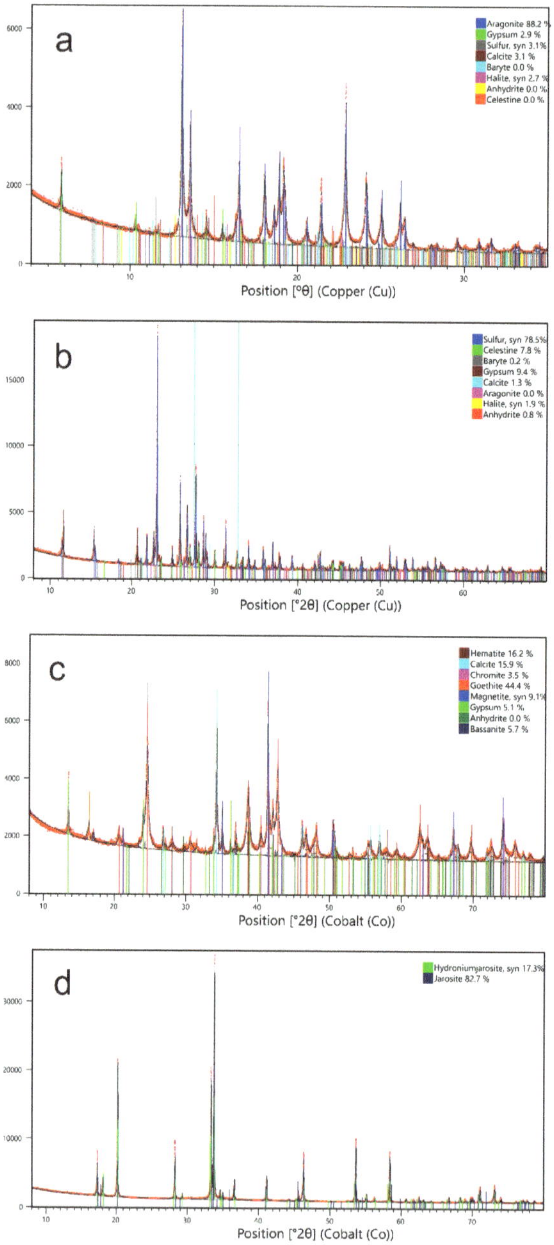

Figure 8. XRD patterns of characteristic samples of Tamiahua Lagoon travertine deposits: (**a**) Sample Tami 1.3; (**b**) Sample Tami 1.6; (**c**) Sample Po 14R; (**d**) Sample Tami 6.2.

Aragonite, elemental sulfur, gypsum, anhydrite, celestine, opal, fluorite, and goethite were identified using scanning electron microscopy coupled to SEM (Figure 9a–d). It is worth mentioning that some authors (e.g., [26]) have also described the presence of fluorite in travertines of deep origin. Quartz (<5 µm) is scarce and has been identified as detrital origin (Figure 9a). Celestine crystals (~10µm) were found to be associated with aragonite by the bright hue of the image due to the presence of Sr (Figure 9a). Measurement with EDS allowed us to determine the existence of 1–3% of Sr in the aragonite structure. Furthermore, the presence of opal particles—rounded in shape, without crystalline faces, and with a diameter of approximately 60 μm—was also identified (Figure 9c), as well as aggregates of native sulfur of poor crystallinity (Figure 9d).

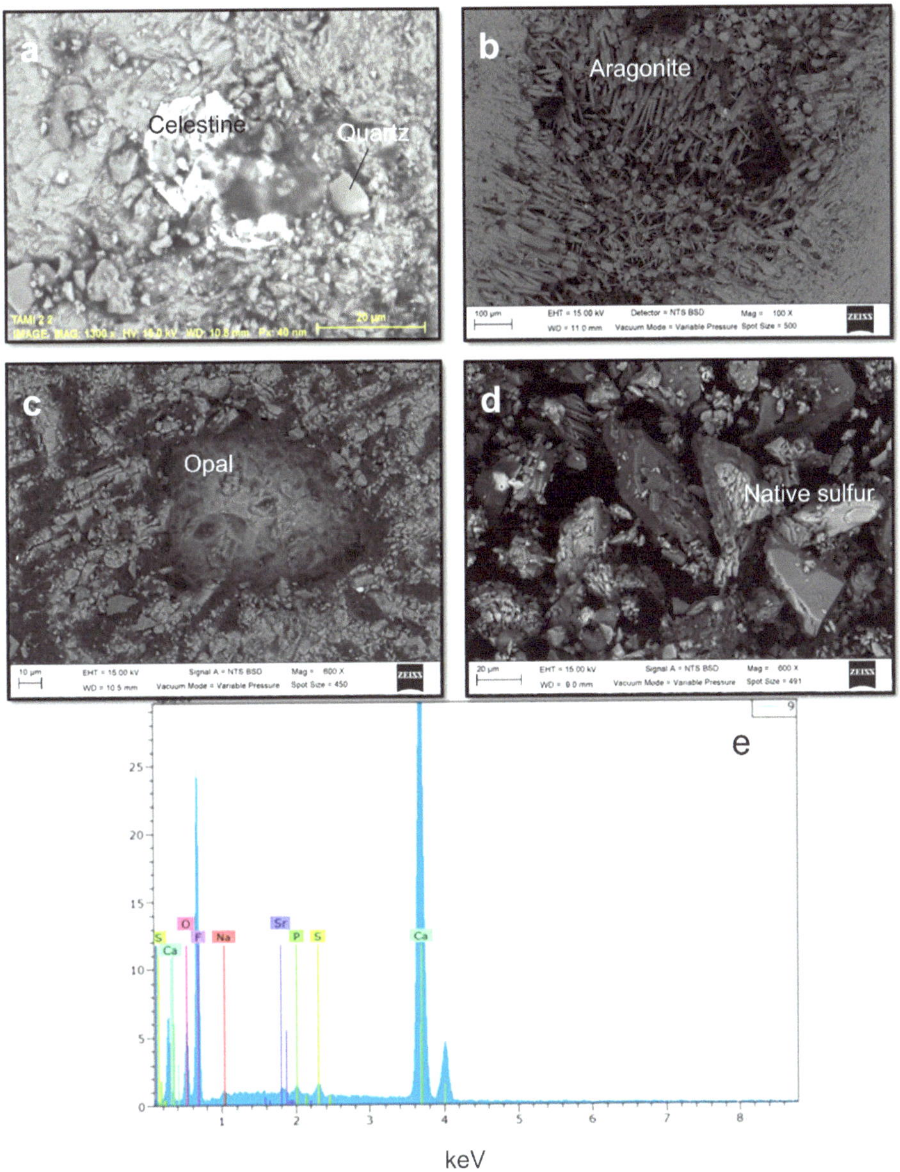

Figure 9. SEM Images. (**a**) Detrital quartz and authigenic celestine crystals (top left image). (**b**) Aragonite prismatic crystals (top right image). (**c**) Rounded opal inclusion (lower left image). (**d**) Native sulfur crystals (lower right image). (**e**) EDS spectrum of fluorite crystal (lower center). The identity of these minerals was identified by SEM-EDS and all of them, except fluorite, also by XRD.

4.1.4. Geochemistry and Isotopic Composition of the Rock

Table 2 and Figure 10 show the concentrations of trace elements measured by ICP in four Tamiahua Lagoon samples with pure aragonite crystals and one sample of native sulfur. The sample labeled as Tami AC has a dark blue coloration and contains the highest concentration of Y, Zr, U, and of all Rare Earth Elements (REE), while the samples Tami BC and Tami 2.2, of white coloration, have a slightly higher concentration of Ni. All aragonite outcrop samples have a high content of Sr (>1%), whereas the barium content in aragonite samples is low (17–82 ppm) [13].

Table 2. Trace element composition of travertine samples including lanthanide concentrations of four aragonite samples and one native sulfur sample (Tami 2AS) normalized to chondrite-type meteorites [52].

Sample	V	Cr	Co	Ni	Cu	Zn	Ga	Ge	As	Rb	Sr	Y	Zr	Nb	
DL	5	20	1	20	10	30	1	0.5	5	1	2	0.5	1	0.2	
TAMI 6.2	<5	<20	<1	<20	<10	<30	<1	<0.5	<5	<1	> 10,000	<0.5	1	<0.2	
TAMI 2.2	<5	<20	<1	50	<10	<30	<1	<0.5	<5	<1	9350	<0.5	<1	<0.2	
TAMI 2AS	<5	<20	<1	<20	<10	<30	<1	<0.5	<5	<1	2370	<0.5	<1	<0.2	
TAMI AC	<5	<20	3	<20	10	<30	<1	<0.5	<5	<1	>10,000	3.5	2	<0.2	
TAMI BC	<5	<20	3	40	<10	<30	<1	<0.5	<5	<1	>10,000	<0.5	<1	<0.2	
Sample	**Mo**	**Ag**	**In**	**Sn**	**Sb**	**Cs**	**Ba**	**Hf**	**Ta**	**W**	**Tl**	**Pb**	**Bi**	**Th**	**U**
DL	2	0.5	0.1	1	0.2	0.1	3	0.1	0.01	0.5	0.05	5	0.1	0.05	0.01
TAMI 6.2	<2	<0.5	<0.1	<1	<0.2	<0.1	35	<0.1	<0.01	<0.5	<0.05	<5	<0.1	<0.05	<0.01
TAMI 2.2	<2	<0.5	<0.1	<1	<0.2	<0.1	17	<0.1	<0.01	0.8	<0.05	<5	<0.1	<0.05	0.01
TAMI 2AS	<2	<0.5	<0.1	<1	<0.2	<0.1	12	<0.1	<0.01	<0.5	<0.05	<5	<0.1	<0.05	<0.01
TAMI AC	<2	<0.5	<0.1	<1	<0.2	<0.1	70	<0.1	<0.01	<0.5	0.06	<5	<0.1	<0.05	1.33
TAMI BC	<2	<0.5	<0.1	<1	<0.2	<0.1	82	<0.1	0.02	0.5	<0.05	<5	0.1	<0.05	0.02
Sample	**La**	**Ce**	**Pr**	**Nd**	**Sm**	**Eu**	**Gd**	**Tb**	**Dy**	**Ho**	**Er**	**Tm**	**Yb**	**Lu**	**ΣREE**
DL	0.05	0.05	0.01	0.05	0.01	0.005	0.01	0.01	0.01	0.01	0.01	0.005	0.01	0.002	
TAMI 6.2	<0.05	<0.05	<0.01	<0.05	< 0.01	<0.005	<0.01	<0.01	0.01	<0.01	<0.01	<0.005	<0.01	<0.002	0.01
TAMI 2.2	<0.05	<0.05	<0.01	<0.05	0.01	<0.005	0.01	<0.01	0.01	<0.01	<0.01	<0.005	<0.01	<0.002	0.03
TAMI AS	<0.05	<0.05	<0.01	<0.05	< 0.01	<0.005	<0.01	<0.01	<0.01	<0.01	<0.01	<0.005	<0.01	<0.002	0
TAMI AC	2.57	3.51	0.4	1.53	0.3	0.057	0.45	0.07	0.42	0.08	0.25	0.034	0.21	0.035	9.916
TAMI BC	0.09	0.17	0.02	0.1	0.01	0.007	0.01	<0.01	0.01	<0.01	0.01	<0.005	0.01	0.002	0.439

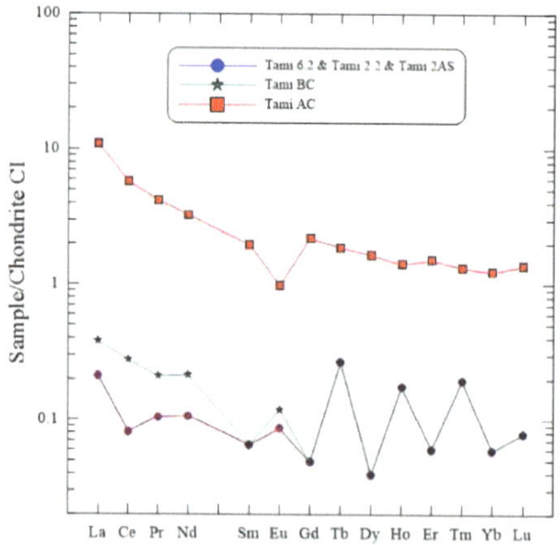

Figure 10. Lanthanide concentrations of four aragonite samples and one sample of native sulfur normalized to chondrite CI meteorites (Wasson and Kallemeyn 1988) [52].

Regarding the lanthanide content (Table 2 and Figure 10), there is a notable difference between the concentrations normalized with chondrite [52]. The Tami AC sample has the highest abundance of lanthanides (∑REE = 9.9 ppm); while light-colored aragonite samples (Tami 6.2 and Tami 2.2) have the lowest (∑REE < 0.5 ppm).

The $\delta^{13}C$ and $\delta^{18}O$ values obtained for the six samples of aragonite (Table 3) are similar; the $\delta^{13}C$ range is of +1.75 to +2.37‰, while for $\delta^{18}O$ it is of +1.70 to −0.78‰.

Table 3. Isotopic composition of travertine deposits of Tamiahua Lagoon (carbonates, sulfates, and sulfur) and solid hydrocarbon samples. Cc—calcite, Arg—aragonite, Gy—gypsum.

Carbon and Oxygen Isotopic Composition of Travertines					
Sample Number	**Sample Code**	**XRD Mineralogy**	**$\delta^{13}C_{VPDB}$ (‰)**	**$\delta^{18}O_{VPDB}$ (‰)**	**$\delta^{18}O_{VSMOW}$ (‰)**
1	Cima 1	31% Ar, 9% Cc, 60% Gy	+2.37	−0.78	+30.11
2	Capa 2	100% Ar	+1.83	−1.39	+29.47
3	Banda 3.1	78% Ar, 28% Gy	+1.93	−1.52	+29.34
4	Banda 3.2	100% Ar	+1.83	−1.60	+29.26
5	Banda 3.3	100% Ar	+2.02	−1.46	+29.40
6	Tami 7	100% Ar	+1.75	−1.70	+29.16
Sulfur Isotopic Composition of Travertines					
Sample Number	**Sample Code**	**XRD Mineralogy**	**$\delta^{34}S$ (‰)**		
1	Tamiahua 2	Gypsum-anhydrite	−4.0		
2	Tamiahua 1	Gypsum-anhydrite	−0.6		
3	Tamiahua 5	Gypsum-anhydrite	−0.8		
4	M 1.2 Cristal	Native sulfur	−0.1		
5	Tamiahua 2R Sup	Native sulfur	−3.7		
6	M 5	Native sulfur	+1.2		
7	M 5.2	Native sulfur	−2.3		
8	M6	Native sulfur	−1.5		
9	M 8.2 Sup	Native sulfur	−0.6		
10	M 13	Native sulfur	+0.3		
Sulfur Isotopic Composition of Hydrocarbons					
Sample Number	**Sample Code**	**$\delta^{13}C_{VPDB}$ (‰)**	**% Sulfur**	**$\delta^{34}S$ (‰)**	
1	MX0242	−26.85	2.16	7.64	
2	MX0243	−26.44	3.02	7.50	

Table 3 shows the $\delta^{34}S$ results for the ten analyzed samples of native sulfur and gypsum. The range of $\delta^{34}S$ is of −4.0‰ to +1.2‰. The isotopic values obtained for native sulfur (+1.2‰ to −3.7‰) are very similar but slightly more positive than the sulfur isotopic values obtained for sulfates (−0.8‰ to −4.0‰).

Table 3 also shows the results for $\delta^{13}C$ and $\delta^{34}S$ for the hydrocarbon asphaltene-like samples (*chapopote*). Isotopic values do not vary significantly; values between −26.85 and −26.44‰ are obtained for carbon, whereas values from +7.65 to +7.50‰ are acquired for sulfur. The extremely negative carbon values of the hydrocarbon are due to its organic origin, and the range of sulfur isotopic values are compatible with a thermochemical ($\delta^{34}S$: 5–15‰) or bacterial ($\delta^{34}S$: 0–10‰) sulfate reduction genesis [53].

4.2. Water System

A Durov diagram [54] was constructed to visualize the chemistry of major ions and to compare it with pH and total dissolved solids (TDS). The pH of the hydrothermal emanation water in the aragonite terraces is slightly acidic (Figure 11a), approaching neutral, as is the water from the well (meteoric water). In comparison, the water from the river and the lagoon is slightly alkaline (Table 4). The temperature recorded in the three-point intake (well, river, and lagoon) is similar, being between 27 °C and 34 °C (the average annual maximum temperature recorded at the climatological station of Majagual in Tamiahua is 29 °C; https://smn.conagua.gob.mx/es/informacion-climatologica-por-estado?estado=ve (accessed 15 June 2022), while the minimum surface temperature of the aragonite terraces was registered as 70 °C, similar to the outer part of the fumaroles in which steam emanates. Table 4 contains the physicochemical data taken and recorded in the field of the water samples, except for the condensate from the fumarole, which was not determined due to the limited amount of water collected.

Table 4. The chemical and physicochemical properties of the water samples collected at different points in the Tamiahua Lagoon area. The isotopic composition of stable elements in the water results is also presented.

Sample	Description	Surficial Temperature (°C)	pH	Electrical Conductivity (mS/cm)	$\delta^{18}O_{VSMOW2}$	$\delta^{2}H_{VSMOW2}$
a	Terraces water	70	6.1	118	+7.94	−5.39
b	Well water	27	6.4	3	−3.82	−18.87
c	Lagoon water	34	8.1	45	+3.73	+23.75
d	River water	32	7.9	0.06	+3.58	+22.49
e	Fumarole water	Not measured	8.4	3	+3.53	−9.54
f	Duplicate of a	70	6.1	Not measured	+7.77	−5.50
g	Duplicate of b	27	6.4	Not measured	−3.76	−18.85

We determined whether the aragonite is in isotopic equilibrium with the water from which it precipitates. Among the numerous equations of carbonate–water isotopic equilibrium, two equations were chosen. The first (Equation (1) is from a classical source [55], and the second (Equation (2)) is one of the most recent recalibrations [56]. First, the calculation uses a mean value of $\delta^{18}O$ = +7.86 from the two available water measurements (Table 4, analyses **a** and **f**). Second, for each carbonate (Table 3), its oxygen value was used ($\delta^{18}O$ = −1.30 to −1.70), except for sample Cima 1, which had much calcite and gypsum besides aragonite. Finally, both oxygen values for each sample (aragonite, water) were used to calculate $\alpha = (1000 + \delta^{18}O_{aragonite})/(1000 + \delta^{18}O_{water})$, which is then input into the equations, and the temperature was calculated by solving the equations iteratively. The error in T obtained from Equation (2) is from the ± stated in the equation [56].

$$1000 \ln \alpha_{carbonate\text{-}water} = 0.9521\,(10^6/T^2) + 11.59\,(10^3/T) - 21.56 \quad (1)$$

$$1000 \ln \alpha_{aragonite\text{-}water} = 17.88\,[\pm 0.13]\,(10^3/T) - 31.14\,[\pm 0.46] \quad (2)$$

The temperature calculated using Equation (1) [55] is 65 °C and using Equation (2) [56] is 69 ± 4 °C, both similar to the water temperature measured in aragonite terraces (70 °C). Based on this result, we can consider that the aragonite precipitation occurred not too far from the chemical equilibrium [55,56].

The ionic balance was carried out with the cation and anion concentration from the different Tamiahua Lagoon water samples (Table 5). The allowable range in ion balance for geothermal systems is ± 10%, therefore the only water sample outside the limit corresponds to the well sample.

The analytical results were plotted on the Durov diagram (Figure 11A), indicating that all water samples can be classified as chloride–sodium water type and are related to subsurface seawater invasion, given their proximity to the shoreline sedimentary basin. The bicarbonates and lithium content are low for almost all the samples, except for the terraces sample, which presents the highest values with 730 mg/L of bicarbonate and 20 mg/L of lithium. The terraces, river, and lagoon samples have similar concentrations in some anions, such as fluorides, chlorides, sulfates, and cations including sodium, magnesium, and calcium. However, the terraces sample contains almost double the chloride (31,013 mg/L) and sodium (15,953 mg/L) content and has half the sulfate content of the other two marginal samples. The lagoon sample with 104 mg/L of fluoride, has almost twice the amount of this element as the other samples. The potassium content is similar for the lagoon and the river samples, which show the highest amount, while the other samples show low concentrations. A ternary Na-K-Mg diagram (Figure 11B) [57,58] was created using the water sample values: these are major elements only for water from terraces, which presents the highest surface temperature, and the fumarole, since it represents the condensate of the vapor and gas that comes from the hydrothermal vent. The condensate sample from the fumarole is almost at the equilibrium line, which corresponds to a temperature of around 200 °C in the reservoir.

Table 5. The concentration of ions and trace elements from water samples in mg/L. DL—detection limit; QL—quantification limit; NA—not analyzed.

Sample	Al	As	B	Ba	Fe	Cs	Mn	SiO_2	Sr	V
Terraces	0.02	0.13	32.27	0.18	<DL	8	<DL	38	73	<DL
River	0.02	≤DL	2.96	0.1	<DL	2.06	0.16	6.6	4.55	<DL
Fumarole	NA	NA	NA	NA	NA	NA	NA	NA	NA	NA
Lagoon	0.42	≤DL	3.11	0.07	0.36	2.88	0.03	6.8	4.61	<DL
Well	0.04	≤DL	0.07	0.08	<DL	1.13	0.5	67.4	0.87	0.01
DL (mg/L)	NA	0.08	0.03	0.003	0.002	NA	0.001	0.15	NA	0.002

Sample	F^-	Cl^-	SO_4^{2-}	HCO_3^-	Na^+	Mg^{2+}	Ca^{2+}	K^+	Li^+
Terraces	58	31,014	1288	730	15,953	1121	410	0.1	20
River	59	17,830	2864	45	9236	1023	452	339	0.03
Fumarole	0.2	907	151	15	634	0.1	40	44	5
Lagoon	104	16,836	2548	45	9762	1080	446	347	0.03
Well	13	864	655	231	366	49	147	15	0.03
DL (mg/L)	0.2	0.3	0.3	NA	0.1	0.1	0.08	0.1	0.03

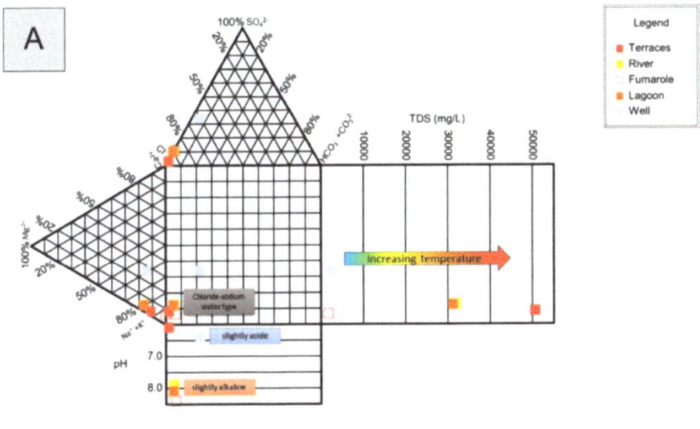

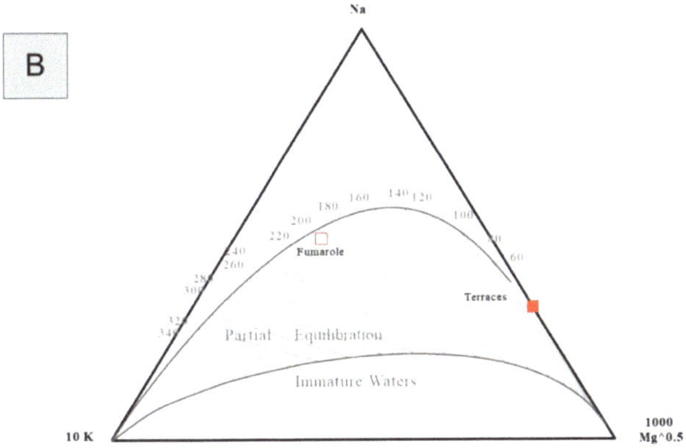

Figure 11. (**A**) Durov diagram showing the dominant ions in relation to pH and Total Dissolved Solids (TDS) of the water samples from Tamiahua Lagoon. The plot was created with the AqQA (version 1.2.0). (**B**) Na-K-Mg diagram showing the chemical equilibrium of the waters of Tamiahua Lagoon. The plot was created with the Liquid Analysis v3 program from Powell and Cumming 2010 [58].

In Table 5, it can be observed that the water labeled as terraces sample presents the highest contents of B, Ba, As, Cs, Li, and Sr with values of 32.27, 0.18, 0.13, 8.0, 9.21, and 20 mg/L, respectively. On the other hand, the well sample has the highest value of SiO_2 with 67.4 mg/L. Although it is of near-alkaline conditions, the abundance of silica in well water can be related to the dissolution of silicate minerals at higher temperatures and depths, conditions that do not exist around the lagoon and surficial terraces where we only found carbonates and sulfates.

The analytical results obtained for trace elements indicate low values, even below the instrument's detection limit, as is the case for As, Fe, Sb, Pb, and V.

The diagram proposed by Arnórsson and Andrésdóttir [59] was replicated for this study in order to represent the relationship between Cl and B in the waters and to infer the origin of these elements either derived from being located in the coastal zone or arising from hydrothermal activity at depth (Figure 12A). Generally, the boron content in groundwater is less than 1 mg/L; concentrations above this may be due to the presence of hydrothermalism [60], evaporative processes, seawater intrusion, evaporite dissolution [60], mineral weathering [60], anthropogenic contamination [61,62], and sorption and desorption processes on mineral surfaces [63,64].

The mineral saturation index was calculated with the software The Geochemist's Workbench (version 11.0.8; module SpecE8) and the results are plotted in Figure 12B.

Isotopic δ^2H values for each type of water are quite different, with negative $\delta^{18}O$ values (−4.0 to −4.1‰) for well water and positive for terrace water (+7.4 to +7.6‰), river water (+3.2%), water from the lagoon (+3.6‰), and fumarole (+3.3‰) (Table 4). The values for the isotopic pairs δ^2H and $\delta^{18}O$ were plotted in Figure 13. The duplicates (aragonite water and well water) yielded similar values to the original samples (Samples 1 and 5).

As can be seen in Figure 13, the well water is found above the meteoric water line, while the other water samples tend to be away from this line; this can be attributed to the fact that water from the terraces and the fumarole may be the product of a surficial mixture between meteoric waters and ascending oil-field derived brines. In addition, the water from Tamiahua Lagoon presents anomalous high values of δ^2H that can be linked to the high coastal environment evaporation rates.

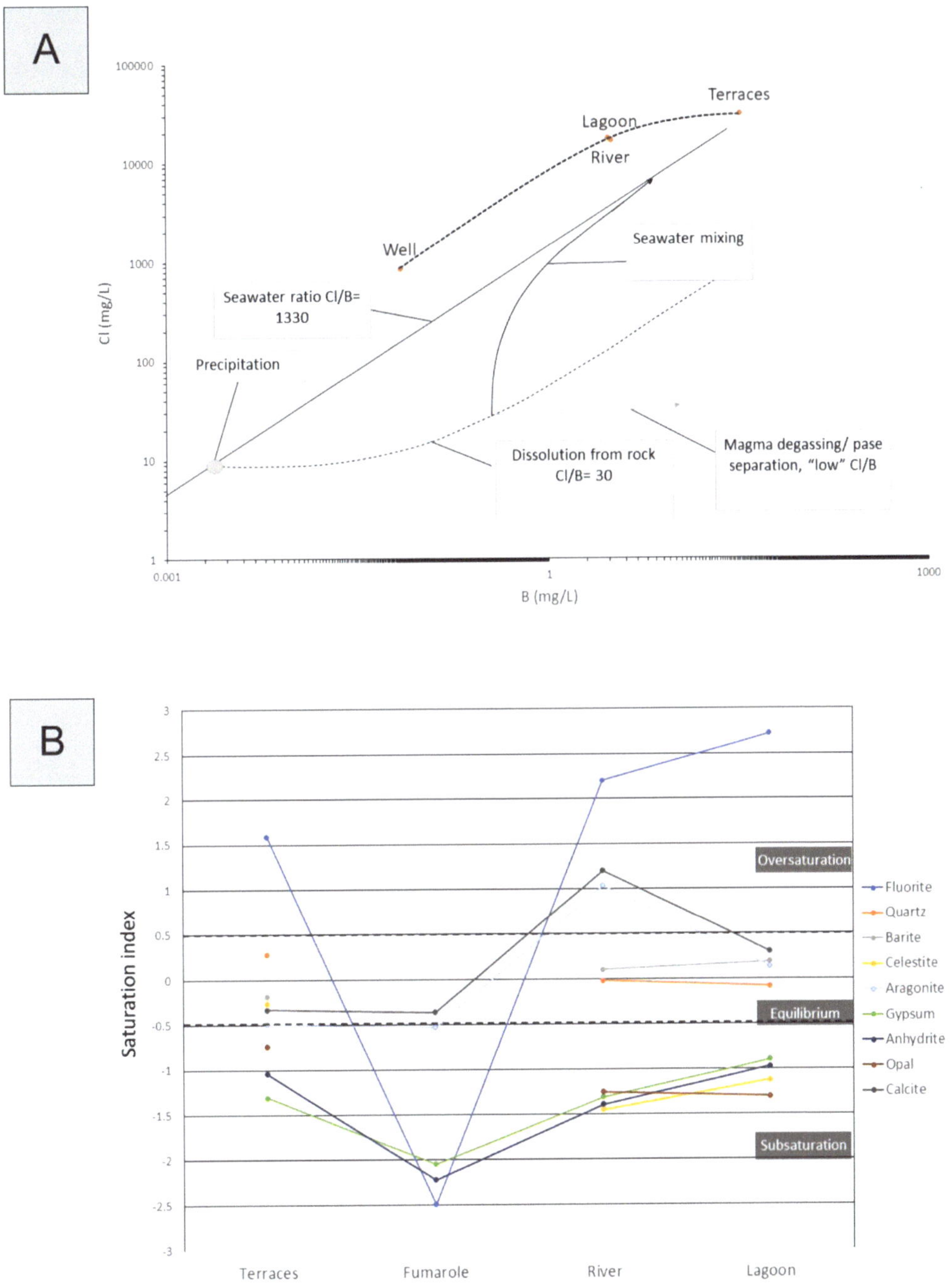

Figure 12. (**A**) Schematic diagram (Arnórsson and Andrésdóttir, 1995) [59] showing the Cl/B ratio of the water samples from Tamiahua Lagoon. (**B**) Saturation index (SI) from the water chemistry of some minerals identified by XRD analysis in rock samples from Tamiahua Lagoon. The SI was calculated with module SpecE8 of the Geochemist's Workbench.

5. Discussion

5.1. Origin of the Hydrothermal System

All the water samples collected at the different points of Tamiahua Lagoon were classified as chloride–sodium water type, related to seawater or sedimentary brines. When analyzing the ion content in the water, it was found that Tamiahua Lagoon exerts a strong influence on the hydrothermal system, mainly by providing Cl^-, SO_4^{2-}, Na^+, and Mg^{2+} as the most abundant ions. Therefore, if the influence of marine sodium chloride is eliminated, the water from the aragonite terraces could be classified as magnesium sulfate since the sulfate is above 1200 mg/L and the magnesium itself is above 1100 mg/L. According to Herman et al. [65], the B/Cl ratio in the terraces sample (0.00104) is indicative of a deep fossil brine input. The other samples (well, lagoon, and river samples) have a ratio outside the range established for a fossil brine (0.0099–0.00056) but have a strong influence from seawater (Figure 12A). Well water also contains similar ions such as Cl^-, SO_4^{2-}, and Na+, possibly due to the proximity to the coastline.

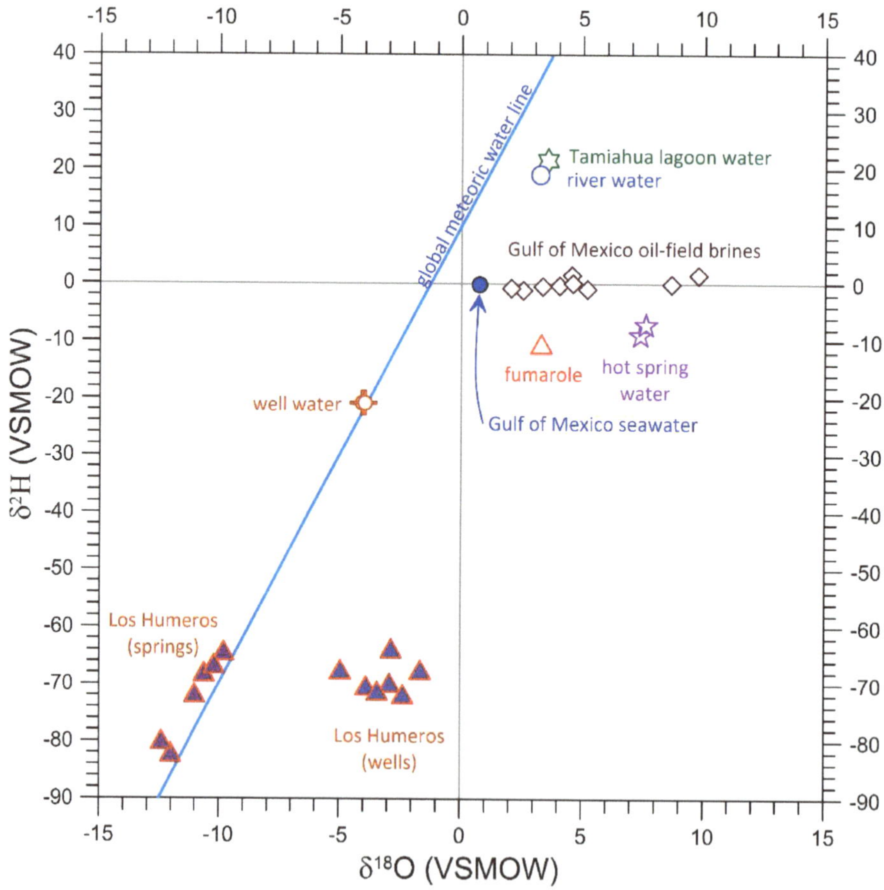

Figure 13. Isotopic composition of different types of waters in Tamiahua Lagoon. Values for the waters from Los Humeros springs and wells were obtained from Martínez-Serrano [66].

As the discharge temperature for the terraces sample was around 70 °C, the corresponding temperature at the depth of this sample point and the fumarole was estimated using the Giggenbach diagram. Before the estimation, the existence of the equilibrium state had to be considered to apply the cationic geothermometry. According to Figure 12B, the terraces sample is coherent with the surface temperature, but the fumarole chemistry reveals a temperature of approximately 200 °C (Figure 12B) at depth, which corresponds to the starting stage of catagenesis, the hydrocarbon generation window. An objective of the present work was to determine the origin of the fluids; therefore, the isotopic composition of hydrogen and oxygen obtained from the water of the aragonite platforms was compared with the nearby water sources: well water, water from Tamiahua Lagoon, and water from a small river that flows into the lagoon and is in contact with the aragonitic formation.

The δ^2H and $\delta^{18}O$ values of the well water sample indicate that it is dominantly meteoric water, since they are projected above the global meteoric water line (Figure 13). Isotopic values are similar, for example, to those measured for the waters of the distant hot springs of the Los Humeros geothermal field [66], but the latter displays more negative values due to latitude and sea distance differences between both areas. On the other hand, isotopic values of fluids sampled at aragonite terraces (δ^2H: −7 to −9‰; $\delta^{18}O$: +7.4 to +7.6‰) are close to those published for the waters of the oil brines of the Gulf of Mexico [67] but with slightly more negative δ^2H. The fumarole sample's isotopic values are similar to those in the brines, but slightly different from those of the terraces sample; these differences could result from a condensate generating an isotopic fractionation or that the vapors come from a deeper source.

The terraces sample that gives rise to the aragonite terraces has two components; the first is a meteoric component given by well water (sample on meteoric global line) with δ^2H at −21‰ and $\delta^{18}O$ at −4.1‰. In contrast, the second component is related to the oil brines of the Gulf of Mexico (average δ^2H = −0.18‰ and $\delta^{18}O$ = +4.75‰).

Tamiahua Lagoon presents an anomaly in the δ^2H of the lagoon sample being highly enriched in deuterium (δ^2H = 21.5‰). These high values have been previously reported in meteoric waters in the Tamiahua Lagoon area, ranging from +17‰ to +35‰; these are observed as an annual trend in the winter seasons, mainly in January and February [68]. Evaporation tends to be significant for the isotopic record only in closed basins of relatively arid regions, where lake waters may be deuterium-enriched by 10–15‰ [69]. Therefore, the high values reported for the waters of Tamiahua Lagoon indicate the presence of an evaporative/hypersaline system. Continuous evaporation of water from Tamiahua Lagoon is an ongoing process and could contribute to an enrichment of the heavy isotopes of the water (^2H and ^{18}O). The increment in $\delta^{18}O$ in basinal brines that comes with a temperature increase is primarily due to its effect on the isotopic and chemical exchange rate between the water and the enclosing rock. Sedimentary rocks have very high $\delta^{18}O$ and the water becomes enriched due to exchange with the surrounding host rocks during rising thermal waters. Furthermore, the loss of H by gases such as H_2S, H_2, or CH_4—associated with hydrocarbon emanations—also provokes deuterium enrichment in water relative to the rocks and precipitates [70]. It is important not to discard the possible influence of deep brines associated with the hydrocarbon layers that were mixed with the surface water in the lagoon area due to the spill that occurred approximately 110 years ago.

5.2. *Origin and Classification of Mineral Travertine System*

5.2.1. Aragonite Precipitation and Its Typologies

Calcite occurs only in Tamiahua travertines as a minor phase of low crystallinity. This point could be verified in the XRD patterns by the punctual presence of very small and wide peaks in the 2theta positions characteristic of this mineral. The calcite saturation indices for the terrace area show the balance of the fluid with this mineral (Figure 12B). The theories that explain the precipitation and formation of aragonite layers instead of calcite precipitation are diverse [71–73]. Three main factors that favor aragonite precipitation have been identified in the water from the Tamiahua aragonite terraces. The first factor is the temperature being greater than 70 °C; when the temperature is higher than 40 °C, aragonite will precipitate regardless of the fluid's composition [74]. The second factor is the Mg/Ca molar ratio, which is greater than 1:1 [75]; magnesium molar content is greater than calcium molar content in the water. The third and last factor is the content of Sr present in the water and the aragonite structure; the strontium content in aragonite is relatively high since it is found as a major element (>1%). This composition has been observed in aragonite formed in hot springs from non-marine sources such as the travertines of Rapolano Terme, Italy [45], where aragonite incorporated approximately 1% strontium into its structure (Figure 13). The Ca^{2+} ion has a coordination number of nine in the aragonite structure, while calcite has a coordination number of six [76]. The Ca^{2+} distribution in calcite has a cubic packing, unlike the Ca^{2+} in aragonite that has an almost hexagonal packing. Sr^{2+} is slightly larger than Ca^{2+} when forming a bond with $CO_3{}^{2-}$; it tends to form a coordination of nine—greater than that formed with Ca^{2+}—that transforms the packing from cubic to hexagonal [77]. When the aragonite nucleates, the structure's growth will be controlled if the concentration of Sr and the temperature do not change. Aragonite does not precipitate exclusively in the zone of higher temperature, and the saturation indices of calcite and aragonite present parallel trends and are very similar to each other in the different parts of the deposit (Figure 12B). Consequently, the high concentration of some ions (mainly Mg, $SO_4{}^{2-}$, and Sr) can be the key factor that favors, by inhibiting the growth of calcite, aragonite precipitation in the Tamiahua zone.

The most evident micro-factory in the different typologies described for Tamiahua is the fibrous laminations of aragonite, formed by aggregates of acicular crystals. These microfabrics, associated with CO_2 degassing and high carbonate precipitation rate, have been preferentially defined in thermogene travertines and are characterized by presenting fine, sequential, or rhythmic lamination, with low permeability and low, mainly fenestral, porosity [78]. Genetically they are related to the inorganic processes that predominate in the formation of the travertines of the Tamiahua coast (e.g., [43,79,80]).

Another micro-factory found in travertines are the so-called "rafts" structures formed at the water–air interface that later, when sinking, accumulate at the bottom of terraces or pools where the flow of water is very slow [46].

A less frequent micro-factory is the formation of "bushes" or dendrites; they appear as fans of fibrous aragonite with undulating extinction (Figure 6d. Whether the origin of these microfabrics is inorganic or biogenic continues to be discussed by various authors; however, the prevailing conditions (for example, temperature ~70 °C) in the Tamiahua area and the predominance of aragonite allow us to link their origin to a dominantly inorganic process [79]. The porous and less consolidated textures seem to be associated with lower flow conditions [20].

5.2.2. Geochemistry, Isotopic Composition, and Travertine Classification

A factor to consider is that the aragonitic laminations present different colorations, especially white and blue in the light and dark tones. When comparing the rare earth content (lanthanides) of the samples with contrasting coloration (white and dark blue), it was possible to verify a notable difference in the samples of distinct colors. The bluish coloration seems to be related to a higher lanthanide content, mainly in light rare earth (Table 2). The colorless (white) aragonite has the lowest $\sum$REE (sum of all lanthanides) content (<0.5 ppm), while the $\sum$REE of dark blue aragonite has higher values (9.9 ppm). The sample Tami BC has a higher abundance in all rare earth than other samples and shows a negative anomaly in Eu (Figure 10); this anomaly is very common in the sediments of the upper continental crust and surface waters [81,82]. The darker bands with the most abundant rare earth content can be related to hydrothermal pulses more enriched in these elements, and it is possible to interpret that the pattern of lanthanides in aragonite is analogous to that of the initial fluids [83].

The stable isotopes of carbon and oxygen in aragonite do not present significant variations, suggesting that the fluid's chemical composition remains constant throughout the crystallization of this mineral. The carbonate dissolved in the fluid that gives rise to the mineralization seems to originate from the dissolution of Mesozoic marine carbonates of the El Abra formation, as they form the carbonate platforms characteristic of the oil fields of the Tampico-Misantla Province. Nonetheless, we cannot rule out a small carbon component from hydrocarbon generating rocks. Although the high contents of strontium in the fluid also seem to indicate that the dissolution of the Mesozoic carbonates may be the fundamental process in the formation of the carbonates of the travertines, a contribution from other rocks rich in calcium (mainly magmatic) cannot be ruled off. Therefore, in a new sampling campaign, we plan to carry out isotopic analyses of water on a more regional scale to better understand the hydrological cycle—for example, possible processes of mixing and isotopic fractionation due to CO_2 degassing—of the area that, due to the presence of the lagoon, the sea, the deep oil system, and anthropogenic activity, results in a quite complex system.

Oxygen isotopic values are very close to zero and the obtained isotopic temperatures, using Equations (1) and (2), seem to indicate equilibrium precipitation conditions despite evaporation and degassing [84]. Additionally, the carbon isotopes of aragonite suggest that the CO_2 degassing is associated with the decarbonization of limestones. The waters are very hot; therefore, the effect of evaporation is probably the dominant $\delta^{18}O$ controlling factor [85].

The classification of the travertines as thermogene was based on the stable isotopes of carbon from aragonite, as well as the microfabric observed during the petrographic study. Travertines of thermogenic origin are characterized by having more positive values of $\delta^{13}C$ (enriched in ^{13}C, −3 to +8‰) than meteogene [6,79]; thus, they are more consistent with the isotopic values obtained for aragonite ($\delta^{13}C$ from +1.75‰ to +2.37‰) in Tamiahua. According to different authors [5], the ^{18}O signature of aragonite is strongly linked to meteoric waters and therefore does not allow to define the origin of the travertines correctly.

According to Teboult et al., (2016) [13], the relation between Ba and Sr allows for discrimination between epigean and hypogean travertines and can help to discriminate between different source rocks. For example, limestones, evaporites, and dolomites exhibit low barium (>100 ppm) and high strontium (>400 ppm) contents [13]. In Tamiahua, the low barium (70 to 82 ppm) and high strontium (>1%) contents of aragonite, and mainly the presence of powerful carbonate layers with levels of evaporites in the subsoil, suggest the fluids' hypogean nature and point to a source rock constituted mainly by limestones and evaporite [13] (Figure 14).

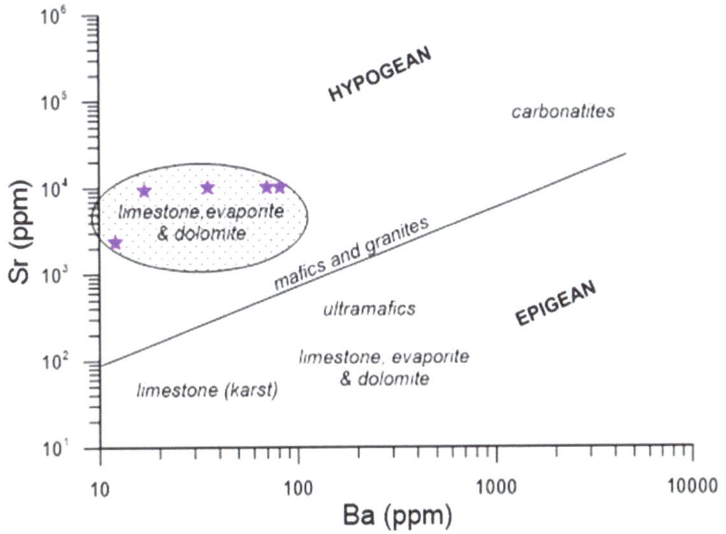

Figure 14. Comparison of Tamiahua Ba/Sr values (purple stars) with the travertine and tufa literature data (dashed area), modified from Teboult et al., (2016) [13].

The El Abra Formation, composed mainly of limestone, contains abundant layers of anhydrite [86], which was thought to be in origin related to the sulfates found interspersed in the layers of carbonates. The thermal springs in the Tamiahua area contain H_2S, giving them the characteristic smell of sulfur. H_2S can likewise be related to hydrocarbons from oil wells [87]. In the presence of hydrocarbons and a reducing environment, bacterial activity consumes H^+ and can release H_2S into the system [88–90]; this reaction (Equation (3)) implies that sulfates can be reduced in the presence of hydrocarbons, producing H_2S.

$$SO_4{}^{2-} + 2(CH_2O)n + 2H^+ \rightarrow 2nCO_2 + 2H_2O + H_2S \quad (3)$$

The isotopic data concluded that the elemental sulfur that covers the carbonate formation is attributed to the mainly inorganic oxidation of the deep-source H_2S [91], probably following Equation (4) and consequently precipitating elemental sulfur and carbonate in one step.

$$Ca\,(HCO_3)_2 + 2H_2S \rightarrow CH_2O + CaCO_3\downarrow + 2S\downarrow + 2H_2O \quad (4)$$

However, the isotopic data do not allow us to rule out the possibility that there is a deep sulfur component of volcanic origin.

5.3. Geotectonic Context of Formation

The rise of hydrocarbons mixed with the carbonate formations can be attributed to a regional fault and fracture system generated by deformation events during the Cretaceous–Paleogene. The first event occurred due to the magmatism generated during the Laramide deformation, which determined tectonic structures with NW-SE orientation. Later, this was followed by a lateral regime that predominated during the Cenozoic [92]. These dextral transpressive slip fault and fracture systems also had an NW-SE orientation [93], affecting the Chicontepec Formation and originating new migration routes for hydrocarbons. These events share the same NW-SE direction, which is in accordance with the fractures identified at different points of the carbonate formations (N05° W to N17° W).

These old fractures and faults are how fluids rise to the surface, allowing waters from different sources to mix during the ascent. However, it is still unclear why the temperature is ~70 °C in the aragonite terraces, while it is less than 35°C in the other nearby water sources. The fractures operate as fluid conduits, permitting the flow of groundwaters to the surface [6]. Different authors such as Berriguete et al., (2017) [94], have determined that the flow of water and its physicochemical characteristics can be greatly affected by diagenetic modification (dissolution, cementation, recrystallization) of the underlying rocks from the substrate. Unfortunately, in the case of the studied travertines and their location at sea level on the coast of Tamiahua lagoon, it is impossible to observe the travertine deposit roots without perforating the carbonated body. Therefore, it is unattainable to verify whether these diagenetic processes modified the water flow in depth [94].

Where the extensional zone is located, and the conditions in which the fluids rise through faults and fractures with NW-SE orientation, suggest that the heat source may come from a deep zone, which can be linked to a regime dominated by convection in a zone extension [30].

6. Conclusions

Based on the analyses carried out on the collected samples, the representative mineralogy of the hydrothermal formations is constituted by aragonite, sulfur, gypsum, anhydrite, celestine, barite, calcite, quartz, opal, hematite, jarosite, and halite.

Aragonite is the main mineral identified in the hydrothermal formation; it has high crystallinity and a mainly inorganic origin. The precipitation of this carbonate is due to several factors, including a high content of Mg in solution, a high content of Sr (>1%) in the aragonite structure, and a temperature of ~70 °C. On the other hand, the scarce calcite crystals found are of low crystallinity and, in some cases, a biogenic origin is attributed to them.

According to the results obtained for the mineral microfabric, the stable isotopes of carbon and oxygen and the temperature measured at the surface, the aragonitic formations of the Tamiahua coast can be classified as travertines mainly of thermogenic (hypogean) origin.

The dominant microfabric is of fibrous laminations formed by acicular crystals of aragonite and fine rhythmic laminations attributed to an inorganic origin. The dendritic textures found in aragonite indicate rapid precipitation.

Native sulfur seems to be related to the emission of H_2S present in hydrocarbons from oil wells, which generates the precipitation of native sulfur and calcium carbonate under suitable physicochemical conditions.

Based on the isotopic results, the water analyzed in the aragonite terraces must have a mixed origin, including two components: a first one of meteoric origin, represented by well water; and a second component isotopically related to the oil-field waters of the Gulf of Mexico. Deep brines associated with the hydrocarbon layers were mixed with the surface water in the lagoon area due to the spill approximately 110 years ago. These basin waters are, in turn, mixtures of meteoric water and evolved seawaters.

The mixing of meteoric water and brackish water is carried out at depth, where it passes through sedimentary sequences in which there are hydrocarbons typical of warehouse rocks of petroleum systems such as the El Abra Formation; these formations contain interspersed evaporites typical of marine environments. Finally, the fluids are heated at depth and rise by convection through faults and fractures generated by extension events; during this process, there occurs the dissolution of carbonates from the sedimentary units through which they rise, as well as the isotopic exchange of carbon, oxygen, and sulfur between fluids and dissolved components.

Once on the surface, the variations in pH, oxidation, the very limited bacterial activity, and the escape of gases result in the precipitation of different minerals such as elemental sulfur—which oxidizes to sulfates—and aragonite, as well as the release of H_2S and CO_2 gases.

Supplementary Materials: The following supporting information can be downloaded at: https://www.mdpi.com/article/10.3390/min12070822/s1, Table S1: Mineralogic composition by XRD of representative samples of travertine. Mineral abundance (%) by the Rietveld method.

Author Contributions: Conceptualization, T.P.-P., R.E.V.-E., M.A.R.-R. and J.S.; methodology, I.P.-T., T.P.-P. and R.E.V.-E.; formal analysis, I.P.-T., T.P.-P. and R.E.V.-E.; investigation, I.P.-T., T.P.-P., R.E.V.-E., M.A.R.-R. and J.S.; resources, T.P.-P., R.E.V.-E. and J.S.; writing—original draft preparation, T.P.-P.; writing—review and editing, I.P.-T., T.P.-P., R.E.V.-E., M.A.R.-R. and J.S.; All authors have read and agreed to the published version of the manuscript.

Funding: This research was partially funded by CONACYT: Laboratorios Nacionales, No. 299087, and, P02 of CeMIE-Geo. Fondos Sectoriales SENER-CONACYT.

Acknowledgments: We especially thank Margarita Reyes Salas and Blanca Sonia Ángeles García for their support using the LANGEM Scanning Electron Microscope, as well as Francisco Javier Otero Trujano and Edith Cienfuegos from the Stable Isotope Laboratory of LANGEM for the isotopic analyses of C and O in carbonates. To the work team of the laboratory of the Geochemistry of Geothermal Fluids Unit of the Institute of Geophysics, UNAM, and Carlos Uriel Miguel Morales, we offer thanks for the support in the field, and Águeda E. Ceniceros Gómez from the Laboratory of Physical and Chemical Analysis of the Environment (LAFQA) of the Institute of Geography, UNAM, for the determination of trace elements in water samples. We are grateful to the NVCLEVS Stable Isotope Laboratory of the University of Salamanca, Spain, and to its work team: Antonio M. Álvarez Valero, Félix García García, Raquel Sáez Ayuso, and Margarita Sotelo Martínez for the isotopic analyses of sulfur. We acknowledge Mireia Solé Pi for the revision of the English. Finally, we are grateful to three anonymous reviewers whose comments helped to improve the manuscript.

Conflicts of Interest: The authors declare no conflict of interest.

References

1. Ford, T.D.; Pedley, H.M. A review of tufa and travertine deposits of the world. *Earth Sci. Rev.* **1996**, *41*, 117–175. [CrossRef]
2. Gandin, A.; Capezzuoli, E. Travertine versus calcareous tufa: Distinctive petrologic features and stable isotopes signatures. *Quat. Ital. J. Quat. Sci.* **2008**, *21*, 125–136.
3. Gandin, A.; Capezzuoli, E. Travertine: Distinctive depositional fabrics of carbonates from thermal spring systems. *Sedimentology* **2014**, *61*, 264–290. [CrossRef]
4. Capezzuoli, E.; Gandin, A.; Pedley, M. Decoding tufa and travertine (freshwater carbonates) in the sedimentary record: The state of the art. *Sedimentology* **2014**, *61*, 1–21. [CrossRef]
5. Matera, P.F.; Ventruti, G.; Zucchi, M.; Brogi, A.; Capezzuoli, E.; Liotta, D.; Yu, T.-L.; Shen, C.-C.; Huntington, K.W.; Rinyu, L.; et al. Geothermal Fluid Variation Recorded by Banded Ca-Carbonate Veins in a Fault-Related, Fissure Ridge-Type Travertine Depositional System (Iano, southern Tuscany, Italy). *Geofluids* **2021**, *2021*, 8817487. [CrossRef]
6. Pentecost, A. *Travertine*; Springer: Berlin/Heidelberg, Germany, 2005; pp. 11–240.
7. Kano, A.; Okumura, T.; Takashima, C.; Shiraishi, F. *Geomicrobiological Properties and Processes of Travertine with a Focus on Japanese Sites*; Springer Nature: Singapore, 2019. [CrossRef]
8. Luo, L.; Wen, H.; Li, Y.; You, Y.; Luo, X. Mineralogical, crystal morphological, and isotopic characteristics of smooth slope travertine deposits at Reshuitang, Tengchong, China. *Sediment. Geol.* **2019**, *381*, 29–45. [CrossRef]
9. Pentecost, A.; Viles, H.A. A review and reassessment of travertine classification. *Géogr. Phys. Quat.* **1994**, *48*, 305–314. [CrossRef]
10. Jones, B.; Peng, X. Signatures of biologically influenced $CaCO_3$ and Mg–Fe silicate precipitation in hot springs: Case study from the Ruidian geothermal area, western Yunnan Province, China. *Sedimentology* **2014**, *61*, 56–89. [CrossRef]
11. Rodríguez-Berriguete, A.; Alonso-Zarza, A.M. Controlling factors and implications for travertine and tufa deposition in a volcanic setting. *Sediment. Geol.* **2019**, *381*, 13–28. [CrossRef]
12. Brogi, A.; Alçiçek, M.C.; Liotta, D.; Capezuoli, E.; Zucchi, M.; Matera, P.F. Step-over fault zones controlling geothermal fluid-flow and travertine formation (Denizli Basin, Turkey). *Geothermics* **2021**, *89*, 101941. [CrossRef]
13. Teboult, P.-A.; Durlet, C.; Gaucher, E.C.; Virgone, A.; Girard, J.-P.; Curie, J.; Lopez, B.; Camoin, G.F. Origins of elements building travertine and tufa: New perspectives provided by isotopic and geochemical tracers. *Sediment. Geol.* **2016**, *334*, 97–114. [CrossRef]
14. Pola, M.; Gandin, A.; Tuccimei, P.; Soligio, M.; Deianas, R.; Fabbri, P.; Zampieri, D. A multidisciplinary approach to understanding carbonate deposition under tectonically controlled hydrothermal circulation: A case study from a recent travertine mound in the Euganean hydrothermal system, northern Italy. *Sedimentology* **2014**, *61*, 172–199. [CrossRef]
15. Bougeault, C.; Vennin, E.; Durlet, C.; Muller, E.; Mercuzot, M.; Chavez, M.; Gérard, E.; Ader, M.; Virgone, A.; Gaucher, E.C. Biotic-abiotic influences on modern Ca-Si-rich hydrothermal spring mounds of the Pastos Grandes volcanic caldera (Bolivia). *Minerals* **2019**, *9*, 380. [CrossRef]
16. Chafetz, H.S.; Patrick, F.R.; Utech, N.M. Microenvironmental controls on mineralogy and habit of $CaCO_3$ precipitates: An example from an active travertine system. *Sedimentology* **1991**, *38*, 107–126. [CrossRef]
17. Chafetz, H.S.; Guidry, S.A. Bacterial shrubs, crystal shrubs, and ray-crystal shrubs: Bacterial vs. abiotic precipitation. *Sediment. Geol.* **1999**, *126*, 57–74. [CrossRef]
18. Koban, C.G.; Schweigert, G. Microbial Origin of Travertine Fabrics: Two Examples from Southern Germany (Pleistocene Stuttgart Travertines and Miocene Riedöschingen Travertine). *Facies* **1993**, *29*, 251–264. [CrossRef]
19. Okumura, T.; Takashima, C.; Shiraishi, F.; Nishida, S.; Yukimura, K.; Naganuma, T.; Koike, H.; Arp, G.; Kano, A. Microbial Processes Forming Daily Lamination in an Aragonite Travertine, Nagano-yu Hot Spring, Southwest Japan. *Geomicrobiol. J.* **2011**, *28*, 135–148. [CrossRef]
20. Okumura, T.; Takashima, C.; Shiraishi, F.; Akmaluddin; Kano, A. Textural transition in an aragonite travertine formed under various flow conditions at Pancuran Pitu, Central Java, Indonesia. *Sediment. Geol.* **2012**, *265–266*, 195–209. [CrossRef]
21. Okumura, T.; Takashima, C.; Shiraishi, F.; Nishida, S.; Kano, A. Processes forming daily lamination in a microbe-rich travertine under low flow condition at the Nagano-yu hot spring, southwestern Japan. *Geomicrobiol. J.* **2013**, *30*, 910–927. [CrossRef]
22. Okumura, T.; Takashima, C.; Kano, A. Textures and processes of laminated travertines formed by unicellular cyanobacteria in Myoken hot spring, southwestern Japan. *Isl. Arc* **2013**, *22*, 410–426. [CrossRef]
23. Shiraishi, F.; Eno, Y.; Nakamura, Y.; Hanzawa, Y.; Asada, J.; Bahniuk, A.M. Relative influence of biotic and abiotic processes on travertine fabrics, Satono-yu hot spring, Japan. *Sedimentology* **2019**, *66*, 459–479. [CrossRef]
24. Takashima, C.; Kano, A. Microbial processes forming daily lamination in a stromatolitic travertine. *Sediment. Geol.* **2008**, *208*, 114–119. [CrossRef]
25. Takashima, C.; Okumura, T.; Nishida, S.; Koike, H.; Kano, A. Bacterial symbiosis forming laminated iron-rich deposits in Okuoku-hachikurou hot spring, Akita Prefecture Japan. *Isl. Arc* **2011**, *20*, 294–304. [CrossRef]
26. Tatarinov, A.V.; Yalovik, L.I.; Kashkak, E.S.; Danilova, E.V.; Khromova, E.A.; Khakhinov, V.V.; Namsaraev, B.B. Mineralogical and geochemical features of bacterial mats and travertines of the Khoito-Gol thermal spring (East Sayan). *Russ. Geol. Geophys.* **2017**, *58*, 47–58. [CrossRef]
27. Folk, R.L.; Chafetz, H.S.; Tiezzi, P.A. Bizarre forms of depositional and diagenetic calcite in hotspring travertines, central Italy. In *Carbonate Cements*; Special Publication, 36, Scheidemann, N., Harris, P., Eds.; Society of Economic Paleontologists and Mineralogists: Tulsa, OK, USA, 1985; pp. 349–369.
28. Fabré, G.; Fiche, J.L. Les concretionnements de l'Aqueduct Romain de Nîmes. *Mediterranée* **1986**, *1–2*, 129–130.

29. Riding, R. (Ed.) Classification of Microbial Carbonates. In *Calcareous Algae and Stromatolites*; Springer: Berlin/Heidelberg, Germany, 1991. [CrossRef]
30. Prol-Ledesma, R.M.; Morán-Zenteno, D.J. Heat flow and geothermal provinces in Mexico. *Geothermics* **2019**, *78*, 183–200. [CrossRef]
31. Ferrari, L.; Tagami, T.; Eguchi, M.; Orozco-Esquivel, M.T.; Petrone, C.M.; Jacobo-Albarrán, J.; López-Martínez, M. Geology, geochronology and tectonic setting of late Cenozoic volcanism along the southwestern Gulf of Mexico: The Eastern Alkaline Province revisited. *J. Volcanol. Geotherm. Res.* **2005**, *146*, 284–306. [CrossRef]
32. Maldonado Lee, J.M.; Rosales Franco, E.; Hernández Loredo, A.; Serrano Gómez, S.J. *Carta Geológico Minera Tamiahua*; F14-9, Veracruz, 1:250,000; Servicio Geológico Mexicano: Pachuca, Mexico, 2004.
33. Santos-Llorente, J.; Uribe Cruz, M.; Benítez-Juárez, M.; Zavala, R.; Olvera Rivera, A. El petróleo en Veracruz. In *Episodios Petroleros*; 50 Aniversario; Petróleos Mexicanos: Mexico City, Mexico, 1988.
34. Révész, K.M.; Landwehr, J.M.; Keybl, J. *Measurement of $\delta^{13}C$ and $\delta^{18}O$ Isotopic Ratios of $CaCO_3$ Using a Thermoquest-Finnigan GasBench II Delta Plus XL Continuous Flow Isotope Ratio Mass Spectrometer with Application to Devils Hole Core DH-11 Calcite*; U.S. Geological Survey, Open-File Report; U.S. Geological Survey: Reston, VA, USA, 2001; pp. 1–257.
35. Révész, K.M.; Landwehr, J.M. $\delta^{13}C$ and $\delta 1^{8}O$ isotopic composition of $CaCO_3$ measured by continuous flow isotope ratio mass spectrometry statistical evaluation and verification by application to Devils Hole Core DH-11 Calcite. *Rapid Commun. Mass Spectrom.* **2002**, *16*, 2102–2114. [CrossRef]
36. Coplen, T. Normalization of oxygen and hydrogen isotope data. *Chem. Geol. Isot. Geosci. Sect.* **1988**, *72*, 293–297. [CrossRef]
37. Coplen, T.B.; Brand, W.A.; Gehre, M.; Gröning, M.; Meijer-Harro, A.J.; Toman, B.; Verkouteren, R.M. New Guidelines for $\delta^{13}C$ Measurements: *Anal. Chem.* **2006**, *78*, 2439–2441. [CrossRef]
38. Grinenko, V.A. Preparation of sulfur dioxide for isotopic analysis. *Z. Neorgan. Khim.* **1962**, *7*, 2478–2483.
39. Altunel, E.; Hancock, P.L. Morphology and structural setting of Quaternary travertines at Pamukkale, Turkey. *Geol. J.* **1993**, *28*, 335–346. [CrossRef]
40. Altunel, E.; Hancock, P.L. Structural attributes of Travertine-filled Extnsional Fissures and the Pamukkale Plateau, western Turkey. *Int. Geol. Rev.* **1996**, *38*, 768–777. [CrossRef]
41. Özkul, M.; Varol, B.; Alcicek, M.C. Depositional environments and petrography of the Denizli travertines. *Bull. Miner. Res. Explor.* **2002**, *125*, 13–29.
42. Chafetz, H.S.; Folk, R.L. Travertines: Depositional morphology and the bacterially constructed constituents. *J. Sediment. Res.* **1984**, *54*, 289–316.
43. Fouke, B.W.; Farmer, J.D.; Des Marais, D.J.; Pratt, L.; Sturchio, N.C.; Burns, P.C.; Discipulo, M.K. Depositional facies and aqueous-solid geochemistry of travertine-depositing hot springs (Angel Terrace, Mammoth Hot Springs, Yellowstone National Park, U.S.A.). *J. Sediment. Res.* **2000**, *70*, 565–585. [CrossRef]
44. Guo, L.; Riding, R. Aragonite laminae in hot water travertine crusts, Rapolano Terme, Italy. *Sedimentology* **1992**, *39*, 1067–1079. [CrossRef]
45. Guo, L.; Riding, R. Hot-Spring travertine facies and sequences, Late Pleistocene, Rapolano Terme, Ilaty. *Sedimentology* **1998**, *45*, 163–180. [CrossRef]
46. Jones, B.; Renaut, R.W. Calcareous spring deposits in continental settings. In *Carbonates in Continental Settings. Facies Environments and Processes*; Alonso-Zarza, A.M., Tanner, L.H., Eds.; Elsevier: Amsterdam, The Netherlands, 2010; pp. 177–224.
47. Liu, Z.; Li, H.; You, C.; Wan, N.; Sun, H. Thickness and stable isotopic characteristics of modern seasonal climate-controlled sub-annual travertine laminas in a travertine-depositing stream at Baishuitai, SW China: Implications for paleoclimate reconstruction. *Environ. Geol.* **2006**, *51*, 257–265. [CrossRef]
48. Liu, H.; Zhou, X.; Zhang, Y.; Wang, M.; Tan, M.; Hai, K.; Yu, M.; Huo, D. Hydrochemical characteristics of travertine-depositing hot springs in western of Yunnan, China. *Quat. Int.* **2020**, *54*, 63–74. [CrossRef]
49. Pentecost, A. The formation of travertine shrubs: Mammoth Hot springs Wyoming. *Geol. Mag.* **1990**, *127*, 159–168. [CrossRef]
50. Allen, E.T.; Day, A.L. *Hot Springs of Yellowstone National Park*; Publication 466; Carnegie Institute of Washington: Washington, DC, USA, 1935; pp. 1–525.
51. Bargar, K.E. *Geology and Thermal History of Mammoth Hot Springs, Yellowstone National Park, Wyoming*; Bulletin 1444; U.S. Geological Survey: Reston, VA, USA, 1978; pp. 1–55.
52. Wasson, J.T.; Kallemeyn, G.W. Composition of chondrites. *Philos. Trans. R. Soc. Lond. Ser. A Math. Phys. Sci.* **1988**, *325*, 533–544.
53. Krouse, R.H.; Viau, C.A.; Eluik, L.S.; Ueda, A.; Halas, S. Chemical and isotopic evidence of thermochemical sulfate reduction by light hydrocarbon gases in deep carbonate reservoirs. *Nature* **1988**, *333*, 415–419. [CrossRef]
54. Durov, S.A. Natural waters and graphic representations of their composition. *Dokl. Akad. Nauk SSSR* **1948**, *59*, 87–90.
55. Kim, S.T.; O'Neil, J.R.; Hillaire-Marcel, C.; Mucci, A. Oxygen isotope fractionation between synthetic aragonite and water: Influence of temperature and Mg^{2+} concentration. *Geochim. Cosmochim. Acta* **2007**, *71*, 4704–4715. [CrossRef]
56. Horita, J.; Clayton, R.N. Comment on the studies of oxygen isotope fractionation between calcium carbonates and water at low temperatures by Zhou and Zheng (2003; 2005). *Geochim. Cosmochim. Acta* **2005**, *71*, 3131–3135. [CrossRef]
57. Giggenbach, W.F. Geothermal solute equilibria. Derivation of Na-K-Mg-Ca geoindicators. *Geochim. Cosmochim. Acta* **1988**, *52*, 2749–2765. [CrossRef]

58. Powell, T.; Cumming, W. Spreadsheets for Geothermal Water and Gas Geochemistry. In Proceedings of the 35th Workshop on Geothermal Reservoir Engineering, Stanford, CA, USA, 1–3 February 2010.
59. Arnórsson, S.; Andrésdóttir, A. Processes controlling the distribution of boron and chlorine in natural waters in Iceland. *Geochim. Cosmochim. Acta* **1995**, *59*, 4125–4146. [CrossRef]
60. Hem, J.D. *Study and Interpretation of the Chemical Characteristics of Natural Water*, 3rd ed.; Water-Supply Paper 2254; US Geological Survey: Reston, VA, USA, 1985; 263p.
61. McArthur, J.M.; Ravenscroft, P.; Safiullah, S.; Thirlwall, M.F. Arsenic in groundwater: Testing pollution mechanism for aquifers in Bangladesh. *Water Resour. Res.* **2001**, *37*, 109–117. [CrossRef]
62. Barth, S. Application of boron isotopes for tracing sources of anthropogenic contamination in groundwater. *Water Res.* **1998**, *32*, 685–690. [CrossRef]
63. Vengosh, A.; De Lange, D.J.; Starinsky, A. Boron isotope and geochemical evidence for the origin of Urania and Bannock brines at the eastern Mediterranean: Effect of water-rock interactions. *Geochim. Cosmochim. Acta* **1998**, *62*, 3221–3228. [CrossRef]
64. Ravenscroft, P.; McArthur, J.M. Mechanism pollution of groundwater by boron: The examples of Bangladesh and Michigan, USA. *Appl. Geochem.* **2004**, *19*, 1413–1430. [CrossRef]
65. Hermann, A.G.; Knate, D.; Schneider, J.; Peters, H. Geochemistry of modern seawater and brines from saltpans: Main components and bromine distribution. *Contrib. Mineral. Petrol.* **1973**, *40*, 1–24. [CrossRef]
66. Martínez-Serrano, R.G. Caractérisaton Minéralogique, Géochimique et Isotopique du Champ Géothermique de Los Humeros, Méxique. Ph.D. Thesis, Institut National Polytechnique de Lorraine, Nancy, France, 1993; p. 232.
67. Clayton, R.N.; Friedman, I.; Graf, D.L.; Mayeda, T.K.; Meents, W.F.; Shimp, N.F. The origin of saline formation Waters. *J. Geophys. Res.* **1996**, *71*, 3869–3882. [CrossRef]
68. Cortés, A.; Durazo, J.; Farvolden, R.N. Studies of isotopic hydrology of the basin of Mexico and vicinity: Annotated bibliography and interpretation: *J. Hydrol.* **1996**, *198*, 346–376. [CrossRef]
69. Sachse, D.; Sachs, J.P. Inverse relationship between D/H fractionation in cyanobacterial lipids and salinity in Christmas Island saline ponds. *Geochim. Cosmochim. Acta* **2008**, *72*, 793–806. [CrossRef]
70. Sharp, Z. *Principles of Stable Isotope Geochemistry*, 2nd ed.; The University of New Mexico: Albuquerque, NM, USA, 2017. [CrossRef]
71. Kitano, Y.; Kawasaki, N. Behaviour of strontium in the progress of calcium carbonate separation from a bicarbonate solution. *J. Earth Sci.* **1958**, *6*, 43–74.
72. Kitano, Y.; Park, K.; Hood, D.W. Pure aragonite synthesis. *J. Geophys. Res.* **1962**, *67*, 4873–4874. [CrossRef]
73. Busenberg, E.; Plummer, L.N. *A Comparative Study of the Dissolution and Crystal Growth Kinetics of Calcite and Aragonite: Ln Studies in Diagenesis*; United States Geological Survey Bulletin, 1578; Mumpton, F.A., Ed.; US Geological Survey: Reston, VA, USA, 1986; pp. 139–168.
74. Folk, R.L. Interaction between bacteria, nannobacteria, and mineral precipitation in hot springs in central Italy. *Geogr. Phys. Quat.* **1994**, *48*, 233–246. [CrossRef]
75. Renault, R.W.; Jones, B. Control son aragonite and calcite precipitation in hot spring travertines at Chemurkeu, Lake Bogoria, Kenya. *Can. J. Earth Sci.* **1997**, *34*, 818–881.
76. Hurlbut, C.S., Jr.; Klein, C. *Manual of Mineralogy*; Editorial Reverté: Barcelona, Spain, 1982; ISBN 978-84-291-4606-6.
77. Sunawana, I.; Takahashi, Y.; Imai, H. Strontium and aragonite-calcite precipitation. *J. Mineral. Petrol. Sci.* **2007**, *102*, 174–181. [CrossRef]
78. Ibrahim, K.M.; Makhlouf, I.M.; El Naqah, A.R.; Al-Thawabteh, S.M. Geochemistry and stable isotopes of travertine from Jordan Valley and Dead Sea areas. *Minerals* **2017**, *7*, 82. [CrossRef]
79. Rodríguez-Berriguete, A.; Alonso-Zarza, A.M.; Cabrera, M.C.; Rodriguez-González, A. The Azuaje travertine: An example of aragonite deposition in a recent volcanic setting, N Gran Canaria Island, Spain. *Sediment. Geol.* **2012**, *277–278*, 61–71. [CrossRef]
80. Erthal, M.; Capezzuoli, E.; Mancini, A.; Claes, H.; Soete, J.; Swennen, R. Shrub morpho-types as indicator for the water flow energy—Tivoli travertine case (Central Italy). *Sediment. Geol.* **2017**, *347*, 79–99. [CrossRef]
81. McLennan, S.M. Chapter 7: Rare earth elements in sedimentary rocks: Influence of provenance and sedimentary processes. In *Geochemistry and Mineralogy of Rare Earth Elements*; Lipin, B.R., McKay, G.A., Eds.; De Gruyter: Berlin, Germany, 2018; pp. 169–200. [CrossRef]
82. Smrzka, D.; Zwicker, J.; Bach, W.; Feng, D.; Himmler, T.; Chen, D.; Peckmann, J. The behavior of trace elements in seawater, sedimentary pore water, and their incorporation into carbonate minerals: A review. *Facies* **2019**, *65*, 41. [CrossRef]
83. Lavrushin, V.Y.; Kuleshov, V.N.; Kikvadz, O.E. Travertines of the Northern Caucasus. *Lithol. Miner. Resour.* **2006**, *41*, 137–164. [CrossRef]
84. Desouky, H.E.; Soete, J.; Claes, H.; Özkul, M.; Vanhaecke, F.; Swennen, R. Novel applications of fluid inclusions and isotope geochemistry in unravelling the genesis of fossil travertine systems. *Sedimentology* **2015**, *62*, 27–56. [CrossRef]
85. Chafetz, H.S.; Lawrence, J.R. Stable isotopic variability within modern travertines. *Geogr. Phys. Quat.* **1994**, *48*, 257–273. [CrossRef]
86. Wilson, J.L.; Ward, W.C. Chapter 4: Early Cretaceous Carbonate Platforms of Northeastern and East-Central Mexico. Cretaceous Carbonate Platforms. *AAPG Bull.* **1993**, *56*, 35–49.
87. Karimi, H.; Moore, F. The source and heating mechanism for Ahram, Mirahmad and Garu thermal springs, Zagros Mountains, Iran. *Geothermics* **2008**, *37*, 84–100. [CrossRef]

88. Kempe, A.L.W.; Thode, H.G. The mechanism of bacterial reduction of sulfate and sulfite from isotope fractionation studies. *Geochim. Cosmochim. Acta* **1968**, *32*, 71–91. [CrossRef]
89. Hill, C.A. *Geology of Carlsbad Caverns and other caves of the Guadalupe Mountains, New Mexico and Texas*; Bulletin 117; New Mexico Bureau Mines and Mineral Resources: Socorro, NM, USA, 1987; 150p.
90. Kompani-Zare, M.; Moore, F. Chemical thermometry and origin of the Dalaki mineral springs, Boshehr Province, Iran. *J. Hydrol.* **2001**, *40*, 189–204.
91. Kele, S.; Demény, A.; Siklósy, Z.; Németh, T.; Tóth, M.; Kovács, M.-B. Chemical and stable isotope composition of recent hot-water travertines and associated thermal waters, from Egerszalók, Hungary: Depositional facies and non-equilibrium fractionation. *Sediment. Geol.* **2008**, *211*, 53–57. [CrossRef]
92. Morán-Zenteno, D.; Cerca, M.; Keppie, J.D. La evolución tectónica y magmática cenozoica del suroeste de México: Avances y problemas de interpretación. *Bol. Soc. Geol. Mex.* **2005**, *57*, 319–341. [CrossRef]
93. Aguayo-Camargo, J.E.; Arellano-Gil, J.; Santillán-Piña, N. Prograding low-density turbidite systems and oil traps at the Lower Paleogene Chicontepec Foreland Basin, East-Central Mexico. *Ing. Investig. Tecnol.* **2018**, *19*, 1–12. [CrossRef]
94. Rodríguez-Berriguete, A.; Alonso-Zarza, A.M.; Martín-García, R. Diagenesis of continental carbonate country rocks underlaying surficial travertine spring deposits. *Quat. Int.* **2017**, *437*, 4–14. [CrossRef]

minerals

MDPI

Article

Response of Travertine Dam to Precipitation over the Past 800 Years in Zabuye Salt Lake, Southwestern Tibetan Plateau

Mingming Li [1], Mianping Zheng [2,*], Chuanyong Ye [2,*], Chenguang Wang [3], Xuefei Zhang [2], Xuefeng Wang [4], Yuanyi Zhao [2] and Yanbo Zhang [5]

1 College of Geoscience and Surveying Engineering, China University of Mining and Technology (Beijing), Beijing 100083, China; testlmm@126.com
2 MNR Key Laboratory of Saline Lake Resources and Environments, Institute of Mineral Resources, Chinese Academy of Geological Sciences, Beijing 100037, China; zhangxuefei2000@163.com (X.Z.); yyizhao@126.com (Y.Z.)
3 Hebei Key Laboratory of Strategic Critical Mineral Resources, Hebei GEO University, Shijiazhuang 050031, China; chenguangwangcags@163.com
4 Key Laboratory of Cenozoic Geology and Environment, Institute of Geology and Geophysics, Chinese Academy of Sciences, Beijing 100029, China; xfwang@mail.iggcas.ac.cn
5 Training Base, Army Engineering University, Xuzhou 221004, China; dtzybo@163.com
* Correspondence: zhengmp2010@126.com (M.Z.); yechuanyong@cags.ac.cn (C.Y.)

Abstract: The Tibetan Plateau is known as the core area of the third pole of the Earth and is a key area for global climate change research. This study uses the Zabuye Salt Lake travertine dam as the research object and U–Th dating as the chronological framework and proposes that the carbon and oxygen isotopes of travertine can be used as a precipitation index through the analysis of hydrogen and oxygen isotopes of spring water, and petrology, mineralogy, carbon and oxygen isotopes of travertine. The precipitation records of Zabuye Salt Lake over the last 800 years show a dry condition in 1191–1374 AD (Medieval Warm Period), a humid condition in 1374–1884 AD (Little Ice Age), and a dry condition in 1884–1982 AD (Current Warn Period), indicating a warm–dry/cold–moist climate pattern, which is consistent with precipitation records from many places on the Tibetan Plateau. We preliminarily point out that travertine can record the evolution of paleoprecipitation (paleomonsoon) at least on the decadal–centennial scale. The Indian summer monsoon has been the main factor influencing precipitation change in Zabuye Salt Lake over the past 800 years, and the change in evapotranspiration intensity caused by temperature change driven by solar radiation is also an important factor affecting dry–moist change.

Keywords: travertine; paleoprecipitation; Zabuye Salt Lake; Tibetan Plateau

Citation: Li, M.; Zheng, M.; Ye, C.; Wang, C.; Zhang, X.; Wang, X.; Zhao, Y.; Zhang, Y. Response of Travertine Dam to Precipitation over the Past 800 Years in Zabuye Salt Lake, Southwestern Tibetan Plateau. *Minerals* **2022**, *12*, 916. https://doi.org/10.3390/min12070916

Academic Editors: Francesca Giustini and Mauro Brilli

Received: 26 June 2022
Accepted: 19 July 2022
Published: 21 July 2022

Publisher's Note: MDPI stays neutral with regard to jurisdictional claims in published maps and institutional affiliations.

1. Introduction

The Tibetan Plateau is known as the core region of the Earth's third pole [1]. It is critical to the hemispheric and even the global atmospheric circulation system due to its huge area, geographic location, and high altitude [2], and is a key region of global climate change research [3]. The past 2000 years have been a key period for global climate change research, and a condition in 2000 years ago can be considered as a baseline for current conditions, as the climate has been very similar to the present [4]. The precipitation records of the Tibetan Plateau over the past 2000 years mainly come from tree rings [5–8], ice cores [9,10], and lake sediments [11–14]. Trees are mainly distributed in the eastern, northeastern, and southeastern parts of the plateau where the climate is suitable, and ice is mainly distributed at high altitudes where snowfall can be continuously preserved. Chronological control for nearly all published paleolimnological records from the Tibetan Plateau has been based on radiocarbon dating [15]; however, the radiocarbon age of lake sediments may be subject to reservoir effects due to the input of dead carbon from local bedrock or wetlands within the catchment [16]. The reservoir effect has made it challenging to establish reliable chronologies for lake sediment cores from the Tibetan Plateau [15]. Little attention has been paid to high-resolution travertine sequences from the past 2000 years; however, travertine is a relatively common sedimentary phenomenon on the Tibetan Plateau and has great potential for environmental archives [17].

Travertine is a non-marine calcium carbonate deposited around springs, rivers, lakes, or caves, mainly composed of calcite and aragonite, and widely distributed in terrestrial environments [18–21]. Pentecost et al. divided it into meteogene and thermogene travertine according to the source of CO_2 in the environmental water. The former is also called tufa, and its carbon originates from soil CO_2 and carbonate rock, with $\delta^{13}C$ mostly ranging from −12‰ to −2‰. The carbon of the latter comes from various sources, including hydrolysis and oxidation of reduced carbon and decarbonation of limestone or directly from the upper mantle, with $\delta^{13}C$ usually ranging from −2‰ to 10‰ [20,22]. In Ford and Pedley's classification, tufa corresponds to meteogene travertine and travertine to thermogene travertine [19], and in this paper we use this term for discussion. For a long time, tufa was mainly used for paleoclimate reconstruction [23], and its resolution can reach years, seasons, months, and maybe even weeks [24–26]. Recently, more researchers have emphasized the close relationship between travertine distribution and climate [17,27–30]. Ricketts et al. compiled a global dataset containing 1649 published ages of travertine, which showed that although the deposition of travertine was spatially controlled by crustal faults and fractures, it was temporally regulated by global or regional climate change [31].

In the hinterland of the Tibetan Plateau, previous climate-related research was carried out on ancient travertine, which had ceased to grow and lacked the direct connection between modern spring (lake) water and its deposition [17,32–36]. Wang et al., based on a systematic study and a summary of their research results, proposed that the widely distributed travertine on the Tibetan Plateau could provide a record of paleoclimate (paleomonsoon) evolution at least over the decadal–centennial time scale [37]. Travertine dams are common in caves, springs, and rivers all over the world, ranging in size from millimeters to several meters [38,39].

This paper studied travertine based on U–Th dating, petrology, mineralogy, and carbon and oxygen isotopes as well as hydrogen and oxygen isotopes of spring (lake) water, taking the growing Zabuye Salt Lake travertine dam as the research object. We discussed the significance of travertine to the paleoenvironment in order to provide a basis for better use of travertine in reconstructing the climate of the Tibetan Plateau.

2. Geological Setting

Zabuye Salt Lake (31°14′47″ N–31°33′10″ N, 83°52′34″ E–84°23′47″ E) is located in the southwest of the Tibetan Plateau, at the intersection of the westerlies and the Indian summer monsoon (Figure 1), and is very sensitive to climate change. According to the meteorological data of the Long-Term Observation Station from 1991 to 2020, the annual average temperature in the Zabuye Salt Lake area is 3.1 °C, the annual average precipitation is 168.7 mm, and the annual average evaporation is 2579.1 mm. Precipitation is concentrated in the rainy season from early July to mid-September, which accounts for more than 90% of the total annual precipitation [40], indicating that summer monsoon rainfall dominates annual precipitation in the Zabuye Salt Lake area.

The Carboniferous and Permian strata are mainly distributed in the northern part of the Zabuye Salt Lake area, which consist of clastic and carbonate rocks. The Cretaceous is exposed in the south and southwest and the Paleogene in the east, and both are composed of clastic and volcanic rocks. The Neogene is distributed in the west and is a set of pyroclastic rocks. The Quaternary is mainly distributed around the lake, including residual slope, fluvial, lacustrine, and travertine deposits. The intrusive rocks in the salt lake area are mainly intermediate-acid rocks, ranging from diorite to granite (Figure 1c). The main structures are NW, NE, and nearly NS trending faults [42].

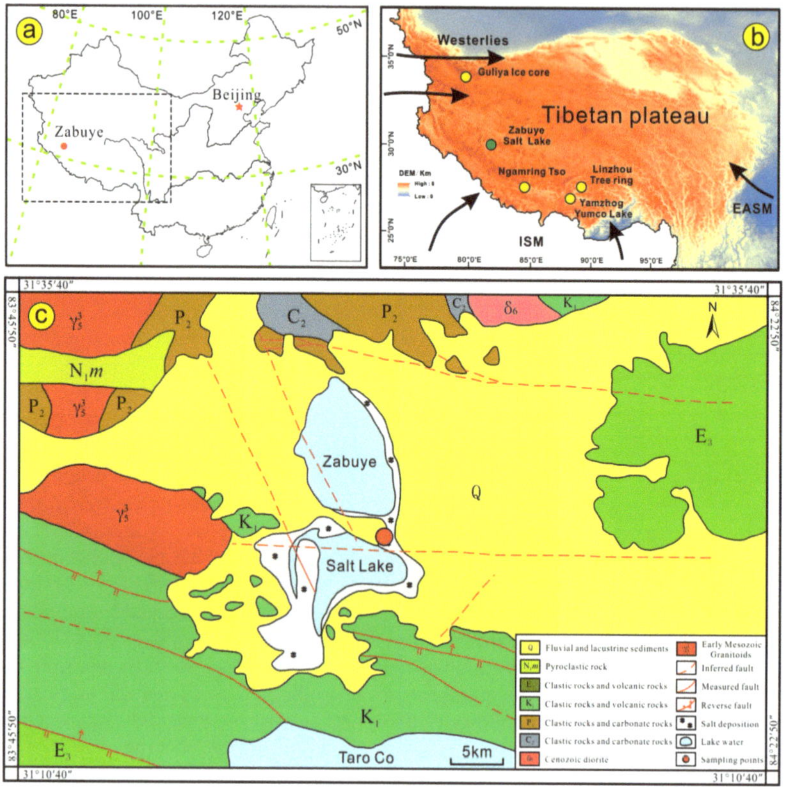

Figure 1. Maps of study site, Zabuye Salt Lake. (**a**,**b**) Zabuye Salt Lake and some related research sites on Tibetan Plateau: Guliya ice core [41], Ngamring Tso [13], Yamzhog Yumco Lake [11], and Linzhou tree rings [7]. EASM stand for the "East Asian summer monsoon", ISM stand for the "Indian summer monsoon", and DEM stand for the "Digital Elevation Model". Black arrows indicate climate systems. (**c**) Simplified geological map of lake area.

Zabuye Salt Lake covers an area of 243 km^2 at an average elevation of about 4421 m. It is a semi-dry salt lake with a combination of surface brine and salt flats. Large-scale ancient travertine accumulation developed in the middle of the lake and formed a travertine island (Figure 1c). From the travertine island as the starting point to the west, a sand embankment was formed that divides Zabuye Lake into south and north lakes, and there is a waterway connecting the two on the east side. The springs around and in the lake are relatively well developed, and those on the travertine island have the largest water inflow [43].

3. Materials and Methods

3.1. Research Materials

There are many springs on Zabuye travertine island. The spring water flows into the salt lake and forms a travertine dam at the junction of the lake water. In this study, samples of spring water, lake water, and travertine were collected from a travertine dam with good topographic conditions (Figure 2). A 23.5 cm section was carved from the travertine dam, and a horizontal travertine bedding was developed (Figure 2e). A total of 24 samples (ZD01-24) were taken from top to bottom. Lake water samples (ZH01) and spring water samples (ZQ01) were taken from both sides of the travertine dam (Figure 2d), and one water sample (ZQ02) was taken from the spring hole. Travertine samples were transported back to the lab and dried in a drying oven at 50 °C.

Figure 2. Outcrops of Zabuye Salt Lake. (**a**) Travertine dam formed by spring. (**b**) Spring outlet. (**c–e**) Close-up of travertine sampling sites, indicated by red circles. Black squares indicate the positions of (**b**,**c**,**e**).

3.2. Analytical Methods

3.2.1. U–Th Dating

Four travertine samples were collected for U–Th dating. Pure and compact calcite was selected as a test sample, and the impurities were washed with alcohol and hydrogen peroxide in an ultrasonic cleaning machine. The procedures for chemical separation and purification of uranium and thorium were similar to those in previous studies [44,45]. The $^{230}Th/^{232}Th$ atomic ratio of Zabuye Salt Lake water as the initial $^{230}Th/^{232}Th$ atomic ratio of ($6.7 \pm 0.67 \times 10^6$) was used to correct the initial ^{230}Th amount. These samples were analyzed on a Thermo Fisher Neptune Plus multi-collector inductively coupled plasma mass spectrometer (MC-ICP-MS) at the Institute of Geology and Geophysics, Chinese Academy of Sciences, Beijing, China.

3.2.2. Petrography and Mineralogy

Petrographic and mineralogic observations of the travertine were conducted using a Leica DM4500P polarizing microscope on polished thin sections, and an FEI Nova NanoSEM 450 scanning electron microscope (with working condition of 20 kV and beam current of 15 μA) on carbon-coated samples. Thin sections were prepared and photographed at the MNR Key Laboratory of Saline Lake Resources and Environments, Beijing, China. The mineralogical composition analysis of travertine was performed by using a Bruker D2-PHASER X-ray diffractometer (Cu Kα, 30 kV, 10 mA, 7°–90° 2θ, 0.02° 2θ step size, 6°/min) at the Sichuan University of Science and Engineering, Zigong, China. Quantified mineral results were analyzed with MDI Jade 6.5 software (Materials Data, Inc., Livermore, CA, USA). The $MgCO_3$ content of the carbonate minerals was calculated from the shift of d-spacing of the (104) reflection peak of calcite from their stoichiometric peak positions in the diffraction spectra [46,47]. Calcite with <5 mol% $MgCO_3$ is classified as low-magnesium calcite (LMC), and calcite with >5 mol% $MgCO_3$ is considered high-magnesium calcite (HMC). These are often denoted simply as calcite and magnesium calcite [48].

3.2.3. Stable Isotopes

Travertine carbon and oxygen isotopes were analyzed by using a Finnigan MAT 253 mass spectrometer. Carbon and oxygen isotopes were analyzed by the 100% phosphoric acid method, and analytical precision was ±0.1‰ for $\delta^{13}C$ and ±0.2‰ for $\delta^{18}O$. The measurement results of carbon and oxygen isotopes were scaled by the V-PDB standard. Spring (lake) hydrogen and oxygen isotopes was measured by isotope ratio mass spectrometry. For the measurement of the $^{18}O/^{16}O$ ratio, the CO_2 equilibration method was employed; for the D/H ratio, H_2 was generated by the Zn-reduction method. Isotope ratios of CO_2 and H_2 were measured using a MAT-253 mass spectrometer, and the results are reported relative to V-SMOW with a standard deviation of ±0.5‰ and ±0.1‰. This work was done at the Analytical Laboratory of the Beijing Research Institute of Uranium Geology, Beijing, China.

4. Results

4.1. Chronology

The U–Th dating results are shown in Table 1 and Figure 3. The $^{230}Th/^{232}Th$ activity ratio of all samples is lower than 20, indicating that they are contaminated by ^{230}Th debris, so the age error has great uncertainty [49], and debris correction must be carried out [50]. The $^{230}Th/^{232}Th$ in Zabuye Lake water can represent the average value of detrital $^{230}Th/^{232}Th$ of the terrigenous detrital material transported into the lake from the periphery of the catchment basin. In this study, $^{230}Th/^{232}Th$ in Zabuye Salt Lake ($6.7 \pm 0.67 \times 10^6$) is used as the initial value of travertine for correction. ZD02 did not get effective age correction. Although the errors of ZD10, ZD17, and ZD24 age values are large, they have a good age sequence of lower old and higher new. By fitting the ages of ZD10, ZD17, and ZD24 samples to make the time-depth trend line of the travertine profile (Figure 3), the following formula can be obtained: Age = 3.6305 × D + 2.678 (R^2 = 0.9766), where D is depth (mm) and R^2 is the square of the correlation coefficient. In this study, samples were collected in the field in 2020, and the top (0 mm) of the travertine profile was deposited, with a theoretical age of 0 years (−20 yr BP). According to the trend formula, the age of the top (0 mm) of the travertine profile was calculated to be 2.678 yr BP, with a difference of only about 23 yr. This indicates that U–Th dating can represent the age of travertine to a certain extent.

Table 1. ^{230}Th dating results for profiles in study area (±2σ).

Sample	^{238}U	^{232}Th	$^{230}Th/^{232}Th$	$d^{234}U$	$^{230}Th/^{238}U$	^{230}Th Age (yr)	^{230}Th Age (yr)	$d^{234}U_{Initial}$	^{230}Th Age (yr BP)
	(ppb)	(ppt)	(Atomic × 10^{-6})	(Measured)	(Activity)	(Uncorrected)	(Corrected)	(Corrected)	(Corrected)
ZD10	4518 ± 16	379,893 ± 7712	7.6 ± 0.2	748.8 ± 4.1	0.0389 ± 0.0003	2452 ± 17	304 ± 308	749 ± 4	283 ± 308
ZD17	3660 ± 11	469,369 ± 9489	8.1 ± 0.2	751.9 ± 3.8	0.0628 ± 0.0004	3972 ± 25	689 ± 470	753 ± 4	668 ± 470
ZD24	3759 ± 13	669,952 ± 13,565	7.9 ± 0.2	746.7 ± 3.9	0.0851 ± 0.0005	5430 ± 36	830 ± 660	748 ± 4	809 ± 660

$d^{234}U = ([^{234}U/^{238}U]_{activity} - 1) \times 1000$. $d^{234}U_{initial}$ was calculated based on ^{230}Th age (T), i.e., $d^{234}U_{initial} = d^{234}U_{measured} \times e^{\lambda 234 \times T}$. Corrected ^{230}Th ages assume the initial $^{230}Th/^{232}Th$ atomic ratio of $6.7 \pm 0.67 \times 10^{-6}$. Those are the values for the modern lake water. The errors are arbitrarily assumed to be 10%. B.P. stands for "Before Present", where the "Present" is defined as the year 2000 AD.

The trend line formula was used to calculate the ages at 0 mm of travertine section. Combined with the measured ages of ZD10, ZD17, and ZD24, the deposition rate of each section was calculated (2.4–4.4 mm/yr). The ages of other samples were obtained by linear interpolation (Figure 3).

4.2. Petrography and Mineralogy

Among the travertine lithotypes proposed by Guo and Riding [51], the crystalline crust was recognized in Zabuye Salt Lake travertine. The travertine was formed from abiotic feather dendrite, radiating dendrite, micrite, and intraclast (Figure 4). The inner clasts are composed of travertine clasts, quartz, and feldspar grains. Travertine stratification is good, with overall density and few voids. The interaction between micrite and microsparry is more reflective of the difference in water environment. The formation of sparry calcite was under hydrodynamic conditions of high flow velocity, and the formation of micrite calcite was under hydrodynamic condition of low flow velocity.

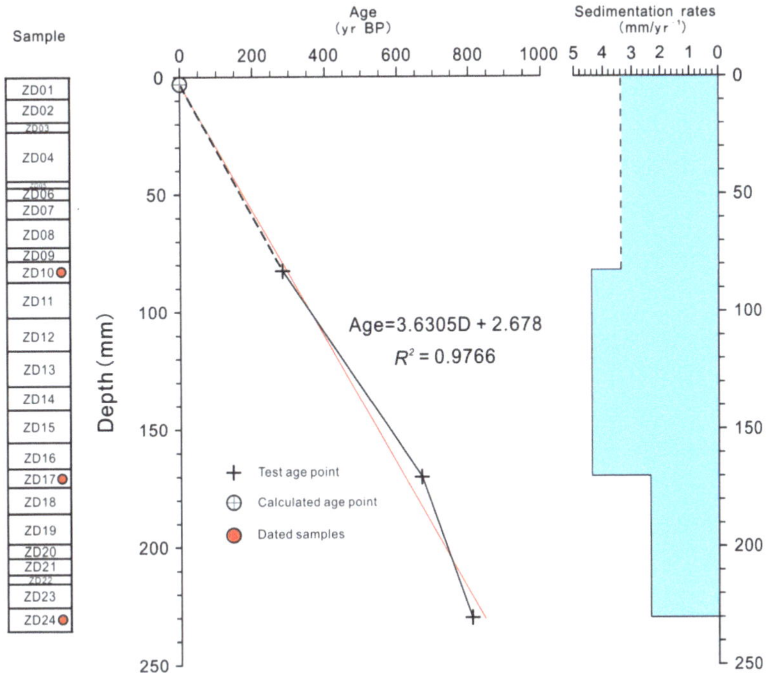

Figure 3. Age-depth profile based on U–Th dating from Zabuye Salt Lake travertine dam. The doted lines represent the value calculated using the age formula.

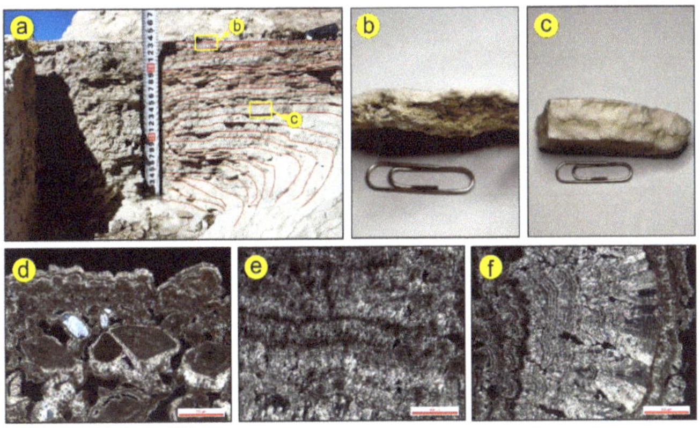

Figure 4. Textures of travertine of Zabuye Salt Lake. (**a**) Travertine profile with good stratification. (**b**) Sample at top of profile (ZD01) with higher porosity. (**c**) Dense middle section sample (ZD12). (**d**) Clastic nuclei are composed of travertine clasts, quartz, and feldspar grains. (**e**) Generation growth relationship exists between base of microsparry and micrite. (**f**) Crystalline dendrite textures showing wavy and banded internal zonations. The length of the red line segment is 500 μm.

XRD analysis of 24 samples shows that the mineral composition of the travertine profile is simple, mainly calcite, and some samples contain a small amount of quartz and feldspar. The calcite content ranges from 95.2% to 100%, with an average of 98.3%. Quartz content varies from 0.5% to 3.8%, with an average of 1.6%. Feldspar content ranges from 0.5% to 1.6%, with an average of 1.2%; feldspar content is very low, and specific feldspar species can no longer be distinguished. Quartz and feldspar are detrital mineral grains of rocks around the basin carried mainly by wind and rain. The presence of small amounts of authigenic quartz and feldspar also cannot be ruled out. The range of $MgCO_3$ content in calcite (mol%) in the travertine dam is 0.5 to 3.6%, with an average of 1.8%, all of which is LMC (Figure 5).

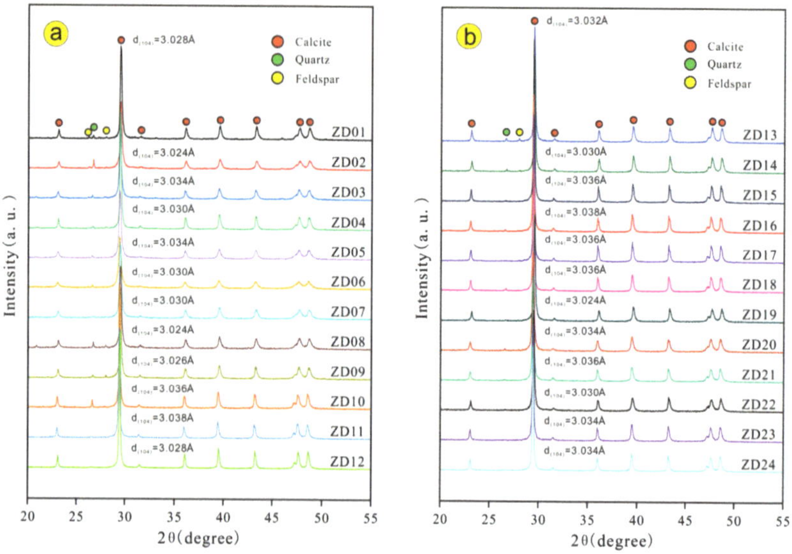

Figure 5. X-ray diffraction patterns of travertine samples. (**a**) X-ray diffraction patterns of samples ZD01-ZD12. (**b**) X-ray diffraction patterns of samples ZD13-ZD24.

Since the travertine dam is basically composed of calcite, the presence of aragonite and other calcium carbonate minerals was not observed, and the influence of the change of carbonate mineral facies on the change of carbon and oxygen isotopes of travertine can be excluded.

4.3. Hydrogen and Oxygen Isotopes of the Water Samples

Two spring samples (ZQ01-02) and one lake water sample (ZH01) were used for hydrogen and oxygen isotope analysis. The $\delta D_{V\text{-}SMOW}$ values ranged from 12.5‰ to 15.9‰, $\delta^{18}O_{V\text{-}SMOW}$ from −118.9‰ to −136.6‰ (Table 2). Craig first found a linear correlation between δD and $\delta^{18}O$ in atmospheric precipitation: $\delta D = 8\delta^{18}O + 10$ [52,53], which is also called the global meteoric water line (GMWL) on the graph of the relationship between δD and $\delta^{18}O$. The δD and $\delta^{18}O$ test data of the three water samples collected were combined with the δD and $\delta^{18}O$ data of the two spring waters samples of the travertine island [43] to draw the δD-$\delta^{18}O$ diagram (Figure 6). It can be seen that for the GMWL, the data of the sampling point have a slight ^{18}O drift, but it is not far from the GMWL, indicating that the hot groundwater comes from atmospheric precipitation but has a certain ^{18}O exchange with rocks and minerals during the deep cycle.

Table 2. Hydrogen and oxygen isotope data of Zabuye water samples.

Sample	$\delta D_{V\text{-}SMOW}$ (‰)	$\delta^{18}O_{V\text{-}SMOW}$ (‰)
ZQ-01	−132.9	−14.5
ZQ-02	−136.6	−15.9
ZH-01	−118.9	−12.8

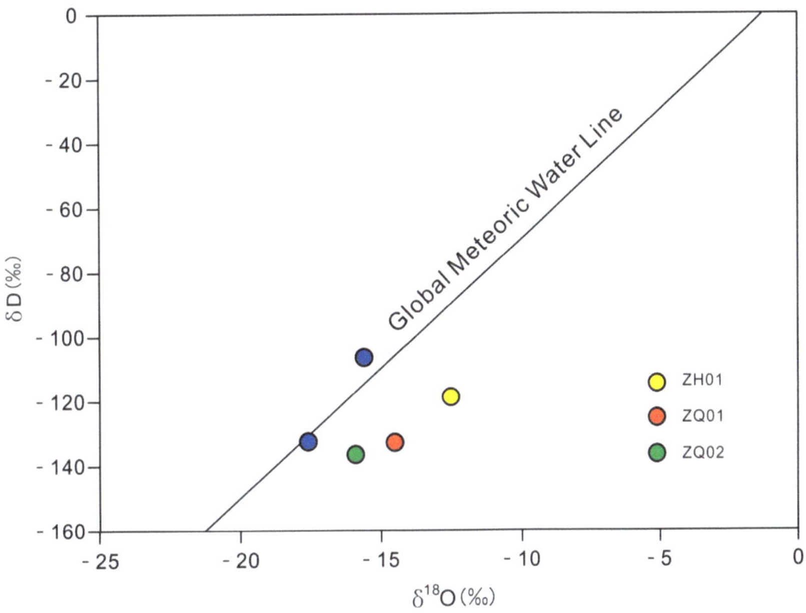

Figure 6. Plot of δD-δ^{18}O of Zabuye spring samples. Blue dot indicates data from [43].

4.4. Carbon and Oxygen Isotopes

Twenty-four travertine samples were used for carbon and oxygen isotope testing. $\delta^{13}C_{V\text{-}PDB}$ values ranged from 0.5‰ to 2.6‰, with an average of 1.0‰, and $\delta^{18}O$ values ranged from −12.4‰ to −6.4‰, with an average value of −10.3‰. There is a good correlation between $\delta^{18}O$ and $\delta^{13}C$ (r = 0.92). All carbon isotope data of travertine are within the range of thermogenic travertine.

The $\delta^{13}C$ value of travertine can be used to calculate the carbon isotopic composition of the parent CO_2 gas, using the empirical equation: $\delta^{13}C_{CO2} = 1.2 \times \delta^{13}C_{travertine} - 10.5$ [54]. Applying this equation to the carbon isotope data of the travertine dam, the obtained $\delta^{13}C_{CO2}$ values ranged from −9.9‰ to −7.4‰, with an average of −9.3‰.

There are three major CO_2 sources for travertine: soil, magma, and limestone decarbonation [39]. The $\delta^{13}C$ of soil CO_2 is controlled by the predominant vegetation type in the region. Globally, the $\delta^{13}C$ of C3 plants varies from −37‰ to −20‰, with an average of −28.7‰ [55]; the $\delta^{13}C$ of C4 plants varies from −15‰ to −9‰, with an average of −13‰ [56]. The $\delta^{13}C$ of CO_2 from the mantle generally ranges from −8‰ to −3‰ [57] or from the $\delta^{13}C$ of magmatic CO_2 generally ranges from −7‰ to −5‰ [58]. The $\delta^{13}C$ of CO_2 from typical limestone metamorphic sources is −1‰ to 2‰ [59]. Comparing the carbon and oxygen isotopes to the data [60] indicates that the CO_2 originated from carbonates or igneous rocks (Figure 7).

Based on the above analysis and taking into account that C4 plants are scarce in high-elevation regions [61], according to the current data, it is suggested that the parent CO_2 of the travertine dam mainly originated from thermal decarbonation of carbonates and intermediate–basic volcanic rocks around the basin, partly from magmatic mantle degassing and soil CO_2. The fault system developed in the Zabuye Salt Lake area, and soil CO_2 in the basin was transferred into the groundwater cycle through atmospheric precipitation and mixed with CO_2 from deep underground sources. Since the $\delta^{13}C$ of soil CO_2 is obviously negative, even a small change in soil CO_2 will cause obvious changes to $\delta^{13}C$ in travertine.

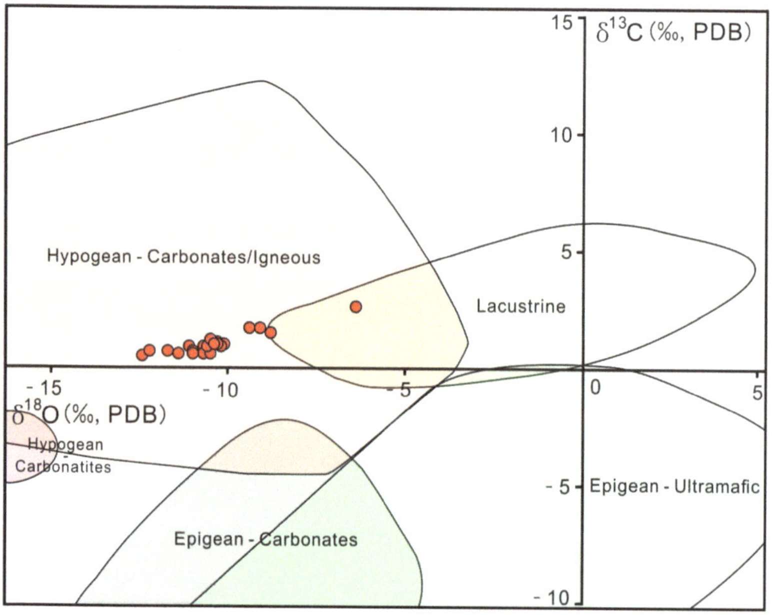

Figure 7. Results of stable isotopes overlain on clusters. All dots plotted within hypogean travertine area of Teboul et al. [60], with CO_2 derived from carbonates or an igneous source.

5. Discussion

5.1. Paleoclimatic Implications

The premise of travertine deposition is flowing water rich in dissolved CO_2 [62]. When high pCO_2 groundwater discharges from the spring, due to the low atmospheric pCO_2, CO_2 will be degassed rapidly, resulting in oversaturation of calcite, then calcite precipitation [63]. Travertine deposits can retain geochemical characteristics inherited from the parent fluid, so it is considered to be a favorable object for reconstructing paleoclimate and paleofluid characteristics [64].

Travertine is a comprehensive product of interactions between underground materials (mantle-derived carbon dioxide, magmatic water, surrounding rocks involved in groundwater circulation), atmospheric precipitation, and surface materials (soil carbon dioxide, major and trace elements). The information carried by travertine can reflect surface information when the subsurface material remains basically unchanged. That is, when the magmatic and tectonic activities in a certain area are stable, the area maintains relatively stable hydrothermal activity. The travertine formed against this background is relatively stable under the influence of mantle-derived CO_2 and CO_2 and magmatic water formed by decarburization of surrounding carbonate rocks. The carbon and oxygen isotope change in travertine is less affected by the above, but more controlled by the change in soil CO_2 supply and atmospheric precipitation, which are directly affected by climatic factors. The amount of groundwater infiltration by atmospheric precipitation is directly related to the rainfall in the region, the soil CO_2 is related to the type and density of vegetation in the region, and the climatic conditions determine the vegetation.

Studies based on the hydrogen and oxygen isotopes of geothermal spring water indicate that geothermal springs are mainly supplied by meteoric precipitation and water from melted glaciers/ice, and most geothermal spring water circulates rapidly deep underground with a cycle time of only 20–40 years [65–67]. The Zabuye Salt Lake area is structurally stable, and no earthquakes of magnitude 5 or greater have been recorded [68]. The average age of travertine samples from the edge stone dam of the lake is about 33 yr.

From the perspective of chronology, travertine can reflect the geological information carried by atmospheric precipitation in the water cycle process in this region.

The $\delta^{13}C$ and $\delta^{18}O$ of travertine show an obvious positive correlation (r = 0.922), indicating that the two are affected by the same or similar factors. In a continuous high-resolution (monthly) study of stable isotopes of $\delta^{13}C$ and $\delta^{18}O$ in travertine in Baishuitai, there is a strong correlation between carbon and oxygen isotopes (r = 0.75); Liu et al. believe that rainwater is the main factor causing the seasonal variation of $\delta^{13}C$ and $\delta^{18}O$ [69]. This indicates that $\delta^{13}C$ and $\delta^{18}O$ of the Zabuye Salt Lake travertine dam can reflect changes in the surface climate environmental system.

Under the condition of relatively stable tectonic movement, CO_2 from deep sources and the hydro-rock interaction of hot springs are relatively fixed. In arid climates, it is difficult to form a suitable plant cover, while in humid climates, $\delta^{13}C$ in travertine shifts to negative values when biomass develops more [70]. The $\delta^{18}O_{H2O}$ value of hot springs depends on the $\delta^{18}O$ value of atmospheric precipitation. Because the temperature of Zabuye spring is stable and less affected by air temperature [43] and the travertine dam is located only a few meters from the spring hole (Figure 2), the effect of water temperature change on its $\delta^{18}O$ during the deposition of travertine is negligible. Therefore, the variation of $\delta^{18}O$ of Zabuye Salt Lake travertine dam can reflect the variation of $\delta^{18}O$ of atmospheric precipitation.

In the transition region between the westerlies and Indian summer monsoon (30°N–35°N) over the Tibetan Plateau, the factors that influence atmospheric precipitation isotope change are complex and are sensitive to the water vapor source and transport process [71]. In this region, the $\delta^{18}O$ in precipitation formed by water vapor from Indian summer monsoon entering the southern plateau is lower, and the stronger monsoon activity lowers the $\delta^{18}O$ in precipitation. The $\delta^{18}O$ values of water vapor from the northern Tibetan Plateau and precipitation formed by local evaporation water vapor are higher [71–73].

Based on the above discussion, in this paper we believe that the $\delta^{13}C$ of travertine dam can be used as an index of precipitation change: lighter (heavier) $\delta^{13}C$ indicates increased (decreased) precipitation; $\delta^{18}O$ mainly indicates variations of monsoon intensity and water vapor source, and lighter (heavier) $\delta^{18}O$ indicates increased (decreased) Indian summer monsoon precipitation.

5.2. Precipitation Changes at Zabuye Salt Lake over the Past 800 Years

With the exception of a warm period in the 20th century, which occurred simultaneously all over the globe, the changes in warm and cold phases around the world before the Industrial Revolution were not synchronized. Based on previous research results [74–77], in this paper we roughly divide the temperature stages of the Tibetan Plateau over the past thousand years into the Medieval Warm Period (MWP, 800–1400 AD), Little Ice Age (LIA, 1400–1900 AD), and Current Warm Period (CWP, 1900–2000 AD).

Based on the changes in $\delta^{13}C$ and $\delta^{18}O$ of travertine in this study, we reconstructed the precipitation records of Zabuye Salt Lake over the past 800 years (1191–1982 AD). The results show a dry condition in 1191-1374 AD (MWP), a humid condition in 1374-1884 AD (LIA), and a dry condition in 1884-1982 AD (CWP), indicating a warm–dry/cold–moist climate pattern (Figure 8). Zhang et al. [74] studied climate change over the past 300 years at Taro Co, 10 km south of Zabuye Salt Lake (Figure 1c), and showed that the climate was humid during 1750–1860 AD and was dry from 1860 AD to the present [78], which is consistent with our reconstruction results.

We compared the precipitation records of Zabuye with other typical precipitation records of the Tibetan Plateau (Figures 1b and 8), including the accumulated records of the Guliya glacier in the northwestern part of the plateau [41], annual mean precipitation based on quantitative pollen reconstruction at Yamzhog Yumco Lake in the southern part of the plateau (MAP) [11], precipitation records reconstructed from tree rings in Linzhou [7], and June to September (JJAS) precipitation based on grain size reconstruction in Ngamring Tso [13]. These records are very similar to the precipitation records of Zabuye Salt Lake, and are also characterized by a warm–dry/cold–moist pattern. In addition, Lugu Lake [79] and

Erhai Lake [80], which are subjected to seasonal risk control in the southeast of the Tibetan Plateau, both showed warm–dry/cold–moist climate characteristics in the MWP/LIA. This climatic pattern appeared not only in areas controlled by the westerlies, but also in wider areas controlled by Indian summer monsoon. In southern Oman, the stalagmite $\delta^{18}O$ record from Qunf Cave shows higher values during the MWP than the LIA, indicating a weakening of Indian summer monsoon intensity during the MWP [81]. Over the past 100–200 years, the Indian summer monsoon has gradually decreased in intensity, and the climate has become drier under warmer conditions [82–84]. In the entire monsoon region, the climate conditions during the MWP and the past 100–200 years were significantly dry and during the LIA were relatively humid, and this pattern was prevalent [79].

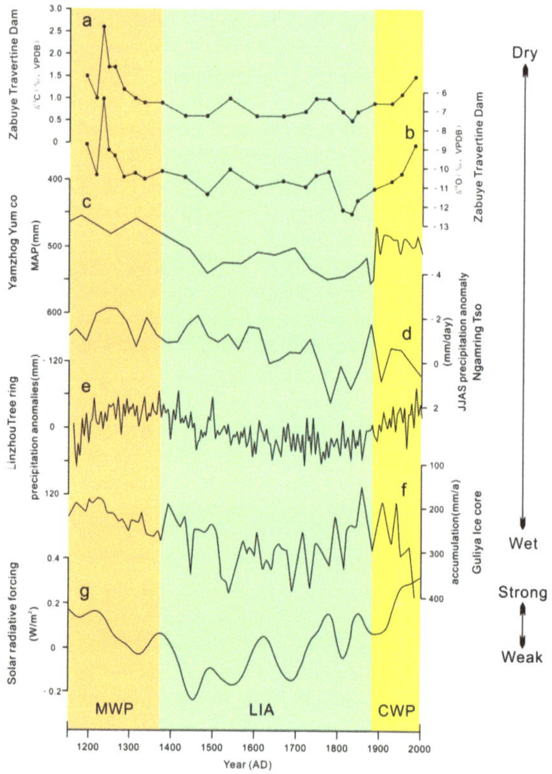

Figure 8. Comparison of climatic records from Zabuye Salt Lake travertine dam with other climatic records. (**a**) Carbon isotopes of carbonate of travertine dam. (**b**) Oxygen isotopes of carbonate. (**c**) Mean annual precipitation (MAP) at Yamzhog Yumco Lake [11]. (**d**) June to September (JJAS) precipitation at Ngamring Tso [13]. (**e**) Tree ring precipitation anomalies at Linzhou [7]. (**f**) Ice core accumulation at Guliya [41]. (**g**) Solar radiative forcing [85]. MWP, LIA, and CWP refer to Medieval Warm Period, Little Ice Age, and Current Warm Period, respectively.

Monsoons and westerlies interact with each other over the Tibetan Plateau at glacial–interglacial, millennial, decadal, and seasonal scales, bringing water vapor to different areas of the plateau [86]. Although the climate characteristics of warm–dry, cold–wet in the Zabuye Salt Lake area are similar to those recorded in the Guliya ice core, the oxygen isotopes of travertine become lighter in humid periods (LIA) and heavier in dry periods (MWP and CWP) (Figure 8a,b), indicating that climate change in the Zabuye Salt Lake area in the past 800 years has mainly been controlled by the influence of the Indian summer monsoon. In addition, the temperature change driven by solar radiation will lead to a change in evaporation intensity, and then affect the dry and wet climate change of the Tibetan Plateau [11]. The precipitation records for the Zabuye Salt Lake area are consistent with the solar radiation (Figure 8g). This study argues that during the MWP, the Zabuye Salt Lake region had higher temperatures, strong evaporation, and less precipitation, resulting in an arid climate during this period; during the LIA, the temperature was low, evaporation was inhibited, and there was more precipitation, so the climate was humid.

6. Conclusions

Based on U–Th dating of a travertine dam, this study establishes the chronology of Zabuye Salt Lake travertine over the past 800 years, preliminarily discusses the carbon and oxygen isotopes of travertine as a precipitation index, and points out that travertine might record the evolution of paleoprecipitation (paleomonsoon) on at least a decadal–centennial scale. The precipitation records of Zabuye Salt Lake over the past 800 years show a dry condition in 1191–1374 AD (MWP), a humid condition in 1374–1884 AD (LIA), and a dry condition in 1884–1982 AD (CWP), indicating a warm–dry/cold–moist climate pattern. The Indian summer monsoon has been the main factor influencing precipitation change at Zabuye Salt Lake over the past 800 years, and the change in evapotranspiration intensity caused by temperature change driven by solar radiation is also an important factor affecting the dry–moist change. It should be pointed out that our precipitation reconstruction index is relatively single and lacks evidence from other proxy indices. Whether the westerlies have an influence and to what extent still needs further study.

Author Contributions: Conceptualization, M.L. and M.Z.; methodology, M.L., M.Z. and C.Y.; software, M.L. and X.W.; validation, M.L. and M.Z.; formal analysis, M.L., X.W. and Y.Z. (Yuanyi Zhao); investigation, M.L., C.Y., X.Z. and Y.Z. (Yuanyi Zhao); resources, M.Z.; data curation, M.Z.; writing—original draft preparation, M.L. and M.Z.; writing—review and editing, M.L., M.Z., C.Y., C.W., X.Z. and Y.Z. (Yanbo Zhang); visualization, M.L. and C.Y.; supervision, M.Z.; project administration, M.Z.; funding acquisition, M.Z. All authors have read and agreed to the published version of the manuscript.

Funding: This work was supported by the National Natural Science Foundation of China (No: 91962219, 41673023).

Data Availability Statement: Not applicable.

Acknowledgments: We thank the anonymous reviewers whose comments have improved the quality of the manuscript.

Conflicts of Interest: The authors declare no conflict of interest.

References

1. Qiu, J. China: The third pole. *Nature* **2008**, *454*, 393–396. [PubMed]
2. Yao, T.; Thompson, L.G.; Mosbrugger, V.; Zhang, F.; Ma, Y.; Luo, T.; Xu, B.; Yang, X.; Joswiak, D.R.; Wang, W.; et al. Third Pole Environment (TPE). *Environ. Dev.* **2012**, *3*, 52–64.
3. Liu, X.; Zheng, H.; Zhang, M.; Liu, C. Identification of dominant climate factor for pan evaporation trend in the Tibetan Plateau. *J. Geogr. Sci.* **2011**, *21*, 594–608.
4. Flantua, S.; Henry, H.; Vuille, M.; Carson, J.; Gosling, W.; Hoyos, I.; Ledru, M.-P.; Montoya, E.; Mayle, F.; Maldonado, A.; et al. Climate variability and human impact in South America during the last 2000 years: Synthesis and perspectives from pollen records. *Clim. Past* **2016**, *12*, 483–523.
5. Shao, X.; Xu, Y.; Yin, Z.Y.; Liang, E.; Zhu, H.; Wang, S. Climatic implications of a 3585-year tree-ring width chronology from the northeastern Qinghai-Tibetan Plateau. *Quat. Sci. Rev.* **2010**, *29*, 2111–2122. [CrossRef]
6. Yang, B.; Qin, C.; Wang, J.; He, M.; Melvin, T.M.; Osborn, T.J.; Briffa, K.R. A 3500-year tree-ring record of annual precipitation on the northeastern Tibetan Plateau. *Proc. Natl. Acad. Sci. USA* **2014**, *111*, 2903–2908.
7. He, M.; Yang, B.; Wang, J.; Wang, Z. Tree-ring-derived millennial precipitation record for the southern Tibetan Plateau and its possible driving mechanism. *Holocene* **2013**, *23*, 36–45.
8. Grießinger, J.; Bräuning, A.; Helle, G.; Thomas, A.; Schleser, G. Late Holocene Asian summer monsoon variability reflected by δ18O in tree-rings from Tibetan junipers. *Geophys. Res. Lett.* **2011**, *38*, L03701. [CrossRef]
9. Yao, T.; Duan, K.; Xu, B.; Wang, N.; Guo, X.; Yang, X. Precipitation record since AD 1600 from ice cores on the central Tibetan Plateau. *Clim. Past* **2008**, *4*, 175–180.

10. Pang, H.; Hou, S.; Zhang, W.; Wu, S.; Jenk, T.M.; Schwikowski, M.; Jouzel, J. Temperature Trends in the Northwestern Tibetan Plateau Constrained by Ice Core Water Isotopes Over the Past 7000 Years. *J. Geophys. Res. Atmos.* **2020**, *125*, e2020JD032560.
11. Guo, C.; Ma, Y.; Meng, H.; Hu, C.; Li, D.; Liu, J.; Luo, C.; Wang, K. Changes in vegetation and environment in Yamzhog Yumco Lake on the southern Tibetan Plateau over past 2000 years. *Palaeogeogr. Palaeoclim. Palaeoecol.* **2018**, *501*, 30–44. [CrossRef]
12. He, Y.; Zhao, C.; Wang, Z.; Wang, H.; Song, M.; Liu, W.; Liu, Z. Late Holocene coupled moisture and temperature changes on the northern Tibetan Plateau. *Quat. Sci. Rev.* **2013**, *80*, 47–57.
13. Conroy, J.L.; Hudson, A.M.; Overpeck, J.T.; Liu, K.-B.; Wang, L.; Cole, J.E. The primacy of multidecadal to centennial variability over late-Holocene forced change of the Asian Monsoon on the southern Tibetan Plateau. *Earth Planet. Sci. Lett.* **2017**, *458*, 337–348.
14. Li, X.; Liang, J.; Hou, J.; Zhang, W. Centennial-scale climate variability during the past 2000 years on the central Tibetan Plateau. *Holocene* **2015**, *25*, 892–899. [CrossRef]
15. Ji, K.; Zhu, E.; Chu, G.; Aquino-López, M.A.; Hou, J. A record of late Holocene precipitation on the Central Tibetan Plateau inferred from varved lake sediments. *J. Paleolimnol.* **2021**, *66*, 439–452. [CrossRef]
16. Hou, J.; D'Andrea, W.J.; Liu, Z. The influence of 14C reservoir age on interpretation of paleolimnological records from the Tibetan Plateau. *Quat. Sci. Rev.* **2012**, *48*, 67–79.
17. Wang, Z.; Yin, J.-J.; Cheng, H.; Ning, Y.; Meyer, M.C. Climatic controls on travertine deposition in southern Tibet during the late Quaternary. *Palaeogeogr. Palaeoclim. Palaeoecol.* **2022**, *589*, 110852. [CrossRef]
18. Viles, H.A.; Tufas, A.G. Travertines and allied carbonate deposits. *Prog. Phys. Geogr.* **1990**, *14*, 19–41. [CrossRef]
19. Ford, T.; Pedley, H. A review of tufa and travertine deposits of the world. *Earth-Sci. Rev.* **1996**, *41*, 117–175.
20. Pentecost, A.; Viles, H. A Review and Reassessment of Travertine Classification. *Geogr. Phys. Quat.* **1994**, *48*, 305–314.
21. Capezzuoli, E.; Gandin, A.; Pedley, M. Decoding tufa and travertine (fresh water carbonates) in the sedimentary record: The state of the art. *Sedimentology* **2014**, *61*, 1–21.
22. Pentecost, A. The quaternary travertine deposits of Europe and Asia Minor. *Quat. Sci. Rev.* **1995**, *14*, 1005–1028.
23. Agliasacchi, E.; Kayseri-Özer, M.S. Multidisciplinary approach for palaeoclimatic signals of the non-marine carbonates: The case of the Sarıkavak tufa deposits (Afyon, SW-Turkey). *Quat. Int.* **2019**, *544*, 41–56.
24. Andrews, J.E.; Brasier, A. Seasonal records of climatic change in annually laminated tufas: Short review and future prospects. *J. Quat. Sci.* **2005**, *20*, 411–421.
25. Andrews, J. Palaeoclimatic records from stable isotopes in riverine tufas: Synthesis and review. *Earth-Sci. Rev.* **2006**, *75*, 85–104.
26. Kano, A.; Kawai, T.; Matsuoka, J.; Ihara, T. High-resolution records of rainfall events from clay bands in tufa. *Geology* **2004**, *32*, 793.
27. Toker, E.; Kayseri-Özer, M.S.; Özkul, M.; Kele, S.; Ariztegui, D. Depositional system and palaeoclimatic interpretations of Middle to Late Pleistocene travertines: Kocabaş, Denizli, south west Turkey. *Sedimentology* **2015**, *62*, 1360–1383.
28. De Filippis, L.; Faccenna, C.; Billi, A.; Anzalone, E.; Brilli, M.; Soligo, M.; Tuccimei, P. Plateau versus fissure ridge travertines from Quaternary geothermal springs of Italy and Turkey: Interactions and feedbacks between fluid discharge, paleoclimate, and tectonics. *Earth-Sci. Rev.* **2013**, *123*, 35–52.
29. Rodríguez-Berriguete, Á.; Alonso-Zarza, A.M.; Martín-García, R.; Cabrera, M.d.C. Sedimentology and geochemistry of a human-induced tufa deposit: Implications for palaeoclimatic research. *Sedimentology* **2018**, *65*, 2253–2277.
30. Faccenna, C.; Soligo, M.; Billi, A.; De Filippis, L.; Funiciello, R.; Rossetti, C.; Tuccimei, P. Late Pleistocene depositional cycles of the Lapis Tiburtinus travertine (Tivoli, Central Italy): Possible influence of climate and fault activity. *Glob. Planet. Chang.* **2008**, *63*, 299–308.
31. Ricketts, J.W.; Ma, L.; Wagler, A.E.; Garcia, V.H. Global travertine deposition modulated by oscillations in climate. *J. Quat. Sci.* **2019**, *34*, 558–568.
32. Su, J.; Tan, H.; Chen, X. The groundwater deep circulation and large-scale geothermal deposition in response to the extension of the Yadong–Gulu rift, South Tibet, China. *J. Volcanol. Geotherm. Res.* **2020**, *395*, 106836.
33. Wang, Z.; Meyer, M.C.; Hoffmann, D.L. Sedimentology, petrography and early diagenesis of a travertine–colluvium succession from Chusang (southern Tibet). *Sediment. Geol.* **2016**, *342*, 218–236.
34. Wang, Z.; Meyer, M.C.; Gliganic, L.A.; Hoffmann, D.L.; May, J.-H. Timing of fluvial terrace formation and concomitant travertine deposition in the upper Sutlej River (Tirthapuri, southwestern Tibet) and paleoclimatic implications. *Quat. Sci. Rev.* **2017**, *169*, 357–377.
35. Zentmyer, R.; Myrow, P.M.; Newell, D.L. Travertine deposits from along the South Tibetan Fault System near Nyalam, Tibet. *Geol. Mag.* **2008**, *145*, 753–765.
36. Gao, J.; Zhou, X.; Fang, B.; Li, T.; Tang, L. U-series dating of the travertine depositing near the Rongma hot springs in northern Tibet, China, and its paleoclimatic implication. *Quat. Int.* **2012**, *298*, 98–106.
37. Wang, Z.J.; Yin, J.; Yuan, D.X. Possibilities and problems associated with travertines and tufas in Quaternary studies: A case of the Tibetan Plateau. *Chin. Sci. Bull.* **2018**, *63*, 1012–1023, (In Chinese with English Abstract).
38. Hammer, Ø.; Dysthe, D.K.; Jamtveit, B. The dynamics of travertine dams. *Earth Planet. Sci. Lett.* **2007**, *256*, 258–263.
39. Pentecost, A.; Travertine, A. (Eds.) *Pentecost;* Springer: Dordrecht, The Netherlands, 2005; p. 446.
40. Ding, T.; Zheng, M.; Nie, Z.; Ma, L.; Ye, C.; Wu, Q.; Zhao, Y.; Yang, D.; Wang, K. Impact of Regional Climate Change on the Development of Lithium Resources in Zabuye Salt Lake, Tibet. *Front. Earth Sci.* **2022**, *10*, 865158.

41. Thompson, L.G.; Mosley-Thompson, E.; Davis, M.E.; Lin, P.N.; Dai, J.; Bolzan, J.F.; Yao, T. A 1000 year climate ice-core record from the Guliya ice cap, China: Its relationship to global climate variability. *Ann. Glaciol.* **1995**, *21*, 175–181.
42. Liu, X.F.; Zheng, M.P.; Qi, W. Sources of Ore-Forming Materials of the Superlarge B and Li Deposit in Zabuye Salt Lake, Tibet, China. *Acta Geol. Sin.* **2007**, *81*, 1709–1715, (In Chinese with English Abstract).
43. Zheng, M.P.; Xiang, J.; Wei, X.J.; Zheng, Y. *Saline Lake on the Qinghai-Xizang (Tibet) Plateau*; Science Press: Beijing, China, 1989; pp. 1–431, (In Chinese with English Abstract).
44. Cheng, H.; Lawrence Edwards, R.; Shen, C.-C.; Polyak, V.J.; Asmerom, Y.; Woodhead, J.; Hellstrom, J.; Wang, Y.; Kong, X.; Spötl, C.; et al. Improvements in 230Th dating, 230Th and 234U half-life values, and U–Th isotopic measurements by multi-collector inductively coupled plasma mass spectrometry. *Earth Planet. Sci. Lett.* **2013**, *371–372*, 82–91.
45. Edwards, R.L.; Chen, J.; Wasserburg, G. 238U234U230Th232Th systematics and the precise measurement of time over the past 500,000 years. *Earth Planet. Sci. Lett.* **1987**, *81*, 175–192.
46. Lumsden, D.N. Discrepancy Between Thin-Section and X-Ray Estimates of Dolomite in Limestone. *J. Sediment. Res.* **1979**, *49*, 429–435.
47. Goldsmith, J.R.; Graf, D.L.; Chodos, A.A.; Joensuu, O.I.; McVicker, L.D. Relation between lattice constants and composition of Ca-Mg carbonates. *Am. Mineral.* **1958**, *43*, 84–101.
48. Haese, R.R.; Smith, J.; Weber, R.; Trafford, J. High-Magnesium Calcite Dissolution in Tropical Continental Shelf Sediments Controlled by Ocean Acidification. *Environ. Sci. Technol.* **2014**, *48*, 8522–8528.
49. Auler, A.S.; Smart, P.L. Late Quaternary Paleoclimate in Semiarid Northeastern Brazil from U-Series Dating of Travertine and Water-Table Speleothems. *Quat. Res.* **2001**, *55*, 159–167.
50. Martínez-Aguirre, A.; Alcaraz-Pelegrina, J.M.; Rodríguez-Vidal, J. U/Th dating of impure carbonates: 230Th/232Th activity ratios in detrital material. *J. Radioanal. Nucl. Chem. Artic.* **2019**, *321*, 71–81.
51. Guo, L.; Riding, R. Hot-spring travertine facies and sequences, Late Pleistocene, Rapolano Terme, Italy. *Sedimentology* **1998**, *45*, 163–180.
52. Craig, H. Isotopic Variations in Meteoric Waters. *Science* **1961**, *133*, 1702–1703.
53. Craig, H. Standard for Reporting Concentrations of Deuterium and Oxygen-18 in Natural Waters. *Science* **1961**, *133*, 1833–1834. [PubMed]
54. Panichi, C.; Tongiorgi, E. Carbon isotopic composition of CO_2 from springs, fumaroles, mofettes and travertines of central and southern Italy: A preliminary prospection method of geothermal areas. In Proceedings of the 2nd U.N. Symposium on the Development and Use of Geothermal Energy, San Francisco, CA, USA, 20–29 May 1975.
55. Kohn Matthew, J. Carbon isotope compositions of terrestrial C3 plants as indicators of (paleo)ecology and (paleo)climate. *Proc. Natl. Acad. Sci. USA* **2010**, *107*, 19691–19695.
56. Wang, G.; Feng, X.; Han, J.; Lp, Z.; Wb, T.; Su, F. Paleovegetation reconstruction using δ13C of Soil Organic Matter. *Biogeosciences* **2008**, *5*, 1325–1337.
57. Minissale, A.; Kerrick, D.M.; Magro, G.; Murrell, M.T.; Paladini, M.; Rihs, S.; Sturchio, N.C.; Tassi, F.; Vaselli, O. Geochemistry of Quaternary travertines in the region north of Rome (Italy): Structural, hydrologic and paleoclimatic implications. *Earth Planet. Sci. Lett.* **2002**, *203*, 709–728.
58. Hoefs, J. *Stable Isotope Geochemistry*; Springer: Berlin/Heidelberg, Germany, 1997; p. 201.
59. Rollinson, H.R. *Using Geochemical Data: Evaluation, Presentation, Interpretation. Mineralogical Magazine*; Longman Scientific and Technical: London, UK, 1993; Volume 58, 352p.
60. Teboul, P.A.; Durlet, C.; Gaucher, E.C.; Virgone, A.; Girard, J.P.; Curie, J.; Lopez, B.; Camoin, G.F. Origins of elements building travertine and tufa: New perspectives provided by isotopic and geochemical tracers. *Sediment. Geol.* **2016**, *334*, 97–114.
61. Boom, A.; Mora, G.; Cleef, A.M.; Hooghiemstra, H. High altitude C4 grasslands in the northern Andes: Relicts from glacial conditions? *Rev. Palaeobot. Palynol.* **2001**, *115*, 147–160.
62. Rodrigo-Naharro, J.; Herrero, M.J.; Delgado-Huertas, A.; Granados, A.; Pérez del Villar, L. Current travertines precipitation related to artificial CO2 leakages from a natural reservoir (Gañuelas-Mazarrón Tertiary Basin, SE Spain). *J. Hydrol.* **2019**, *577*, 123997.
63. Mancini, A.; Frondini, F.; Capezzuoli, E.; Mejia, E.G.; Lezzi, G.; Matarazzi, D.; Brogi, A.; Swennen, R. Evaluating the geogenic CO_2 flux from geothermal areas by analysing quaternary travertine masses. New data from western central Italy and review of previous CO_2 flux data. *Quat. Sci. Rev.* **2019**, *215*, 132–143.
64. Kokh, S.N.; Shnyukov, Y.F.; Sokol, E.V.; Novikova, S.A.; Kozmenko, O.A.; Semenova, D.V.; Rybak, E.N. Heavy carbon travertine related to methane generation: A case study of the Big Tarkhan cold spring, Kerch Peninsula, Crimea. *Sediment. Geol.* **2015**, *325*, 26–40.
65. Fang, B.; Yang, Y.J.; Wang, G.H.; Zhou, X.; Chen, S.N. Characteristics and resource evaluation of the Jiwa geothemal field in central Qiangtang, northern Tibet, China. *Geol. Bull. China* **2009**, *28*, 1335–1341, (In Chinese with English Abstract).
66. Zhou, L. Characteristics of the Typical Hot Springs in the Central Tibet. Master's Thesis, China University of Geosciences (Beijing), Beijing, China, 2012. (In Chinese with English Abstract).
67. Chen, C.J. Provenance Analysis of Xiqin Geothermal Field, in Lazi County, Tibet. Master's Thesis, Hebei GEO University, Shijiazhuang, China, 2019.

68. Chevalier, M.L.; Tapponnier, P.; Woerd, J.; Leloup, P.H.; Wang, S.; Pan, J.; Bai, M.; Kali, E.; Liu, X.; Li, H. Late Quaternary Extension Rates Across the Northern Half of the Yadong-Gulu Rift: Implication for East-West Extension in Southern Tibet. *J. Geophys. Res. Solid Earth* **2020**, *125*, e2019JB019106.
69. Liu, Z.; Li, H.; You, C.; Wan, N.; Sun, H. Thickness and stable isotopic characteristics of modern seasonal climate-controlled sub-annual travertine laminas in a travertine-depositing stream at Baishuitai, SW China: Implications for paleoclimate reconstruction. *Environ. Earth Sci.* **2006**, *51*, 257–265. [CrossRef]
70. Prado-Pérez, A.J.; Delgado Huertas, A.; Crespo, M.; SÁNchez, A.; Villar, L. Late Pleistocene and Holocene mid-latitude palaeoclimatic and palaeoenvironmental reconstruction: An approach based on the isotopic record from a travertine formation in the Guadix-Baza basin, Spain. *Geol. Mag.* **2013**, *150*, 602–625.
71. Yao, T.; Masson-Delmotte, V.; Gao, J.; Yu, W.; Yang, X.; Risi, C.; Sturm, C.; Werner, M.; Zhao, H.; He, Y.; et al. A review of climatic controls on δ18O in precipitation over the Tibetan Plateau: Observations and simulations. *Rev. Geophys.* **2013**, *51*, 525–548.
72. Tian, L.; Yao, T.; Numaguti, A.; Sun, W. Stable Isotope Variations in Monsoon Precipitation on the Tibetan Plateau. *J. Meteorol. Soc. Jpn. Ser. II* **2001**, *79*, 959–966. [CrossRef]
73. Yu, W.; Yao, T.; Tian, L.; Ma, Y.; Ichiyanagi, K.; Wang, Y.; Sun, W. Relationships between δ18O in precipitation and air temperature and moisture origin on a south–north transect of the Tibetan Plateau. *Atmos. Res.* **2008**, *87*, 158–169. [CrossRef]
74. Feng, X.; Zhao, C.; D'Andrea, W.J.; Liang, J.; Zhou, A.; Shen, J. Temperature fluctuations during the Common Era in subtropical southwestern China inferred from brGDGTs in a remote alpine lake. *Earth Planet. Sci. Lett.* **2019**, *510*, 26–36.
75. Ge, Q.; Hao, Z.; Zheng, J.; Shao, X. Temperature changes over the past 2000 yr in China and comparison with the Northern Hemisphere. *Clim. Past* **2013**, *9*, 1153–1160. [CrossRef]
76. Bao, Y.; Braäuning, A.; Yafeng, S. Late Holocene temperature fluctuations on the Tibetan Plateau. *Quat. Sci. Rev.* **2003**, *22*, 2335–2344. [CrossRef]
77. Li, X.; Zhang, Y.; Wang, M.; Yan, J.; Fan, B.; Xing, W.; He, Y.; Hou, J. Centennial-Scale Temperature Change During the Common Era Revealed by Quantitative Temperature Reconstructions on the Tibetan Plateau. *Front. Earth Sci.* **2020**, *8*, 360. [CrossRef]
78. Zhang, X.L.; Xu, B.Q.; Li, J.L.; Xi, Y.; Gao, S.P.; Wang, M. Climatic and environmental changes over the past about 300 years recorded by lake sediments in Taro Co, southwestern Tibetan Plateau. *J. Earth Sci. Environ.* **2012**, *34*, 79–90, (In Chinese with English Abstract).
79. Sheng, E.; Yu, K.; Xu, H.; Lan, J.; Liu, B.; Che, S. Late Holocene Indian summer monsoon precipitation history at Lake Lugu, northwestern Yunnan Province, southwestern China. *Palaeogeogr. Palaeoclim. Palaeoecol.* **2015**, *438*, 24–33. [CrossRef]
80. Xu, H.; Zhou, X.; Lan, J.; Liu, B.; Sheng, E.; Yu, K.; Cheng, P.; Wu, F.; Hong, B.; Yeager, K.M.; et al. Late Holocene Indian summer monsoon variations recorded at Lake Erhai, Southwestern China. *Quat. Res.* **2015**, *83*, 307–314. [CrossRef]
81. Fleitmann, D.; Burns Stephen, J.; Mudelsee, M.; Neff, U.; Kramers, J.; Mangini, A.; Matter, A. Holocene Forcing of the Indian Monsoon Recorded in a Stalagmite from Southern Oman. *Science* **2003**, *300*, 1737–1739. [CrossRef]
82. Duan, K.; Yao, T.; Thompson, L.G. Low-frequency of southern Asian monsoon variability using a 295-year record from the Dasuopu ice core in the central Himalayas. *Geophys. Res. Lett.* **2004**, *31*, L16209. [CrossRef]
83. Chu, G.; Sun, Q.; Yang, K.; Li, A.; Yu, X.; Xu, T.; Yan, F.; Wang, H.; Liu, M.; Wang, X.; et al. Evidence for decreasing South Asian summer monsoon in the past 160 years from varved sediment in Lake Xinluhai, Tibetan Plateau. *J. Geophys. Res. Earth Surf.* **2011**, *116*, D02116. [CrossRef]
84. Xu, H.; Hong, Y.; Hong, B. Decreasing Indian summer monsoon intensity after 1860 AD in the global warming epoch. *Clim. Dyn.* **2012**, *39*, 2079–2088, Erratum in *Clim. Dyn.* **2012**, *39*, 2089. [CrossRef]
85. Mann, M.; Cane, M.; Zebiak, S.; Clement, A. Volcanic and Solar Forcing of the Tropical Pacific over the Past 1000 Years. *J. Clim.* **2005**, *18*, 447–456. [CrossRef]
86. Sun, Z.; Yuan, K.; Hou, X.; Ji, K.; Li, C.-G.; Wang, M.; Hou, J. Centennial-scale interplay between the Indian Summer Monsoon and the Westerlies revealed from Ngamring Co, southern Tibetan Plateau. *Holocene* **2020**, *30*, 1163–1173. [CrossRef]

minerals MDPI

Article

Non-Linear Clumped Isotopes from DIC Endmember Mixing and Kinetic Isotope Fractionation in High pH Anthropogenic Tufa

Chris Holdsworth [1], John MacDonald [1] and Cedric John [2,*]

[1] School of Geographical and Earth Sciences, University of Glasgow, Glasgow G12 8QQ, UK
[2] Department of Earth Science and Engineering, Imperial College London, London SW7 2AZ, UK
* Correspondence: cedric.john@imperial.ac.uk

Abstract: Clumped isotope values (Δ_{47}) of carbonates forming in high pH conditions do not correspond to mineral precipitation temperatures due to certain effects including kinetic isotope fractionation and dissolved inorganic carbon (DIC) endmember mixing. Field-based archives of these carbonate environments are needed to evaluate and quantify these effects accurately. In this study, we measure the clumped isotope values of anthropogenic carbonates for the first time. Tufa layers were analyzed from samples precipitating in a high pH (>10) stream that drains a major slag heap in north east England. Δ_{47} values are 0.044‰–0.183‰ higher than expected equilibrium values. Non-linear distribution of clumped isotope data is diagnostic of DIC endmember mixing, rather than partial equilibration of DIC. Episodic dilution of hydroxide-rich stream waters by equilibrated rainfall surface runoff provides the mechanism by which mixing occurs. Δ_{47} values are ~0.010‰–0.145‰ higher than linear clumped isotope mixing profiles, suggesting that the majority of Δ_{47} increase results from a combination of endmember non-linear mixing effects and an atmosphere-hydroxide sourcing of DIC. The diagnostic trends and variation in clumped isotope values present in these results demonstrates the potential of anthropogenic carbonate systems as a useful archive for studying and quantifying kinetic effects in clumped isotopes.

Keywords: tufa; anthropogenic carbonates; clumped isotopes; CO_2 hydroxylation; kinetic isotope fractionation; isotope disequilibrium

Citation: Holdsworth, C.; MacDonald, J.; John, C. Non-Linear Clumped Isotopes from DIC Endmember Mixing and Kinetic Isotope Fractionation in High pH Anthropogenic Tufa. *Minerals* **2022**, *12*, 1611. https://doi.org/10.3390/min12121611

Academic Editors: Francesca Giustini and Mauro Brilli

Received: 1 November 2022
Revised: 2 December 2022
Accepted: 10 December 2022
Published: 14 December 2022

Publisher's Note: MDPI stays neutral with regard to jurisdictional claims in published maps and institutional affiliations.

1. Introduction

1.1. Clumped Isotope Palaeothermometry

Clumped isotope analysis uses the abundance of ^{13}C-^{18}O bonds in carbonate anions to measure mineral precipitation temperature. For carbonates forming in a system where water and composite dissolved inorganic carbon (DIC) pool are close to isotopic equilibrium, the occurrence of this heavy isotope clumping is inversely proportional to mineral precipitation temperature [1–6]. Clumped isotope values (Δ_{47}) correspond to excess mass 47 CO_2 measured during acid digestion of carbonates. These data are calibrated against carbonates precipitated at known temperatures [1,7–14], with increasing Δ_{47} equating to decreasing mineral precipitation temperature. The ability to extract precipitation temperatures directly from carbonate mineral chemistry without any prior knowledge of parent water chemistry gives the clumped isotope method a significant advantage over conventional oxygen isotope palaeothermometry. As a result, there are now numerous examples of clumped isotope studies from across the earth sciences, covering topics including paleoclimate reconstruction [15–20], diagenetic histories [21–25], and fluid flow [26–28].

1.2. Non-Equilibrium Precipitation

There are certain carbonate-forming environments where equilibrium is not reached prior to mineral precipitation. Examples of this have been recorded at extremes of pH [29–34], temperature [16,35–38] and precipitation rates [5,39,40]. In these settings, $\delta^{13}C$, $\delta^{18}O$ and Δ_{47} record disequilibrium effects; isotopic and kinetic processes that relate to solution chemistry and carbonate-forming reactions, rather than mineral precipitation temperatures [5,6,39–42].

Kinetic Isotope Effects

Carbonates precipitating rapidly in alkaline waters (pH > 10) at low temperatures (<30 °C) are likely to record kinetic isotope effects due to water-DIC equilibration times increasing with pH and decreasing with temperature [5]. In high pH conditions, CO_2 is rapidly draw in to the water column and converted into DIC via CO_2 hydroxylation [31]. This kinetic process preferentially selects lighter isotopes (^{12}C and ^{16}O), depleting DIC in ^{13}C and ^{18}O [42–44]. Near-instantaneous carbonate precipitation prevents any isotopic equilibration between DIC and water, therefore preserving kinetic depletions in mineral bulk isotope signatures that are significantly below expected equilibrium values [31,32,34,45]. Our understanding of how these kinetic effects impact clumped isotopes is still in its infancy [6,46]. Depletions of ^{13}C and ^{18}O in high pH carbonates in Oman have corresponding enrichments in Δ_{47} [33], with authors concluding CO_2 hydroxylation also to be the cause of this. In this study, we investigate human-made (anthropogenic) carbonate deposits that form in similarly alkaline conditions and, in this instance, even lower temperatures (<20 °C). The independence of Δ_{47} from growth rate above pH 10 means only kinetic disequilibrium effects should be recorded in these carbonates if the DIC speciation Δ_{47} offset is corrected for [5,6,35,39,40,47–51].

1.3. Anthropogenic Carbonates

Anthropogenic carbonates typically form in the low temperature, high pH conditions described above. Weathering of human-sourced materials such as legacy industrial waste results in the liberation of divalent cations (e.g., Ca, Mg, and Fe) and hydroxide (OH^-) into surface waters. These alkaline leachates promote rapid in-gassing and hydroxylation of atmospheric CO_2, followed by near-instantaneous carbonate precipitation [52,53]. A similar process of carbonate formation occurs through the weathering of natural silicate minerals in mafic and ultramafic rocks [33,34,54]. Accumulations of anthropogenic carbonates are recorded at banana plantations [55], quarries [56,57], landfill sites [58,59], furnace slag heaps [53,60–65] urban soils [52,66,67], and on cement [68,69] or concrete structures [70,71].

1.4. Purpose of Study

Our understanding of how kinetic effects impact Δ_{47} is particularly hindered by a lack of field case study datasets from high pH, hydroxylation-related carbonate archives. This study tests anthropogenic carbonates as a useful archive for evaluating the impact of kinetic fractionations, such as CO_2 hydroxylation, on Δ_{47}. These clumped isotope values will be the first taken from human-derived carbonates.

Samples were collected from a high pH (>10) stream that drains a legacy slag heap on the outskirts of Consett in northeast England. Carbonate layers from up- and down-stream samples were analysed and compared to historic environmental data (water temperature, flow, chemistry, etc. in [53,63,65,72]) to quantify the extent of clumped isotope disequilibrium. We expect Δ_{47} values to record a kinetic enrichment from CO_2 hydroxylation, combined with the product of any isotope endmember mixing processes.

2. Materials and Methods

2.1. Site of Study

Carbonate samples for this study were collected from the Howden Burn, a ~500 m stream that drains the main slag heap of the now-closed Consett Iron and Steel Works in northeast England (Figure 1A). The works was operational between the mid-19th century and 1980, producing ~120 Mt of steel and iron and >20 Mt of slag materials over its lifespan [53]. Slag was dumped in heaps surrounding the site (Figure 1B), as is captured by the westerly progression of heap deposits seen between 1880 and 1960 on historic map records (Supplementary Materials Figures S1–S7). The main heaps are underlain by alluvium and glacial till resting on top of Carboniferous mudstones [53]. Following the closure of the works the heaps were landscaped and redeveloped in the late 20th century. Recent estimates suggest the heaps reach a maximum thickness of 45 m and have a total volume of ~16 Mm^3 [53]. The constituents of the heaps vary substantially, with steel and iron slag, spoils, flue dusts, pot ashes and building demolition materials all being recorded as present [73]. Average compound composition of blast furnace and steel slag are reported in [63]. 314 tons of Ca is calculated to have leached from the main heap between 1980 and 2017 [53]. Based on CaO and MgO concentrations of heap constituents, this represents as little as 0.004% of the Ca present.

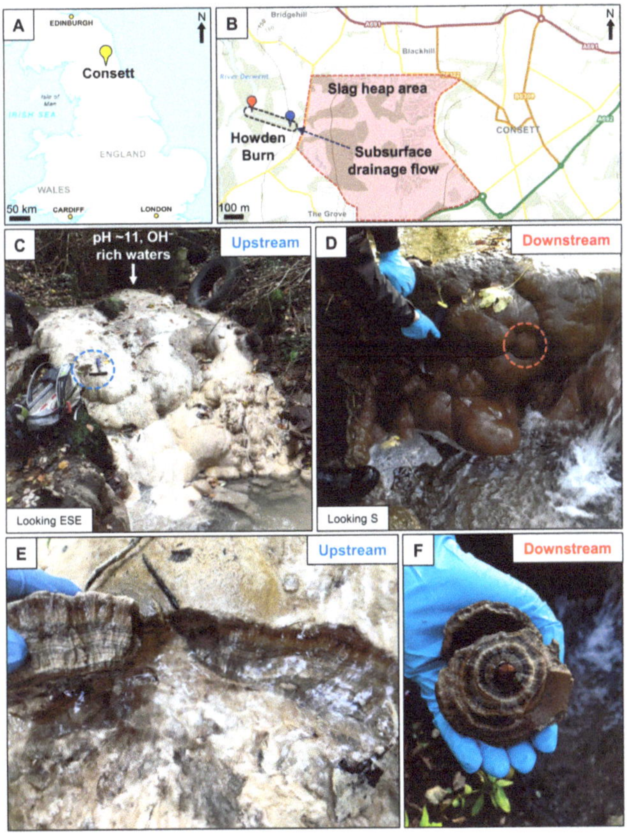

Figure 1. (**A**) Location of Consett in northeast England. (**B**) Area where slag deposition occurred during the lifespan of the steel and iron works. Red dashed area shows the extent of slag deposition in Consett. Black dashed circle indicates the extent of the Howden Burn. Blue and red markers indicate the locations of upstream (blue) and downstream (red) samples on the Howden Burn. (**C**) Upstream tufa barrage located immediately after the Howden Burn emergence from slag heap subsurface drainage. Stream waters here are hyperalkaline (pH~11), rich in OH^-, Ca and slag-derived metals. Upstream sampling point indicated by dashed blue circle. (**D**) Downstream tufa barrage located ~50 m before River Derwent confluence. Deposits are darker and harder than those found upstream. Downstream sample in-situ location indicated by dashed red circle. (**E**) In-situ picture of upstream sampling point showing mm-scale tufa layering; (**F**) Picture of downstream tufa sample showing tufa layer accumulation around a small stick.

2.1.1. Howden Burn

The Howden Burn drains the interior of the largest heap at Consett, emerging from a culverted subsurface drainage outflow on the western side of Pemberton Road. Historic map records document the direct infilling of the Howden Burn profile with slag between 1880 and the steelworks closure (Supplementary Materials Figures S1–S7). The culverted section from which the burn emerges follows the 1880 burn profile for ~150 m into the heap before ceasing on current drainage map records. The burn also receives leachate discharge from a surface water manhole cover which overflows to the immediate south east of the culvert emergence. Total burn catchment is measured at 0.68 km^2 and surface waters flow for ~500 m before reaching a confluence with the River Derwent [53].

The main hydrochemical characteristics of the Howden Burn are summarized in Table 1. Waters are hydroxide enriched and contain high loadings of Ca and metals leached from the heap contents [63,65]. Water temperature records from 1978–1999 range from 4 °C to 20 °C, with a median value of 10.3 °C [53]. Subsequent water temperature measurements record average values of 13.0 °C upstream and 12.8 °C downstream [72]. Burn pH records from 1978 to 1999 range from 6.5 to 12.7, with a median value of 11.0 [53]. Lower pH measurements (<9) present in this range correspond to measurements taken before slag heap redevelopment in the late 1990s and during periods when the burn was being actively dosed with acid to reduce surface water pH [53]. Following heap redevelopment non-slag leaching waters were redirected away from the burn via a combined sewer overflow pipeline [65]. This led to a change in burn catchment and a relative concentration of high pH (>10) slag leachate waters [65]. Subsequent water pH measurements throughout the 2000s show this higher pH has remained [65], with the most recent records showing average values of 11.6 upstream and 11.3 downstream [72].

Table 1. Major hydrochemical characteristics of Howden Burn discharge from [53] unless marked with * [72]. Median values are shown with data ranges in brackets to highlight the concentration of high pH slag leachate in stream waters following heap redevelopment in the late 1990s.

Temperature (°C)	10.3 (4.4–20.0)
pH	11.0 (6.5–12.7)
Flow (l/s)	5.0 (2.5–10.0)
Major Ions (mg/l)	
Al	0.1 (<0.1–5)
Ca	150 (45–315)
Fe	1.6 (<0.1–4.7)
Mg	8.5 (3.3–13.9)
K	412 (143–688)
Na	172 (45–287)
Cl	67 (2–550)
SO_4	755 (42–2700)
Total Alkalinity	162 (93–315)
OH^- *	29 upstream, 13 downstream
CO_3 *	139 upstream, 20 downstream
HCO_3 *	9 upstream, 3 downstream

2.1.2. Carbonate Classification

Enrichment of OH^- and Ca in burn waters leads to substantial accumulations of $CaCO_3$, which precipitates via the formation of (bi)carbonate from the hydroxylation of atmospheric CO_2 [53]. Precipitation rates have been calculated to range between 8 and 259 $g/m^2/day$ by dividing the difference in Ca mass loss at the leachate source and downstream by the sampling area (see [53] for full details). This is estimated to produce 11–117 kg of $CaCO_3$ per day and 4–42 tons per year [53]. Such substantial volumes of precipitation results in extensive smothering of the stream channel and banks, increased opacity of the water column and accumulations of carbonate terraces, barrages and mini-cascade systems. The largest of these barrages is found immediately after the first emergence of the culverted watercourse (Figure 1C). Barrage size decreases downstream corresponding to changes in carbonate color and morphology.

Previous XRD analysis of Howden Burn carbonates indicates $CaCO_3$ polymorphs aragonite and vaterite are present upstream but not downstream [63]. Precipitates are white-pale yellow upstream and resemble a soft, unlithified sediment. Downstream color darkens through orange to red-brown, and hardness increases towards that of a lithified carbonate rock (Figure 1C,D). Carbonates precipitating in the Howden Burn share characteristics indicative of tufas and travertines. The climatic and hydrological characteristics of the Howden Burn combined with the millimeter-scale laminations and cascade/barrage morphology of carbonates dictate that a tufa characterization is most appropriate for Howden Burn carbonates [74–77]. Tufa laminations vary from mm to cm scale, with intermittent dark clay layers and some darker carbonate layers, possibly indicating incorporation of local organic material (e.g., leaves from Howden Wood).

2.1.3. Samples

Tufa samples were taken from the Howden Burn in October 2017. For the purposes of this study five distinctive layers were analyzed from an upstream (CHS5; Figure 1E) and downstream (CHS16; Figure 1F) tufa sample. Sample CHS5 was taken from the largest tufa barrage in the Burn, located immediately after the emergence of leachate waters from the culvert. Sample CHS16 was taken from the last major tufa barrage before the Howden Burn-River Derwent confluence. Analyzing one tufa from either end of the Burn profile allows for a relative comparison between clumped isotope compositions upstream and downstream, whilst analyzing differing layers within each sample provides a relative measure of changes in clumped isotope compositions through the growth history of the sample (i.e., time). The five layers sampled from each tufa were labelled A–E, with 'A' corresponding to the youngest and outer-most layer, and 'E' corresponding to the oldest and inner-most layer (Supplementary Materials Figures S8 and S9). Due to the lack of absolute dating of tufa layers, adjacent layers were not always chosen for analysis. Instead, layers of differing color and morphology were chosen to assess for any impact these factors have on layer clumped isotope values.

2.2. *Data Acquisition*

2.2.1. Stable Isotope Measurements

$\delta^{13}C$, $\delta^{18}O$ VPDB and Δ_{47} clumped isotope measurements were made in the Qatar Stable Isotope Laboratory at Imperial College London in July 2018. A minimum of three replicates were analyzed from each tufa layer, using aliquots of 5–7 mg powdered sample in a randomized order. Replicates were analyzed used the automated Imperial Batch EXtraction (IBEX) system. Individual samples were dropped into a vacuum-sealed common acid bath for reaction with 105% orthophosphoric acid for 10 min at 90 °C [78]. Acid digestion produces CO_2 gas, which is then fed through a series of water, silver and Porapak-Q traps to remove water, sulfur and hydrocarbons respectively using He carrier gas [14,78–80]. This removes potential contaminants from CO_2 gas, which is crucial given the nature abundance of mass 47 CO_2 is only 44 ppm [3,80]. CO_2 is captured in a second water trap that is maintained at liquid nitrogen temperature, before being automatically loaded for analysis on a dual-inlet Thermo Fisher (Bremen, Germany) MAT 253 isotope ratio mass spectrometer [78,80].

Analytical protocols for the Imperial College lab are described in full in Dale et al. [21] and generally follow those outlined in Huntington et al. [79] and Dennis et al. [81]. The MAT 253 measures masses 44 to 49 of CO_2 simultaneously using modified collector arrays and uses an Oztech reference gas standard [80]. Each replicate measurement comprises eight acquisitions, with seven cycles per acquisition, each with an integration time of 26 s [78,80,82]. Measurements are repeated seven times per replicate in two different MAT

253 instruments, resulting in an analysis time of two hours per replicate on each mass spectrometer [78].

2.2.2. Data Analysis and Reduction

Clumped and bulk isotope data were corrected using the free software Easotope [82]. A pressure baseline correction, as described in Bernasconi et al. [83], was performed and Δ_{47} data were then projected into the absolute reference frame, Carbon Dioxide Equilibrated Scale (CDES), of Dennis et al. [81]. This was carried out using a secondary transfer function based on measurements of ETH1, ETH2, ETH3, ETH4, and Carrara Marble (ICM) carbonate standards [14,84,85]. For this study, one standard was analyzed per three samples. Only samples with Δ_{48} and δ_{48} values that fall within 1‰ of the heated gas line (Δ_{48} offset) and had a 49 parameter of <0.2 were accepted [14,82]. These signals were used as indicators for the presence of hydrocarbons, chlorocarbons and sulfur-bearing contaminants [78,79,86,87]. An acid fractionation factor of +0.082‰ [8] was added to the corrected Δ_{47} data to bring data into the 25 °C scale used in calibrations. Results were then converted into temperature using the calibration of Davies and John [14], which is based on recalculated values from Kluge et al. [9] and new inorganic precipitates. This calibration accounts for updated ^{17}O parameters [88], is within analytical error of Bonifacie et al. [12] and the universal clumped isotope calibration of Kelson et al. [11], and includes the recent community inter-comparison efforts of Petersen et al. [13] regarding the ^{17}O correction and clumped isotope calibration. This study is mainly concerned with the impact of CO_2 hydroxylation on Δ_{47}, therefore exact temperature calibration is not of paramount importance.

Data for each tufa layer corresponds to the average of replicate analyses, of which there were a minimum of three per layer. All Δ_{47} values are reported in the CO_2 equilibrated scale [81] with 1σ as error, whilst measured $\delta^{13}C$ and $\delta^{18}O$ values are reported in the Vienna PeeDee Belemnite (VPDB) scale, also with 1σ as error. Average values for $\delta^{13}C$ and $\delta^{18}O$ VPDB replicates of standards fell within error of quoted values [84,85], except for $\delta^{13}C$ for Carrara Marble (by 0.05‰) and $\delta^{18}O$ for ETH4 (by 0.02‰). Average values for Δ_{47} replicates of standards fell within error of quoted values for Carrara Marble and ETH3, but not ETH1 (by 0.007‰), ETH2 (by 0.022‰) or ETH4 (by 0.012‰). Full standards and samples datasets can be found in Research Data files deposited in the British Geological Survey (BGS) National Geoscience Data Centre (NGDC) at [89]. Clumped isotope temperature calculations are given in degrees centigrade with one standard error. A 95% confidence level for temperature values is also available in research data files. Source fluid $\delta^{18}O$ values were calculated using the carbonate–water equilibrium fractionation factors of Kim and O'Neil [90] and are reported in the Vienna Standard Mean Ocean Water (VSMOW) scale with same errors as calculated temperatures. Due to the high pH of the Howden Burn these calculated fluid $\delta^{18}O$ VSMOW values are not expected to provide an accurate estimate of parent solution $\delta^{18}O$. They are included to provide a comparison against regional seasonal $\delta^{18}O_{water}$ data [91,92] and highlight the disequilibrium of carbonate precipitation in this environment.

2.3. *Non-Linear Mixing Model*

The non-linear mixing curves used to interpret clumped isotope data were derived using the non-linear mixing equation of Guo [6], formerly discussed by Defliese and Lohmann [93] and Eiler and Schauble [86]:

$$\Delta_{47,mix} \approx (1-f) \times \Delta_{47,A} + f \times \Delta_{47,B} + f \times (1-f) \times (\delta^{13}C_A - \delta^{13}C_B) \times (\delta^{18}O_A - \delta^{18}O_B) \quad (1)$$

where the *A* and *B* components are the equilibrium (lithogenic or biogenic) and disequilibrium (hydroxylation) endmembers respectively and *f* is the mixing ratio of endmember *B* component. Expected equilibrium clumped isotope values were calculated by reversing the Δ_{47}-temperature conversion [14] for the historic water temperature range of the Howden Burn (4–20 °C; [53]), to give an expected equilibrium clumped isotope range of

0.708‰–0.763‰. Equilibrium carbonate mineral $\delta^{18}O_{mineral}$ VPDB values under were calculated using regional meteoric rainfall values of Darling et al. [91], along with the carbonate-water equilibrium fractionation factor [90]. The disequilibrium used in the model corresponds to a carbonate sample from the hyperalkaline spring waters of the Semail ophiolite in Oman [33]. The near-identical bulk isotope values of this endmember sample ($\delta^{13}C$ VPDB = −27.0‰; $\delta^{18}O$ VPDB = −16.7‰) to the Dietzel et al. [32] hydroxylation bulk isotope values ($\delta^{13}C$ = −25.3‰; $\delta^{18}O$ = −17.6‰) makes this appropriate for the purposes of this study.

3. Results

3.1. Carbonate $\delta^{13}C$ and $\delta^{18}O$ Isotope Values

$\delta^{13}C$ and $\delta^{18}O$ VPDB values of downstream tufa layers are generally lower than upstream layers (Figure 2). Layer 5A in the exception to this, with the lowest $\delta^{13}C$ and $\delta^{18}O$ VPDB values of the data set of −19.49‰ (±0.07‰) and −12.54‰ (±0.13‰) respectively. Upstream layer values range from −14.03‰ (±0.04‰) to −19.49‰ (±0.07‰) for $\delta^{13}C$ VPDB and −6.85‰ (±0.21‰) to −12.54‰ (±0.13‰) for $\delta^{18}O$ VPDB. Downstream values range from −15.94‰ (±0.02‰) to −18.65‰ (±0.25‰) for $\delta^{13}C$ VPDB and −9.93‰ (±0.08‰) to −12.06‰ (±0.30‰) for $\delta^{18}O$ VPDB.

3.2. Carbonate Δ_{47} Clumped Isotope Values

Trends in clumped isotope data generally correspond to those present in bulk isotope data, meaning layers with lower $\delta^{13}C$ and $\delta^{18}O$ VPDB values have higher Δ_{47} values. Exceptions to this trend are layers 16C (0.881 ± 0.011‰) and 16D (0.897 ± 0.009‰). Layer 5A has the highest $\mathbf{\Delta_{47}}$ value of the dataset (0.900 ± 0.009‰), whilst 5D the lowest (0.827 ± 0.011‰). Upstream values range between 0.827‰ (±0.011‰) and 0.900‰ (±0.009‰) and downstream values range from 0.828‰ (±0.021‰) to 0.897‰ (±0.009‰).

3.3. Calculated Temperatures and $\delta^{18}O_{fluid}$

Calculated temperature values for upstream and downstream data points correspond to the trends seen in Δ_{47} values (Figure 2). Upstream layer temperatures range from −25.39 °C (±0.99 °C) to −10.96 °C (±1.42 °C) and downstream temperatures range from −11.18 °C (±2.69 °C) to −25.02 °C (±0.91 °C). $\delta^{18}O_{fluid}$ VSMOW values derived from $\delta^{18}O_{mineral}$ VPDB and Δ_{47} show ranges of −12.54‰ (±0.50‰) to −22.41‰ (±0.37‰) for upstream layers and −15.78‰ (±0.75‰) to −21.15‰ (±0.32‰) for downstream.

Upstream (CHS5)

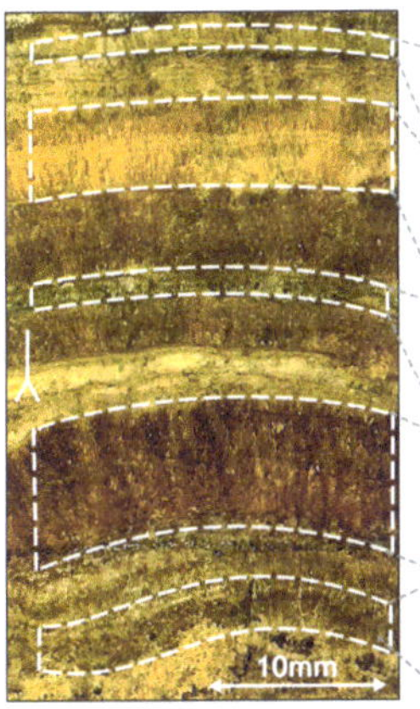

	$\delta^{13}C_{mineral}$ VPDB [‰] (± 1SD)	$\delta^{18}O_{mineral}$ VPDB [‰] (± 1SD)	Δ_{47} [‰] (± 1SD)	T [°C] (± 1SE)	$\delta^{18}O_{fluid}$ VSMOW [‰] (± 1SE)
5A	−19.49 (± 0.07)	−12.54 (± 0.13)	0.900 (± 0.009)	−25.39 (± 0.99)	−22.41 (± 0.37)
5B	−16.20 (± 0.07)	−8.75 (± 0.10)	0.869 (± 0.007)	−19.70 (± 0.70)	−16.88 (± 0.25)
5C	−16.00 (± 0.21)	−8.51 (± 0.29)	0.852 (± 0.035)	−16.25 (± 4.22)	−15.68 (± 1.31)
5D	−14.46 (± 0.14)	−6.85 (± 0.21)	0.827 (± 0.011)	−10.96 (± 1.42)	−12.54 (± 0.50)
5E	−14.03 (± 0.04)	−7.34 (± 0.03)	0.836 (± 0.017)	−12.83 (± 2.18)	−13.55 (± 0.59)

Figure 2. *Cont.*

Downstream (CHS16)

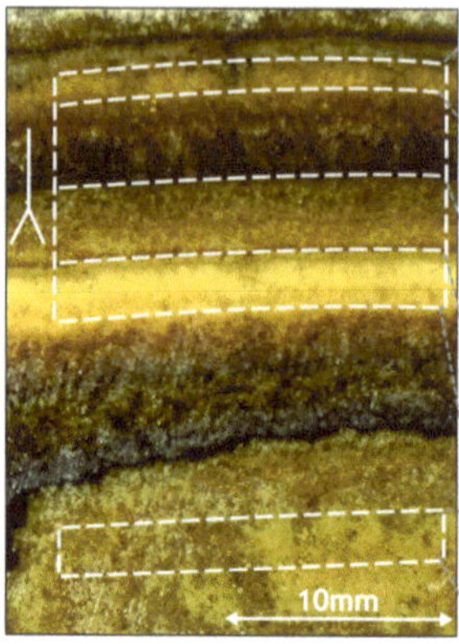

	$\delta^{13}C_{mineral}$ VPDB [‰] (± 1SD)	$\delta^{18}O_{mineral}$ VPDB [‰] (± 1SD)	Δ_{47} [‰] (± 1SD)	T [°C] (± 1SE)	$\delta^{18}O_{fluid}$ VSMOW [‰] (± 1SE)
16A	−16.86 (± 0.05)	−10.02 (± 0.07)	0.856 (± 0.019)	−17.00 (± 2.23)	−17.43 (± 0.65)
16B	−15.94 (± 0.02)	−9.93 (± 0.08)	0.828 (± 0.021)	−11.18 (± 2.69)	−15.78 (± 0.75)
16C	−18.65 (± 0.25)	−12.06 (± 0.30)	0.881 (± 0.011)	−21.88 (± 1.29)	−20.90 (± 0.54)
16D	−17.86 (± 0.10)	−11.41 (± 0.09)	0.897 (± 0.009)	−25.02 (± 0.91)	−21.15 (± 0.32)
16E	−17.20 (± 0.15)	−10.61 (± 0.10)	0.859 (± 0.028)	−17.61 (± 3.27)	−18.22 (± 0.95)

Figure 2. Bulk $\delta^{13}C$ and $\delta^{18}O$ VPDB isotope and Δ_{47} clumped isotope results with corresponding temperature and $\delta^{18}O_{fluid}$ VSMOW calculations for upstream (**top**) and downstream (**bottom**) tufa layers. Sample numbers 5A–5E and 16A–16E correspond to the hand specimen number from field sampling (5 and 16) and the youngest and outer-most layer (A), and the oldest and inner-most layer (E). White dashed lines indicate the layer and area from which each respective data point (e.g. 5A, etc.) was acquired. Errors are given in brackets and correspond to 1 standard deviation for bulk and clumped isotope data and 1 standard error for temperature and $\delta^{18}O_{fluid}$ VSMOW data.

4. Discussion

Carbonate clumped isotope temperatures calculated in this study are ~20–35 °C colder than the Howden Burn median water temperature of 10 °C (Table 1). This substantial negative offset corresponds to higher Δ_{47} clumped isotope and lower bulk $\delta^{13}C$ and $\delta^{18}O$ VPDB values, relative to those expected for carbonates precipitating at Howden Burn water temperatures (4–20 °C). Calculated $\delta^{18}O_{fluid}$ VSMOW values are comparably offset from regional values of −8.2‰ (winter) and −7.0‰ (summer) [91,92]. Using the historic temperature range above and the Δ_{47}-temperature calibration of Davies and John [14], clumped isotope values are 0.044‰–0.183‰ higher than expected. Given that clumped isotopes have provided accurate and reliable reconstructions of carbonate mineral precipitation temperatures in other temperate European [10] and Japanese [94] tufas, isotope data offsets in this study must be caused by hydrochemical characteristics of the Howden Burn that differ to that of the examples above.

4.1. High pH Carbonates

The dominant characteristic of the Howden Burn hydrochemical environment is high pH (Table 1). Measurements regularly exceed pH 11, compared to pH < 9 in the European and Japanese tufa studies [10,94]. High pH values result from the leaching of slag minerals such as larnite (Equation (2)) and portlandite (Equation (3)), which enriches Howden Burn waters in hydroxide (OH^-) and divalent cations (e.g., Ca):

$$Ca_2SiO_4 + 4H_2O \rightarrow 2Ca^{2+} + H_4SiO_4 + 4OH^- \quad (2)$$

$$Ca(OH)_2 \rightarrow Ca^{2+} + 2OH^- \quad (3)$$

This creates a specialized carbonate forming environment that differs from the majority of carbonate archives. OH^- abundance promotes in-gassing and absorption of atmospheric CO_2 into the water column, where newly formed (bi)carbonate then reacts with leached Ca to rapidly precipitate calcium carbonate ($CaCO_3$) [32,51]:

$$CO_{2\,(g)} \rightarrow CO_{2\,(aq)} \quad (4)$$

$$CO_{2\,(aq)} + H_2O \leftrightarrow H_2CO_3 \quad (5)$$

$$H_2CO_3 \leftrightarrow HCO_3^- + H^+ \quad (6)$$

$$HCO_3^- + OH^- \rightarrow H_2O + CO_3^{2-} \quad (7)$$

$$Ca^{2+} + CO_3^{2-} \leftrightarrow CaCO_3 \quad (8)$$

4.1.1. DIC Speciation

There are two significant consequences of this carbonate forming process for clumped isotope values. First, the speciation of composite DIC changes with solution pH [39,41]. Above pH~10.3, dominant speciation transitions from bicarbonate (HCO_3^-) to carbonate (CO_3^{2-}), meaning DIC speciation in high pH environments such as the Howden Burn differs from that of lower pH, 'natural' carbonate precipitating environments (e.g., Kato et al. [94]; Kele et al. [10]). Speciation of DIC is known to produces certain offsets in clumped isotope values [5,39,41]. For a system such as the Howden Burn, which is pH > 10.3 and CO_3^{2-} dominated (see Table 1), Δ_{47} values are offset by up to −0.05‰ [5,39,41]. This is the opposite to the trend seen in Howden Burn tufa data, which suggests DIC speciation is not responsible for higher clumped isotope values in Howden Burn tufas.

4.1.2. CO_2 Hydroxylation

The second consequence of this carbonate forming process for clumped isotope values relates to the source of DIC and the process by which CO_2 moves from the atmosphere into carbonate minerals. The in-gassing of CO_2 produces a kinetic isotope fractionation during the reaction between aqueous CO_2 and OH^- (Equation (7)), known as CO_2 hydroxylation. This produces a stable isotope enrichment factor of −18.8‰ for $\delta^{13}C$ VPDB [32,51,95] and −13‰ for $\delta^{18}O$ VPDB [96,97] in the resultant DIC pool. Abundant supply of cations then allows for near-instantaneous carbonate precipitation (Equation (8)). This, combined with the substantial time requirements for isotopic equilibration at low temperature and high pH (>10^6 min) means tufa layers inherit isotopic signals that correspond to the kinetic fractionation of CO_2 and atmosphere-hydroxide sourcing of DIC, rather than mineral precipitation temperatures [5,6,39,98,99]. These signals are referred to as clumped isotope disequilibrium effects.

Evidence of this isotopic disequilibrium is recorded in previous studies of Consett slag-derived carbonates. Strong depletions of ^{13}C and ^{18}O in Howden Burn tufas [53] and other carbonate morphologies [52] are diagnostic of an atmospheric CO_2 source and hydroxylation disequilibrium effects. Carbonates precipitating in hyperalkaline springs in Oman record comparably negative bulk isotope values [33]. Clumped isotope compositions were also measured for these carbonates and yield values up to 0.200‰ higher than those expected for equilibrium precipitation [33]. This is comparable to the offset of Δ_{47} values seen in this study, which given the differing climates and ambient temperatures of northern England and Oman highlights the lack of temperature correspondence of clumped isotope compositions in high pH carbonates.

4.2. Variation in Disequilibrium

Variation in clumped and bulk isotope values suggests the rate or extent of CO_2 hydroxylation and/or carbonate precipitation in the Howden Burn is changing. In high pH environments such as this, carbonate precipitation rate is defined as:

$$r = C_o \times (D \times k \times [OH^-])^{0.5} \quad (9)$$

where C_o is the CO_2 concentration in solution (mol cm^{-3}), D is the diffusion coefficient of CO_2 through a liquid ($cm^2\ s^{-1}$), k is the rate constant for hydroxylation ($cm^3\ mol^{-1}\ s^{-1}$) and OH^- is the hydroxide concentration in solution (mol cm^{-3}) [32,51,100]. If standard values for D ($10^{6.41}$ $cm^2\ s^{-1}$) and k ($10^{-4.83}$ $cm^3\ mol^{-1}\ s^{-1}$) are assumed, then the primary rate-controlling step of carbonate precipitation is the hydroxylation of dissolved CO_2 gas [52]. Hydroxylation requires the supply of CO_2 and OH^- to stream waters, therefore factors that interrupt or change the availability of either will impact upon the extent of tufa isotopic disequilibrium.

4.2.1. Supply of CO_2

Supply of CO_2 has increased since the onset of tufa precipitation (~1880) due to increasing atmospheric CO_2 concentrations from the burning of fossil fuels [101]. The subsequent increased greenhouse effect and planetary warming decreased atmospheric CO_2 $\delta^{13}C$ by ~1.5‰ in the 20th century [102]. The magnitude of this change is much smaller than the disequilibrium and intra-dataset variation of isotope values seen in this study. Further, if CO_2 supply was responsible for the variation in isotope data, the extent of isotopic disequilibrium should decrease with layer age. In the upstream sample the exact opposite is recorded, with $\delta^{13}C$ and $\delta^{18}O$ VPDB values decreasing and Δ_{47} values increasing along growth (Figure 2). The layers sampled in this tufa are not adjacent either, meaning that this trend may not be representative of the overall trend with layer age. This point is emphasized by the lack of a unidirectional trend in isotope values for the adjacent layers measured in the downstream sample (Figure 2). Together this suggests that increased supply of CO_2 from anthropogenic emissions is not responsible for variation in layer values.

4.2.2. Supply of OH^-

Previous studies of Howden Burn hydrochemistry have highlighted the inverse correlation between stream water pH and carbonate precipitation [63,65]. This suggests that whilst the supply of OH^- has remained relatively constant, its relative concentration in the Howden Burn varies due to dilution from rainfall surface runoff. During periods of high rainfall dilution is at its most extreme, reducing stream pH and thus the capacity for CO_2 hydroxylation and subsequent carbonate precipitation (as per Equation (9)). Mixing of circumneutral meteoric waters also supplies stream waters with non-hydroxylation related DIC that is already equilibrated to ambient temperature conditions. As a result, carbonates that precipitate during periods of higher rainfall and dilution record bulk and clumped isotope signatures trending back towards that of expected equilibrium values [33].

4.3. *Endmember Mixing*

The seasonal, daily, and sometimes hourly fluctuations of rainfall and subsequent dilution of hydroxide in the Howden Burn results in a distribution of tufa isotope data along bulk and clumped isotope mixing trends. This produces a linear spread of $\delta^{13}C$ and $\delta^{18}O$ VPDB data between disequilibrium and equilibrium endmembers (see Figure 3). The disequilibrium endmember corresponds to a hydroxylation endmember ($\delta^{13}C = -25.3$‰, $\delta^{18}O = -17.6$‰) [32]. There are two potential equilibrium endmembers for the Howden Burn; lithogenic ($\delta^{13}C$ = 0‰, $\delta^{18}O$ = 0‰) [53] or biogenic ($\delta^{13}C = -8.0$‰; $\delta^{18}O = -6.8$‰) [103]. Previous studies of Howden Burn carbonates have assumed a lithogenic endmember for calculating CO_2 sequestration rates in the stream [53]. If the same approach is applied to bulk isotope data presented here, tufas contain 55%–77% atmospheric CO_2. This is lower than the 80% average in Mayes et al. [53] but within the overall range, therefore supporting CO_2 sequestration estimates for the stream.

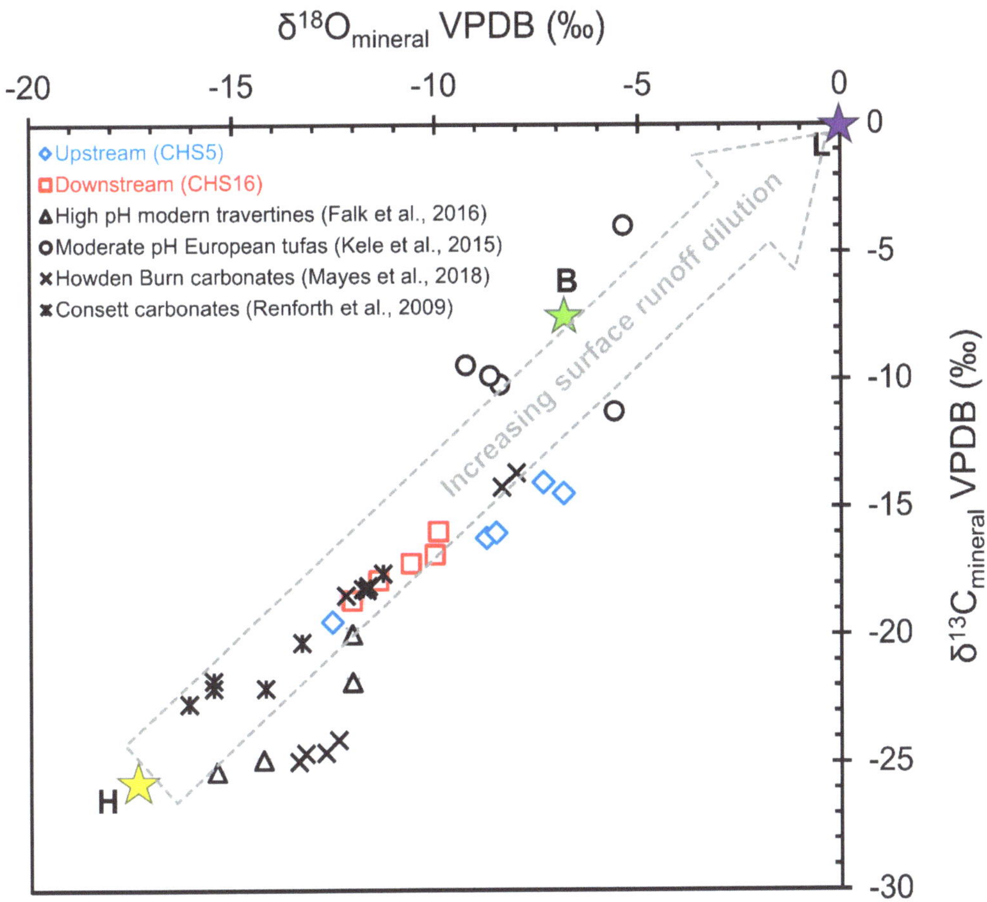

Figure 3. Plot of up- and down-stream tufa layer bulk $\delta^{13}C$ and $\delta^{18}O$ VPDB isotope data. Results from previous Consett carbonates [52,53], hyperalkaline modern travertines [33] and temperate circumneutral tufas [10] are also shown, along with values for hydroxylation (H; yellow star, from [32]), biogenic (B; green star, from [103]) and lithogenic (L; purple star, from [52,56]) isotope endmember values. Arrow shows linear mixing trend of data, which for Howden Burn carbonates corresponds to the variable extents of surface runoff dilution of hyperalkaline leachate waters.

Given there is no significant source of lithogenic carbon in the Howden Burn's catchment and the stream itself is surrounded by woodland, we suggest a biogenic equilibrium endmember may be more suitable for this carbonate system. The proximity of the woodland around the stream means that rainfall runoff and surface waters interact with significant quantities of biogenic carbon prior to entering the burn, particularly in higher rainfall months. Trends in isotope data relating to tufa layer color support this mechanism of carbon sourcing. Darker layers tend to record lesser extents of carbonate isotope disequilibrium. Assuming darker layers correspond to those rich organic materials, this suggests the main source of equilibrium, or non-hydroxylation related carbon, is biogenic rather than lithogenic. In this scenario, the proportion of atmospheric CO_2 in these tufas is $\leq$50%, which implies the CO_2 sequestration estimates of Mayes et al. [53] are towards the upper limit of plausibility.

Non-Linear Mixing

The linear mixing trend of bulk isotopes evident in Figure 3 could alternatively be interpreted as partial equilibration of DIC prior to carbonate precipitation, shifting isotope values back towards those expected from equilibrium precipitation. We know this to be unlikely due to the low temperature, high pH conditions in the Howden Burn, and the records of rapid precipitation rates [53]. However, the point remains that bulk isotope data alone cannot empirically rule out partial equilibration. The distribution of clumped isotope data (Figure 4) can provide this quantitative evidence. If partial equilibration of DIC were responsible for variation in isotope data, one would expect to see a linear mixing trend between equilibrium and hydroxylation disequilibrium endmember clumped isotope values. Instead, we see that mixing is non-linear, plotting between a hydroxylation disequilibrium endmember: Δ_{47} = 0.851‰, δ^{13}C VPDB = −27.0‰ and δ^{18}O VPDB = −16.7‰ [33] and winter (4 °C) and summer (20 °C) equilibrium endmembers for Howden Burn tufas.

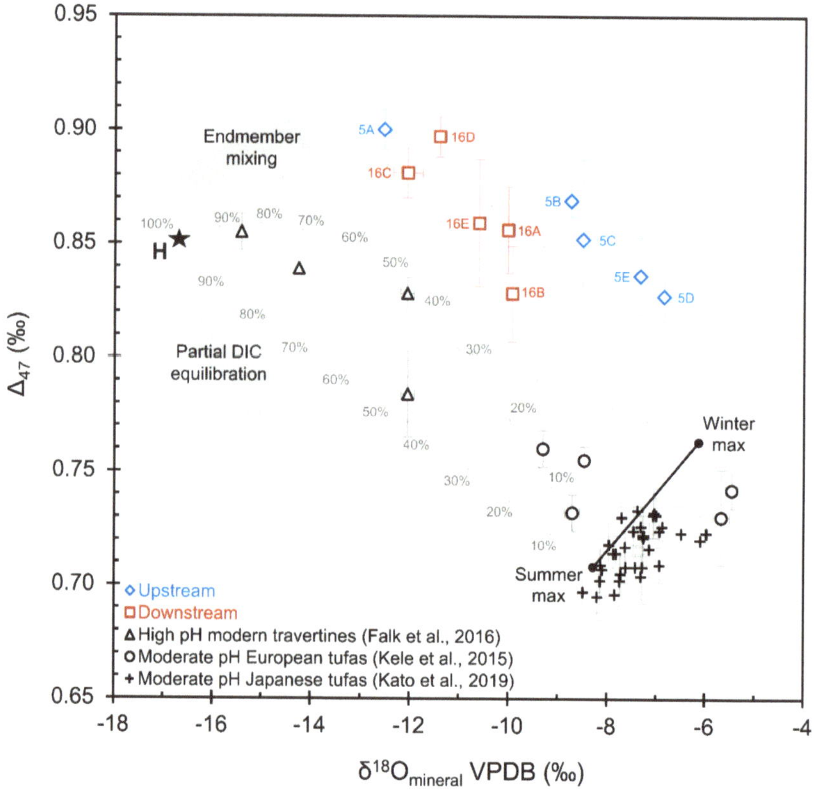

Figure 4. Plot of Δ_{47} and $\delta^{18}O_{mineral}$ VPDB values for up- and down-stream layers, with data from previous carbonate studies plotted for reference [10,94]. Disequilibrium hydroxylation (H; black star, from [33]) and seasonal equilibrium endmember values are shown and were used to construct a non-linear mixing zone for DIC endmember mixing in the Howden Burn system. Winter and summer maximum equilibrium values were calculated using water temperature records [53] and regional meteoric rainfall δ^{18}O VSMOW. Grey % values correspond to the ratio of disequilibrium hydroxylation component for any given point on the mixing zone. Linear mixing from DIC equilibration is also shown. Distribution of tufa layer data points between the equilibrium and disequilibrium endmembers corresponds to differing extents of high pH slag leachate dilution from rainfall surface runoff.

This non-linear spread of clumped isotope data is diagnostic of mixing between two distinct DIC endmembers, rather than any partial equilibration of water DIC [86,93]. Variation in apparent clumped isotope disequilibrium along this mixing curve results from changing ratios of the respective endmember components in the final DIC mix, with the largest deviance from a linear mixing trend coming from a 50:50 endmember mix [93]. This mechanism of clumped isotope disequilibrium conforms to field observations regarding rainfall dilution in the Howden Burn, with episodic rainfall dilution of high pH leachate providing the means for mixing of isotopically distinct endmembers.

4.4. Mixing Model Insights

Only half of the tufa layers measured in this study fall within error of the zone of endmember mixing, with the majority of these being from the downstream sample. There are two possible reasons for this; (1) other processes are increasing Δ_{47} that are separate from DIC endmember mixing, or (2) endmember values used to construct the non-linear mixing zone shown on Figure 4 are not wholly representative of the Howden Burn carbonate system.

4.4.1. pH Effect

We know from previous studies that precipitation rate is unlikely to have contributed any Δ_{47} increase due to the high recorded rates and high pH [5,39,40,53]. However, the speciation effect of high pH is known to produce a Δ_{47} reduction of up to 0.050‰ in CO_3^{2-} dominated solutions (see Table 1 for CO_3^{2-} dominance) [5,39,41]. This offset only reaches its maximum once pH > ~12 [39,41], meaning Δ_{47} decrease may not be as much as 0.050‰ in the Howden Burn system, particularly during high rainfall events. If data in this study are corrected for a full pH speciation effect of 0.050‰, this increases Δ_{47} further and takes all data points above and out of the non-linear mixing zone in Figure 4. This means pH speciation is not the cause of some data points plotting above the calculated zone of non-linear endmember mixing.

4.4.2. Endmember Suitability

This leaves the question of how representative mixing model endmember values are of the Howden Burn system. Equilibrium Δ_{47} calculations use temperature measurements from the Howden Burn and therefore directly correspond to this environment. Despite not being taken as part of this study, the values are sourced from an extensive and long-lasting water temperature archive of the Howden Burn (compiled in [53]) and are therefore deemed to be representative and reliable. Bulk isotope values of the hydroxylation disequilibrium endmember from Falk et al., [33] are near identical to those cited in Dietzel et al. [32], meaning the disequilibrium endmember is also appropriate for modelling this system. That leaves equilibrium $\delta^{18}O_{fluid}$ VSMOW values, which used regional summer and winter $\delta^{18}O_{fluid}$ VSMOW from Keyworth in England [91,92]. These were the most representative data available for the purposes of this study, but given Keyworth is ~150 miles south of Consett, these values may differ slightly from the actual $\delta^{18}O_{fluid}$ VSMOW of meteoric rainfall entering the Howden Burn. Further study of the burn hydrochemistry and isotopic measurements of endmember components would be needed to confirm this.

Accurate endmember bulk isotope values for the non-linear mixing model in Figure 4 are important since a relatively small variation can lead to significant differences in the extent of non-linear Δ_{47} offset. For example, bulk isotope values differing by 10‰ produce a non-linear mixing effect and Δ_{47} increase of up to ~0.025‰ compared to linear mixing, whereas a difference of 15‰ produces up to ~0.055‰ increase [8]. Given the underestimation of the model mixing curve relative to tufa isotope data, equilibrium endmember $\delta^{18}O_{mineral}$ VPDB and therefore $\delta^{18}O_{fluid}$ VSMOW would need to be higher if the mixing zone is to cover where all ten tufa layers plot [93].

4.4.3. Correlation with Environmental Conditions

Despite the slight mismatch between non-linear mixing model zone and measured Δ_{47} values, the distribution of data points on Figure 4 do provide certain notable trends. Higher upstream $\delta^{18}O_{mineral}$ VPDB values (apart from data point 5A) suggest upstream tufa layers correspond closer to winter temperature and/or water chemistry conditions, relative to downstream layers. One possible environmental explanation for this could be that during winter months increased rainfall and surface runoff dilution of leachate limits the extent of high pH carbonate precipitation to upstream areas. In the drier summer months, dilution is reduced, and CO_2 hydroxylation-related precipitation takes place further downstream. As a result, downstream carbonates preferentially record summer conditions relative to upstream deposits. However, it is important to acknowledge that the sample size presented here is relatively small. To quantitatively test if this phenomenon is genuine a larger sample size of tufa taken across various period of a seasonal year would be needed, along with accompanying water chemistry data.

Another notable observation from Figure 4 is that the majority of tufa layers plot closer to equilibrium endmembers, suggesting they have a majority equilibrium (i.e., not hydroxylation) component. This is perhaps unexpected given the recorded dominance of hydroxide in this environment (Table 1), but is likely a consequence of the complex control of rainfall in the Howden Burn system. The dilution effect of surface runoff previously discussed lowers stream water pH, reducing the capacity for CO_2 hydroxylation to occur. At the same time, higher stream discharge increases the precipitation rate of carbonate minerals such as calcite [75,76,104], increasing the volume of carbonate forming during these high rainfall, high flow and lower pH periods. This complexity highlights the importance of understanding and quantifying the controls on carbonate precipitation for interpreting trends in clumped isotope data.

4.5. *Quantifying Δ_{47} Enrichment*

Modelling studies of Defliese and Lohmann [93] and more recently Guo [6] suggest the majority of clumped isotope disequilibrium from kinetic processes such as CO_2 hydroxylation results from non-linear mixing effects rather than any intrinsic fractionation of Δ_{47}. Without a more accurate characterization of Howden Burn DIC endmember components it is not possible to quantitatively conclude whether this is the case for data in this study. Compared to linear Δ_{47} mixing, clumped isotope values in this study are between ~0.010‰ (16B) and ~0.145‰ (5B) higher than winter and summer linear mixing profiles respectively (see Figure 4). Given bulk isotope endmembers differ by 27‰ for $\delta^{13}C$ VPDB and 8‰–10‰ for $\delta^{18}O$ VPDB, these results comparable to the modelling work of Defliese and Lohmann [93], which predict a ~0.050‰ increase in Δ_{47} when endmembers differ by 15‰. The combined effects of this endmember mixing and atmosphere-hydroxide DIC sourcing increases clumped isotope values by 0.044‰–0.183‰ compared to derived equilibrium values. Given pH speciation is known to reduce clumped isotope values by up to 0.050‰ in CO_3^{2-} dominated solutions, overall Δ_{47} enrichment could be as high as 0.233‰. This extent of clumped isotope enrichment bares close resemblance to that recorded in Oman hyperalkaline carbonates, despite the differing climates and water temperatures of the two environments [33].

4.6. *Significance and Future Work*

This study provides the first clumped isotope data from anthropogenic carbonates, adding to the small number of high pH, non-equilibrium carbonate datasets recorded in the carbonate archive. The heterogeneity of tufa layer isotope values demonstrates that there are multiple competing processes controlling carbonate isotope values in high pH, riverine carbonate systems. The magnitude of clumped isotope disequilibrium and diagnostic trends present in Δ_{47} data demonstrates the potential of using anthropogenic carbonates to better understand the impact of kinetic effects on clumped isotope values. Future studies could analyze larger numbers of individual carbonate layers from multiple sites (i.e., >2) across the anthropogenic system. These investigations should be concurrent with monitoring of mineral precipitation rates and the accompanying hydrochemical conditions (e.g., real-time rainfall, stream discharge, temperature, water $\delta^{18}O$) over a full seasonal year. This will quantitatively test which of the environmental variables discussed in this submission (e.g., rainfall, pH, etc.) are the cause and dominant control of clumped isotope disequilibrium variation. Future work should also determine factors such as the organic matter content of carbonates and the consistency of isotope equilibrium endmember signatures.

5. Conclusions

The aim of this study was to investigate the clumped isotope values recorded in anthropogenic tufas and test these human-derived carbonates as an archive for evaluating the impact of kinetic fractionations on Δ_{47}. We found that, as expected, carbonate clumped isotope values do not correspond to mineral precipitation temperatures and instead record enrichments of +0.044‰–0.183‰ relative to expected equilibrium values. This range includes a negative Δ_{47} offset of up to 0.050‰ from the pH speciation effect of CO_3^{2-} dominated solutions, meaning hydroxylation-related Δ_{47} enrichment could be as high as 0.233‰. Non-linear distribution of clumped isotope data indicates DIC endmember mixing is responsible for variable extents of disequilibrium, rather than partial equilibration of DIC. Episodic dilution of hydroxide-rich stream waters by equilibrated rainfall surface runoff provides a plausible mechanism for endmember mixing in this environment.

Increases of ~0.010‰–0.145‰ in Δ_{47} values relative to linear summer and winter mixing profiles indicates that the majority of increase in clumped isotope values results from a combination of non-linear mixing effects and atmosphere-hydroxide sourcing of DIC. Through time, tufa layers inherit clumped isotope signatures that correspond to the mixing of disequilibrium (CO_2 hydroxylation) and equilibrium (surface runoff) endmembers. Layers that form during high rainfall (i.e., high dilution, lower pH) events are expected to record bulk and clumped isotope values closer to expected equilibrium values than those forming during low rainfall (i.e., low dilution, high pH) periods.

The diagnostic trends and variation in clumped isotope values present in this small dataset demonstrates the potential of anthropogenic carbonate systems as a useful archive for studying and quantifying kinetic effects in clumped isotopes.

Supplementary Materials: The following supporting information can be downloaded at: https://www.mdpi.com/article/10.3390/min12121611/s1; Figure S1: 1880 map showing the original, pre-steelworks profile of the Howden Burn. Figure S2: 1890 map showing the initial infilling of the Howden Burn profile. Figure S3: 1920 map showing the infilling of the Howden Burn after 30 years of production. Figure S4: 1930 map showing slag deposition reaching Pemberton Road. Figure S5: 1960 map showing some re-profiling of the Burn under Pemberton Road. Figure S6: 1945 aerial photograph captures the extent of slag deposition in the Howden Burn profile. Figure S7: Present day aerial view of Howden Burn and now-vegetated Consett heaps. Figure S8: Picture of upstream Howden Burn tufa hand specimen CHS5. Figure S9: Picture of downstream Howden Burn tufa hand specimen CHS16.

Author Contributions: Conceptualization, C.H. and J.M.; methodology, C.H. and J.M.; software, C.J.; validation, C.J.; formal analysis, C.H. and C.J.; investigation, C.H.; resources, C.H. and J.M.; data curation, C.H. and C.J.; writing—original draft preparation, C.H.; writing—review and editing, C.H., J.M. and C.J.; visualization, C.H. and J.M.; supervision, C.H. and J.M.; project administration, C.H. and J.M.; funding acquisition, C.H. and J.M. All authors have read and agreed to the published version of the manuscript.

Funding: C.H. was funded by an Angus Mitchell research scholarship and a Sir Alwyn Williams postgraduate research award, both from the University of Glasgow. J.M. acknowledges funding from the SAGES+ Small Grant Scheme for the clumped isotope analysis.

Data Availability Statement: The data presented in this study are openly available in the British Geological Survey (BGS) National Geoscience Data Centre (NGDC) at [89].

Acknowledgments: We would like to thank Editor and the three anonymous reviewers for their guidance and constructive suggestions that helped improve the manuscript. Will Mayes is thanked for showing the authors around the field site and for many fruitful discussions during the work. The late Paul Younger is thanked for introducing J.M. to Consett, anthropogenic tufas, and steel slags more generally.

Conflicts of Interest: The authors declare no conflict of interest. The funders had no role in the design of the study; in the collection, analyses, or interpretation of data; in the writing of the manuscript; or in the decision to publish the results.

References

1. Ghosh, P.; Adkins, J.; Affek, H.; Balta, B.; Guo, W.; Schauble, E.A.; Schrag, D.; Eiler, J.M. ^{13}C–^{18}O Bonds in Carbonate Minerals: A New Kind of Paleothermometer. *Geochim. Cosmochim. Acta* **2006**, *70*, 1439–1456. [CrossRef]
2. Schauble, E.A.; Ghosh, P.; Eiler, J.M. Preferential Formation of ^{13}C–^{18}O Bonds in Carbonate Minerals, Estimated Using First-Principles Lattice Dynamics. *Geochim. Cosmochim. Acta* **2006**, *70*, 2510–2529. [CrossRef]
3. Eiler, J.M. "Clumped-Isotope" Geochemistry—The Study of Naturally-Occurring, Multiply-Substituted Isotopologues. *Earth Planet. Sci. Lett.* **2007**, *262*, 309–327. [CrossRef]
4. Guo, W.; Mosenfelder, J.L.; Goddard, W.A.; Eiler, J.M. Isotopic Fractionations Associated with Phosphoric Acid Digestion of Carbonate Minerals: Insights from First-Principles Theoretical Modeling and Clumped Isotope Measurements. *Geochim. Cosmochim. Acta* **2009**, *73*, 7203–7225. [CrossRef]
5. Kluge, T.; John, C.M.; Boch, R.; Kele, S. Assessment of Factors Controlling Clumped Isotopes and $\delta^{18}O$ Values of Hydrothermal Vent Calcites. *Geochem. Geophys. Geosyst.* **2018**, *19*, 1844–1858. [CrossRef]
6. Guo, W. Kinetic Clumped Isotope Fractionation in the DIC-H_2O-CO_2 System: Patterns, Controls, and Implications. *Geochim. Cosmochim. Acta* **2020**, *268*, 230–257. [CrossRef]
7. Zaarur, S.; Affek, H.P.; Brandon, M.T. A Revised Calibration of the Clumped Isotope Thermometer. *Earth Planet. Sci. Lett.* **2013**, *382*, 47–57. [CrossRef]
8. Defliese, W.F.; Hren, M.T.; Lohmann, K.C. Compositional and Temperature Effects of Phosphoric Acid Fractionation on Δ_{47} Analysis and Implications for Discrepant Calibrations. *Chem. Geol.* **2015**, *396*, 51–60. [CrossRef]
9. Kluge, T.; John, C.M.; Jourdan, A.-L.; Davis, S.; Crawshaw, J. Laboratory Calibration of the Calcium Carbonate Clumped Isotope Thermometer in the 25–250 °C Temperature Range. *Geochim. Cosmochim. Acta* **2015**, *157*, 213–227. [CrossRef]
10. Kele, S.; Breitenbach, S.F.M.; Capezzuoli, E.; Meckler, A.N.; Ziegler, M.; Millan, I.M.; Kluge, T.; Deák, J.; Hanselmann, K.; John, C.M.; et al. Temperature Dependence of Oxygen- and Clumped Isotope Fractionation in Carbonates: A Study of Travertines and Tufas in the 6–95 °C Temperature Range. *Geochim. Cosmochim. Acta* **2015**, *168*, 172–192. [CrossRef]
11. Kelson, J.R.; Huntington, K.W.; Schauer, A.J.; Saenger, C.; Lechler, A.R. Toward a Universal Carbonate Clumped Isotope Calibration: Diverse Synthesis and Preparatory Methods Suggest a Single Temperature Relationship. *Geochim. Cosmochim. Acta* **2017**, *197*, 104–131. [CrossRef]
12. Bonifacie, M.; Calmels, D.; Eiler, J.M.; Horita, J.; Chaduteau, C.; Vasconcelos, C.; Agrinier, P.; Katz, A.; Passey, B.H.; Ferry, J.M.; et al. Calibration of the Dolomite Clumped Isotope Thermometer from 25 to 350 °C, and Implications for a Universal Calibration for All (Ca, Mg, Fe)CO_3 Carbonates. *Geochim. Cosmochim. Acta* **2017**, *200*, 255–279. [CrossRef]
13. Petersen, S.V.; Defliese, W.F.; Saenger, C.; Daëron, M.; Huntington, K.W.; John, C.M.; Kelson, J.R.; Bernasconi, S.M.; Colman, A.S.; Kluge, T.; et al. Effects of Improved ^{17}O Correction on Interlaboratory Agreement in Clumped Isotope Calibrations, Estimates of Mineral-Specific Offsets, and Temperature Dependence of Acid Digestion Fractionation. *Geochem. Geophys. Geosyst.* **2019**, *20*, 3495–3519. [CrossRef]
14. Davies, A.J.; John, C.M. The Clumped ($^{13}C^{18}O$) Isotope Composition of Echinoid Calcite: Further Evidence for "Vital Effects" in the Clumped Isotope Proxy. *Geochim. Cosmochim. Acta* **2019**, *245*, 172–189. [CrossRef]
15. Affek, H.P.; Bar-Matthews, M.; Ayalon, A.; Matthews, A.; Eiler, J.M. Glacial/Interglacial Temperature Variations in Soreq Cave Speleothems as Recorded by 'Clumped Isotope' Thermometry. *Geochim. Cosmochim. Acta* **2008**, *72*, 5351–5360. [CrossRef]
16. Kluge, T.; Affek, H.P.; Zhang, Y.G.; Dublyansky, Y.; Spötl, C.; Immenhauser, A.; Richter, D.K. Clumped Isotope Thermometry of Cryogenic Cave Carbonates. *Geochim. Cosmochim. Acta* **2014**, *126*, 541–554. [CrossRef]
17. Henkes, G.A.; Passey, B.H.; Grossman, E.L.; Shenton, B.J.; Pérez-Huerta, A.; Yancey, T.E. Temperature Limits for Preservation of Primary Calcite Clumped Isotope Paleotemperatures. *Geochim. Cosmochim. Acta* **2014**, *139*, 362–382. [CrossRef]
18. Rodríguez-Sanz, L.; Bernasconi, S.M.; Marino, G.; Heslop, D.; Müller, I.A.; Fernandez, A.; Grant, K.M.; Rohling, E.J. Penultimate Deglacial Warming across the Mediterranean Sea Revealed by Clumped Isotopes in Foraminifera. *Sci. Rep.* **2017**, *7*, 16572. [CrossRef]
19. Evans, D.; Sagoo, N.; Renema, W.; Cotton, L.J.; Müller, W.; Todd, J.A.; Saraswati, P.K.; Stassen, P.; Ziegler, M.; Pearson, P.N.; et al. Eocene Greenhouse Climate Revealed by Coupled Clumped Isotope-Mg/Ca Thermometry. *Proc. Natl. Acad. Sci. USA* **2018**, *115*, 1174–1179. [CrossRef]

20. Henkes, G.A.; Passey, B.H.; Grossman, E.L.; Shenton, B.J.; Yancey, T.E.; Pérez-Huerta, A. Temperature Evolution and the Oxygen Isotope Composition of Phanerozoic Oceans from Carbonate Clumped Isotope Thermometry. *Earth Planet. Sci. Lett.* **2018**, *490*, 40–50. [CrossRef]
21. Dale, A.; John, C.M.; Mozley, P.S.; Smalley, P.C.; Muggeridge, A.H. Time-Capsule Concretions: Unlocking Burial Diagenetic Processes in the Mancos Shale Using Carbonate Clumped Isotopes. *Earth Planet. Sci. Lett.* **2014**, *394*, 30–37. [CrossRef]
22. Shenton, B.J.; Grossman, E.L.; Passey, B.H.; Henkes, G.A.; Becker, T.P.; Laya, J.C.; Perez-Huerta, A.; Becker, S.P.; Lawson, M. Clumped Isotope Thermometry in Deeply Buried Sedimentary Carbonates: The Effects of Bond Reordering and Recrystallization. *Geol. Soc. Am. Bull.* **2015**, *127*, 1036–1051. [CrossRef]
23. MacDonald, J.M.; John, C.M.; Girard, J.-P. Testing Clumped Isotopes as a Reservoir Characterization Tool: A Comparison with Fluid Inclusions in a Dolomitized Sedimentary Carbonate Reservoir Buried to 2–4 Km. *Geol. Soc. Lond. Spec. Publ.* **2018**, *468*, 189–202. [CrossRef]
24. Lawson, M.; Shenton, B.J.; Stolper, D.A.; Eiler, J.M.; Rasbury, E.T.; Becker, T.P.; Phillips-Lander, C.M.; Buono, A.S.; Becker, S.P.; Pottorf, R.; et al. Deciphering the Diagenetic History of the El Abra Formation of Eastern Mexico Using Reordered Clumped Isotope Temperatures and U-Pb Dating. *GSA Bull.* **2018**, *130*, 617–629. [CrossRef]
25. Pagel, M.; Bonifacie, M.; Schneider, D.A.; Gautheron, C.; Brigaud, B.; Calmels, D.; Cros, A.; Saint-Bezar, B.; Landrein, P.; Sutcliffe, C.; et al. Improving Paleohydrological and Diagenetic Reconstructions in Calcite Veins and Breccia of a Sedimentary Basin by Combining Δ_{47} Temperature, δ^{18}Owater and U-Pb Age. *Chem. Geol.* **2018**, *481*, 1–17. [CrossRef]
26. Millán, M.I.; Machel, H.; Bernasconi, S.M. Constraining Temperatures of Formation and Composition of Dolomitizing Fluids In the Upper Devonian Nisku Formation (Alberta, Canada) With Clumped Isotopes. *J. Sediment. Res.* **2016**, *86*, 107–112. [CrossRef]
27. Mangenot, X.; Gasparrini, M.; Rouchon, V.; Bonifacie, M. Basin-Scale Thermal and Fluid Flow Histories Revealed by Carbonate Clumped Isotopes (Δ_{47})–Middle Jurassic Carbonates of the Paris Basin Depocentre. *Sedimentology* **2018**, *65*, 123–150. [CrossRef]
28. MacDonald, J.M.; Faithfull, J.W.; Roberts, N.M.W.; Davies, A.J.; Holdsworth, C.M.; Newton, M.; Williamson, S.; Boyce, A.; John, C.M. Clumped-Isotope Palaeothermometry and LA-ICP-MS U–Pb Dating of Lava-Pile Hydrothermal Calcite Veins. *Contrib. Mineral. Petrol.* **2019**, *174*, 63. [CrossRef]
29. Barnes, I.; O'Neil, J.R. Calcium-Magnesium Carbonate Solid Solutions from Holocene Conglomerate Cements and Travertines in the Coast Range of California. *Geochim. Cosmochim. Acta* **1971**, *35*, 699–718. [CrossRef]
30. Barnes, I.; O'neil, J.R. The Relationship between Fluids in Some Fresh Alpine-Type Ultramafics and Possible Modern Serpentinization, Western United States. *GSA Bull.* **1969**, *80*, 1947–1960. [CrossRef]
31. Clark, I.D.; Fontes, J.-C.; Fritz, P. Stable Isotope Disequilibria in Travertine from High PH Waters: Laboratory Investigations and Field Observations from Oman. *Geochim. Cosmochim. Acta* **1992**, *56*, 2041–2050. [CrossRef]
32. Dietzel, M.; Usdowski, E.; Hoefs, J. Chemical and $^{13}C/^{12}C$- and $^{18}O/^{16}O$-Isotope Evolution of Alkaline Drainage Waters and the Precipitation of Calcite. *Appl. Geochem.* **1992**, *7*, 177–184. [CrossRef]
33. Falk, E.S.; Guo, W.; Paukert, A.N.; Matter, J.M.; Mervine, E.M.; Kelemen, P.B. Controls on the Stable Isotope Compositions of Travertine from Hyperalkaline Springs in Oman: Insights from Clumped Isotope Measurements. *Geochim. Cosmochim. Acta* **2016**, *192*, 1–28. [CrossRef]
34. O'Neil, J.R.; Barnes, I. C^{13} and O^{18} Compositions in Some Fresh-Water Carbonates Associated with Ultramafic Rocks and Serpentinites: Western United States. *Geochim. Cosmochim. Acta* **1971**, *35*, 687–697. [CrossRef]
35. Burgener, L.K.; Huntington, K.W.; Sletten, R.; Watkins, J.M.; Quade, J.; Hallet, B. Clumped Isotope Constraints on Equilibrium Carbonate Formation and Kinetic Isotope Effects in Freezing Soils. *Geochim. Cosmochim. Acta* **2018**, *235*, 402–430. [CrossRef]
36. Clark, I.D.; Lauriol, B. Kinetic Enrichment of Stable Isotopes in Cryogenic Calcites. *Chem. Geol.* **1992**, *102*, 217–228. [CrossRef]
37. Friedman, I. Some Investigations of the Deposition of Travertine from Hot Springs—I. The Isotopic Chemistry of a Travertine-Depositing Spring. *Geochim. Cosmochim. Acta* **1970**, *34*, 1303–1315. [CrossRef]
38. Loyd, S.J.; Sample, J.; Tripati, R.E.; Defliese, W.F.; Brooks, K.; Hovland, M.; Torres, M.; Marlow, J.; Hancock, L.G.; Martin, R.; et al. Methane Seep Carbonates Yield Clumped Isotope Signatures out of Equilibrium with Formation Temperatures. *Nat. Commun.* **2016**, *7*, 12274. [CrossRef]
39. Tripati, A.K.; Hill, P.S.; Eagle, R.A.; Mosenfelder, J.L.; Tang, J.; Schauble, E.A.; Eiler, J.M.; Zeebe, R.E.; Uchikawa, J.; Coplen, T.B.; et al. Beyond Temperature: Clumped Isotope Signatures in Dissolved Inorganic Carbon Species and the Influence of Solution Chemistry on Carbonate Mineral Composition. *Geochim. Cosmochim. Acta* **2015**, *166*, 344–371. [CrossRef]
40. Watkins, J.M.; Hunt, J.D. A Process-Based Model for Non-Equilibrium Clumped Isotope Effects in Carbonates. *Earth Planet. Sci. Lett.* **2015**, *432*, 152–165. [CrossRef]
41. Hill, P.S.; Tripati, A.K.; Schauble, E.A. Theoretical Constraints on the Effects of PH, Salinity, and Temperature on Clumped Isotope Signatures of Dissolved Inorganic Carbon Species and Precipitating Carbonate Minerals. *Geochim. Cosmochim. Acta* **2014**, *125*, 610–652. [CrossRef]
42. Tang, J.; Dietzel, M.; Fernandez, A.; Tripati, A.K.; Rosenheim, B.E. Evaluation of Kinetic Effects on Clumped Isotope Fractionation (Δ_{47}) during Inorganic Calcite Precipitation. *Geochim. Cosmochim. Acta* **2014**, *134*, 120–136. [CrossRef]
43. McConnaughey, T. ^{13}C and ^{18}O Isotopic Disequilibrium in Biological Carbonates: I. Patterns. *Geochim. Cosmochim. Acta* **1989**, *53*, 151–162. [CrossRef]
44. McConnaughey, T. ^{13}C and ^{18}O Isotopic Disequilibrium in Biological Carbonates: II. In Vitro Simulation of Kinetic Isotope Effects. *Geochim. Cosmochim. Acta* **1989**, *53*, 163–171. [CrossRef]

45. Wilson, S.A.; Barker, S.L.L.; Dipple, G.M.; Atudorei, V. Isotopic Disequilibrium during Uptake of Atmospheric CO_2 into Mine Process Waters: Implications for CO_2 Sequestration. *Environ. Sci. Technol.* **2010**, *44*, 9522–9529. [CrossRef]
46. Bajnai, D.; Guo, W.; Spötl, C.; Coplen, T.B.; Methner, K.; Löffler, N.; Krsnik, E.; Gischler, E.; Hansen, M.; Henkel, D.; et al. Dual Clumped Isotope Thermometry Resolves Kinetic Biases in Carbonate Formation Temperatures. *Nat. Commun.* **2020**, *11*, 4005. [CrossRef]
47. Guo, W.; Zhou, C. Patterns and Controls of Disequilibrium Isotope Effects in Speleothems: Insights from an Isotope-Enabled Diffusion-Reaction Model and Implications for Quantitative Thermometry. *Geochim. Cosmochim. Acta* **2019**, *267*, 196–226. [CrossRef]
48. Mills, G.A.; Urey, H.C. The Kinetics of Isotopic Exchange between Carbon Dioxide, Bicarbonate Ion, Carbonate Ion and Water. *J. Am. Chem. Soc.* **1940**, *62*, 1019–1026. [CrossRef]
49. Pinsent, B.R.W.; Pearson, L.; Roughton, F.J.W. The Kinetics of Combination of Carbon Dioxide with Hydroxide Ions. *Trans. Faraday Soc.* **1956**, *52*, 1512. [CrossRef]
50. Turner, J.V. Kinetic Fractionation of Carbon-13 during Calcium Carbonate Precipitation. *Geochim. Cosmochim. Acta* **1982**, *46*, 1183–1191. [CrossRef]
51. Usdowski, E.; Hoefs, J. $^{13}C/^{12}C$ Partitioning and Kinetics of CO_2 Absorption by Hydroxide Buffer Solutions. *Earth Planet. Sci. Lett.* **1986**, *80*, 130–134. [CrossRef]
52. Renforth, P.; Manning, D.A.C.; Lopez-Capel, E. Carbonate Precipitation in Artificial Soils as a Sink for Atmospheric Carbon Dioxide. *Appl. Geochem.* **2009**, *24*, 1757–1764. [CrossRef]
53. Mayes, W.M.; Riley, A.L.; Gomes, H.I.; Brabham, P.; Hamlyn, J.; Pullin, H.; Renforth, P. Atmospheric CO_2 Sequestration in Iron and Steel Slag: Consett, County Durham, United Kingdom. *Environ. Sci. Technol.* **2018**, *52*, 7892–7900. [CrossRef]
54. Kelemen, P.B.; Matter, J. In Situ Carbonation of Peridotite for CO_2 Storage. *Proc. Natl. Acad. Sci. USA* **2008**, *105*, 17295–17300. [CrossRef]
55. Rodríguez-Berriguete, Á.; Alonso-Zarza, A.M.; Martín-García, R.; Cabrera, M.d.C. Sedimentology and Geochemistry of a Human-Induced Tufa Deposit: Implications for Palaeoclimatic Research. *Sedimentology* **2018**, *65*, 2253–2277. [CrossRef]
56. Andrews, J.E.; Gare, S.G.; Dennis, P.F. Unusual Isotopic Phenomena in Welsh Quarry Water and Carbonate Crusts. *Terra Nova* **1997**, *9*, 67–70. [CrossRef]
57. Moyce, E.B.A.; Milodowski, A.E.; Morris, K.; Shaw, S. Herbert's Quarry, South Wales—An Analogue for Host-Rock Alteration at a Cementitious Radioactive Waste Repository? *Mineral. Mag.* **2015**, *79*, 1407–1418. [CrossRef]
58. Ettler, V.; Zelená, O.; Mihaljevič, M.; Šebek, O.; Strnad, L.; Coufal, P.; Bezdička, P. Removal of Trace Elements from Landfill Leachate by Calcite Precipitation. *J. Geochem. Explor.* **2006**, *88*, 28–31. [CrossRef]
59. Manning, D.A.C. Calcite Precipitation in Landfills: An Essential Product of Waste Stabilization. *Mineral. Mag.* **2001**, *65*, 603–610. [CrossRef]
60. Bayless, E.R.; Schulz, M.S. Mineral Precipitation and Dissolution at Two Slag-Disposal Sites in Northwestern Indiana, USA. *Environ. Geol.* **2003**, *45*, 252–261. [CrossRef]
61. Roadcap, G.S.; Kelly, W.R.; Bethke, C.M. Geochemistry of Extremely Alkaline (PH > 12) Ground Water in Slag-Fill Aquifers. *Groundwater* **2005**, *43*, 806–816. [CrossRef] [PubMed]
62. Huijgen, W.J.J.; Comans, R.N.J. Carbonation of Steel Slag for CO_2 Sequestration: Leaching of Products and Reaction Mechanisms. *Environ. Sci. Technol.* **2006**, *40*, 2790–2796. [CrossRef] [PubMed]
63. Mayes, W.M.; Younger, P.L.; Aumônier, J. Hydrogeochemistry of Alkaline Steel Slag Leachates in the UK. *Water Air Soil Pollut.* **2008**, *195*, 35–50. [CrossRef]
64. Hall, C.; Large, D.J.; Adderley, B.; West, H.M. Calcium Leaching from Waste Steelmaking Slag: Significance of Leachate Chemistry and Effects on Slag Grain Mineralogy. *Miner. Eng.* **2014**, *65*, 156–162. [CrossRef]
65. Riley, A.L.; Mayes, W.M. Long-Term Evolution of Highly Alkaline Steel Slag Drainage Waters. *Environ. Monit. Assess.* **2015**, *187*, 463. [CrossRef]
66. Washbourne, C.-L.; Lopez-Capel, E.; Renforth, P.; Ascough, P.L.; Manning, D.A.C. Rapid Removal of Atmospheric CO_2 by Urban Soils. *Environ. Sci. Technol.* **2015**, *49*, 5434–5440. [CrossRef]
67. Washbourne, C.-L.; Renforth, P.; Manning, D.A.C. Investigating Carbonate Formation in Urban Soils as a Method for Capture and Storage of Atmospheric Carbon. *Sci. Total Environ.* **2012**, *431*, 166–175. [CrossRef]
68. Moyce, E.B.A.; Rochelle, C.; Morris, K.; Milodowski, A.E.; Chen, X.; Thornton, S.; Small, J.S.; Shaw, S. Rock Alteration in Alkaline Cement Waters over 15 Years and Its Relevance to the Geological Disposal of Nuclear Waste. *Appl. Geochem.* **2014**, *50*, 91–105. [CrossRef]
69. Martin, L.H.J.; Leemann, A.; Milodowski, A.E.; Mäder, U.K.; Münch, B.; Giroud, N. A Natural Cement Analogue Study to Understand the Long-Term Behaviour of Cements in Nuclear Waste Repositories: Maqarin (Jordan). *Appl. Geochem.* **2016**, *71*, 20–34. [CrossRef]
70. Brocken, H.; Nijland, T.G. White Efflorescence on Brick Masonry and Concrete Masonry Blocks, with Special Emphasis on Sulfate Efflorescence on Concrete Blocks. *Constr. Build. Mater.* **2004**, *18*, 315–323. [CrossRef]
71. Macleod, G.; Fallick, A.E.; Hall, A.J. The Mechanism of Carbonate Growth on Concrete Structures, as Elucidated by Carbon and Oxygen Isotope Analyses. *Chem. Geol. Isot. Geosci. Sect.* **1991**, *86*, 335–343. [CrossRef]

72. Hull, S.L.; Oty, U.V.; Mayes, W.M. Rapid Recovery of Benthic Invertebrates Downstream of Hyperalkaline Steel Slag Discharges. *Hydrobiologia* **2014**, *736*, 83–97. [CrossRef]
73. Harber, A.J.; Forth, R.A. The Contamination of Former Iron and Steel Works Sites. *Environ. Geol.* **2001**, *40*, 324–330. [CrossRef]
74. Ford, T.D.; Pedley, H.M. A Review of Tufa and Travertine Deposits of the World. *Earth-Sci. Rev.* **1996**, *41*, 117–175. [CrossRef]
75. Kano, A.; Matsuoka, J.; Kojo, T.; Fujii, H. Origin of Annual Laminations in Tufa Deposits, Southwest Japan. *Palaeogeogr. Palaeoclimatol. Palaeoecol.* **2003**, *191*, 243–262. [CrossRef]
76. Kawai, T.; Kano, A.; Hori, M. Geochemical and Hydrological Controls on Biannual Lamination of Tufa Deposits. *Sediment. Geol.* **2009**, *213*, 41–50. [CrossRef]
77. Capezzuoli, E.; Gandin, A.; Pedley, M. Decoding Tufa and Travertine (Fresh Water Carbonates) in the Sedimentary Record: The State of the Art. *Sedimentology* **2014**, *61*, 1–21. [CrossRef]
78. Adlan, Q.; Davies, A.J.; John, C.M. Effects of Oxygen Plasma Ashing Treatment on Carbonate Clumped Isotopes. *Rapid Commun. Mass Spectrom.* **2020**, *34*, e8802. [CrossRef]
79. Huntington, K.W.; Eiler, J.M.; Affek, H.P.; Guo, W.; Bonifacie, M.; Yeung, L.Y.; Thiagarajan, N.; Passey, B.; Tripati, A.; Daëron, M.; et al. Methods and Limitations of 'Clumped' CO_2 Isotope (Δ_{47}) Analysis by Gas-Source Isotope Ratio Mass Spectrometry. *J. Mass Spectrom.* **2009**, *44*, 1318–1329. [CrossRef]
80. Cruset, D.; Cantarero, I.; Travé, A.; Vergés, J.; John, C.M. Crestal Graben Fluid Evolution during Growth of the Puig-Reig Anticline (South Pyrenean Fold and Thrust Belt). *J. Geodyn.* **2016**, *101*, 30–50. [CrossRef]
81. Dennis, K.J.; Affek, H.P.; Passey, B.H.; Schrag, D.P.; Eiler, J.M. Defining an Absolute Reference Frame for 'Clumped' Isotope Studies of CO_2. *Geochim. Cosmochim. Acta* **2011**, *75*, 7117–7131. [CrossRef]
82. John, C.M.; Bowen, D. Community Software for Challenging Isotope Analysis: First Applications of 'Easotope' to Clumped Isotopes. *Rapid Commun. Mass Spectrom.* **2016**, *30*, 2285–2300. [CrossRef] [PubMed]
83. Bernasconi, S.M.; Hu, B.; Wacker, U.; Fiebig, J.; Breitenbach, S.F.M.; Rutz, T. Background Effects on Faraday Collectors in Gas-Source Mass Spectrometry and Implications for Clumped Isotope Measurements. *Rapid Commun. Mass Spectrom.* **2013**, *27*, 603–612. [CrossRef] [PubMed]
84. Meckler, A.N.; Ziegler, M.; Millán, M.I.; Breitenbach, S.F.M.; Bernasconi, S.M. Long-Term Performance of the Kiel Carbonate Device with a New Correction Scheme for Clumped Isotope Measurements. *Rapid Commun. Mass Spectrom.* **2014**, *28*, 1705–1715. [CrossRef] [PubMed]
85. Bernasconi, S.M.; Müller, I.A.; Bergmann, K.D.; Breitenbach, S.F.M.; Fernandez, A.; Hodell, D.A.; Jaggi, M.; Meckler, A.N.; Millan, I.; Ziegler, M. Reducing Uncertainties in Carbonate Clumped Isotope Analysis Through Consistent Carbonate-Based Standardization. *Geochem. Geophys. Geosyst.* **2018**, *19*, 2895–2914. [CrossRef]
86. Eiler, J.M.; Schauble, E. $^{18}O^{13}C^{16}O$ in Earth's Atmosphere. *Geochim. Cosmochim. Acta* **2004**, *68*, 4767–4777. [CrossRef]
87. Guo, W.; Eiler, J.M. Temperatures of Aqueous Alteration and Evidence for Methane Generation on the Parent Bodies of the CM Chondrites. *Geochim. Cosmochim. Acta* **2007**, *71*, 5565–5575. [CrossRef]
88. Brand, W.A.; Assonov, S.S.; Coplen, T.B. Correction for the ^{17}O Interference in $\delta(^{13}C)$ Measurements When Analyzing CO_2 with Stable Isotope Mass Spectrometry (IUPAC Technical Report). *Pure Appl. Chem.* **2010**, *82*, 1719–1733. [CrossRef]
89. Holdsworth, C.; MacDonald, J.; John, C. *Clumped isotope data from high-pH anthropogenic tufa in the Howden Burn, Consett, North East England*; NERC EDS National Geoscience Data Centre: Nottingham, UK, 2022. [CrossRef]
90. Kim, S.-T.; O'Neil, J.R. Equilibrium and Nonequilibrium Oxygen Isotope Effects in Synthetic Carbonates. *Geochim. Cosmochim. Acta* **1997**, *61*, 3461–3475. [CrossRef]
91. Darling, W.G.; Bath, A.H.; Talbot, J.C. The O and H Stable Isotope Composition of Freshwaters in the British Isles. 2. Surface Waters and Groundwater. *Hydrol. Earth Syst. Sci.* **2003**, *7*, 183–195. [CrossRef]
92. Darling, W.G.; Talbot, J.C. The O and H Stable Isotope Composition of Freshwaters in the British Isles. 1. Rainfall. *Hydrol. Earth Syst. Sci.* **2003**, *7*, 163–181. [CrossRef]
93. Defliese, W.F.; Lohmann, K.C. Non-Linear Mixing Effects on Mass-47 CO_2 Clumped Isotope Thermometry: Patterns and Implications. *Rapid Commun. Mass Spectrom.* **2015**, *29*, 901–909. [CrossRef]
94. Kato, H.; Amekawa, S.; Kano, A.; Mori, T.; Kuwahara, Y.; Quade, J. Seasonal Temperature Changes Obtained from Carbonate Clumped Isotopes of Annually Laminated Tufas from Japan: Discrepancy between Natural and Synthetic Calcites. *Geochim. Cosmochim. Acta* **2019**, *244*, 548–564. [CrossRef]
95. Létolle, R.; Gégout, P.; Moranville-Regourd, M.; Gaveau, B. Carbon-13 and Oxygen-18 Mass Spectrometry as a Potential Tool for the Study of Carbonate Phases in Concretes. *J. Am. Ceram. Soc.* **1990**, *73*, 3617–3625. [CrossRef]
96. Böttcher, M.E.; Neubert, N.; Escher, P.; von Allmen, K.; Samankassou, E.; Nägler, T.F. Multi-Isotope (Ba, C, O) Partitioning during Experimental Carbonatization of a Hyper-Alkaline Solution. *Geochemistry* **2018**, *78*, 241–247. [CrossRef]
97. Zeebe, R.E. Oxygen Isotope Fractionation between Water and the Aqueous Hydroxide Ion. *Geochim. Cosmochim. Acta* **2020**, *289*, 182–195. [CrossRef]
98. Weise, A.; Kluge, T. Isotope Exchange Rates in Dissolved Inorganic Carbon between 40 °C and 90 °C. *Geochim. Cosmochim. Acta* **2020**, *268*, 56–72. [CrossRef]
99. Uchikawa, J.; Zeebe, R.E. The Effect of Carbonic Anhydrase on the Kinetics and Equilibrium of the Oxygen Isotope Exchange in the CO2–H2O System: Implications for Δ18O Vital Effects in Biogenic Carbonates. *Geochim. Cosmochim. Acta* **2012**, *95*, 15–34. [CrossRef]

100. Danckwerts, P.V.; Lannus, A. Gas-Liquid Reactions. *J. Electrochem. Soc.* **1970**, *117*, 369C. [CrossRef]
101. Blunden, J.; Arndt, D.S. State of the Climate in 2018. *Bull. Am. Meteorol. Soc.* **2019**, *100*, Si-S306. [CrossRef]
102. Francey, R.J.; Allison, C.E.; Etheridge, D.M.; Trudinger, C.M.; Enting, I.G.; Leuenberger, M.; Langenfelds, R.L.; Michel, E.; Steele, L.P. A 1000-Year High Precision Record of $\delta^{13}C$ in Atmospheric CO_2. *Tellus B* **1999**, *51*, 170–193. [CrossRef]
103. Cerling, T.E. The Stable Isotopic Composition of Modern Soil Carbonate and Its Relationship to Climate. *Earth Planet. Sci. Lett.* **1984**, *71*, 229–240. [CrossRef]
104. Dreybrodt, W.; Buhmann, D. A Mass Transfer Model for Dissolution and Precipitation of Calcite from Solutions in Turbulent Motion. *Chem. Geol.* **1991**, *90*, 107–122. [CrossRef]

minerals

MDPI

Article

Controls on Mg/Ca Ratios in Recent Stromatolites: Insights from Fluvial Systems in the Iberian Range (Spain)

Luis F. Auqué [1], M. Cinta Osácar [1,2,*], Concha Arenas [1,2], Neven Cukrov [3], Sonja Lojen [4] and Carlos Sancho [†]

1 Departamento de Ciencias de la Tierra, Universidad de Zaragoza, Pedro Cerbuna 12, 50009 Zaragoza, Spain
2 Institute for Research on Environmental Sciences of Aragón (IUCA) and Geotransfer Group, University of Zaragoza, 50009 Zaragoza, Spain
3 Division for Marine and Environmental Research, Ruder Bošković Institute, 10000 Zagreb, Croatia
4 Department of Environmental Sciences, Jožef Stefan Institute, Jamova cesta 39, 1000 Ljubljana, Slovenia
* Correspondence: cinta@unizar.es
† Deceased.

Abstract: The utility of the Mg/Ca elemental ratio of calcite ($(Mg/Ca)_{calcite}$) as a temperature indicator in continental carbonate deposits is a matter of debate due to the different results obtained by diverse authors. In this study, we aimed to test the reliability of the $(Mg/Ca)_{calcite}$ in fluvial carbonates. We selected the recent tufa stromatolite records of four rivers on the Iberian Peninsula for the trace element analysis based on six-monthly sampling. Previous sedimentary and hydrological studies on these fluvial basins provided the information for this work. The water temperature estimates for the stromatolite $(Mg/Ca)_{calcite}$ substantially differed from the measured water temperatures in most of the studied cases. We thus assessed other factors that participate in the control of the Mg partitioning between water and calcite. The correction of the detrital Mg content yielded water temperatures that matched the measured ones in one of the rivers. The $(Mg/Ca)_{water}$, water discharge and calcite precipitation rates may also occasionally influence the $(Mg/Ca)_{calcite}$. The six-month behaviour of some of these parameters could interfere with the relationship between the $(Mg/Ca)_{calcite}$ and water temperature. According to these results, and their comparison with other non-marine carbonates, the wide variety of parameters that are involved in the $(Mg/Ca)_{calcite}$ limit it as a geochemical thermometer in continental sedimentary environments.

Keywords: recent fluvial stromatolites; Mg partitioning; $(Mg/Ca)_{calcite}$ application; water temperature; geochemical thermometer reliability

Citation: Auqué, L.F.; Osácar, M.C.; Arenas, C.; Cukrov, N.; Lojen, S.; Sancho, C. Controls on Mg/Ca Ratios in Recent Stromatolites: Insights from Fluvial Systems in the Iberian Range (Spain). *Minerals* **2023**, *13*, 57. https://doi.org/10.3390/min13010057

Academic Editors: Francesca Giustini and Mauro Brilli

Received: 30 November 2022
Revised: 22 December 2022
Accepted: 25 December 2022
Published: 29 December 2022

1. Introduction

The use of trace elements in carbonate deposits as climate proxies has been a matter of debate for a long time. In the case of the Mg/Ca ratio in sedimentary nonmarine carbonates, we lack a consensus on its application to the geological record to obtain temperature data because of the varied and usually negative or inconclusive results that researchers have obtained in several studies [1–8]. One important problem is the presence of Mg-bearing clay minerals and other detrital particles (e.g., carbonates) that can be present in small amounts in the carbonate deposits, which is added, in the sample analysis, to the Mg incorporated in the calcite lattice (e.g., [6]) and produce misleading temperature results for the calcite precipitation. The estimation of the contribution of allochthonous Mg to the total amount of Mg in the rock is not straightforward and requires meticulous mineralogical analysis [1]. Moreover, the influence of the biological activity on the Mg partitioning [4,7,8], or the changes in the Mg/Ca ratios of the water that are not related to temperature [2,3], could affect the Mg/Ca ratios in fluvial carbonate deposits. These inconsistencies could be the result of the complex hydrology in carbonate aquifers, the seasonal patterns of the water recharge in some karst systems (e.g., [5]), prior calcite precipitation within the aquifer (e.g., [9]) or the anthropogenic contamination of the water (e.g., [3]). In some cases, the causes of the Mg/Ca variations in water and minerals over time cannot be depicted (e.g., [2]). The use of the isotope compositions of trace metals that coprecipitate with calcite (Mg, U) as a prospective method for estimating the sources of Mg in authigenic carbonates ([10–12] and the references therein) might be an additional way to clarify the sometimes-puzzling Mg/Ca variations in some tufa-precipitating systems.

In this work, we compared the recent fluvial stromatolite records from four river basins in the Iberian Range, Spain. The four basins have similar Mediterranean continental climate conditions. However, the deposition rates, stable isotope composition of the water and carbonate deposits formed in each case are noticeably different. The available data on the depositional rates and isotopic, hydrochemical and climatic parameters for each basin taken every six months, over a period from 3 to 13 years, as well as the corresponding recent stromatolite records, offer an excellent opportunity to decipher whether—or under which conditions—geochemical proxies, such as the Mg/Ca ratio, are reliable indicators of the climatic conditions in sedimentary carbonates.

Therefore, in this work, we aimed to test the reliability of the Mg/Ca ratio in fluvial tufa calcite as the depositional (detrital sediment input), hydrological (discharge, in-aquifer residence time of water) and climate (temperature, precipitation, evaporation) indicators of past environments. For this purpose, we analysed the Mg/Ca ratios of the recent six-month fluvial stromatolite records and their water counterparts obtained from 1999 to 2012 for four river valleys in the Iberian Range, Spain. We compared the calculated temperature values with the corresponding measured water temperatures and the water temperatures calculated from the tufa $\delta^{18}O_{calcite}$ in the same records by the authors of [13]. We analysed the influence of the different parameters on the tufa Mg content, and we compared the results with similar outcomes for other fluvial systems. According to our findings, the use of the Mg content in calcite as a temperature proxy is difficult because of the complex interplay of the other acting parameters (e.g., the water discharge and calcite precipitation rates). Specifically, the influence of the Mg derived from detrital minerals may be responsible for the generally higher temperatures yielded by this proxy.

2. Location, Geological Context, Climate and Hydrology

The Añamaza, Piedra, Mesa and Ebrón Rivers are located along a 210 km long, nearly N–S transect across the Iberian Range (Figure 1). The Iberian Range is an Alpine intraplate mountain range that was formed as the result of the convergence of the Eurasian and Iberian plates [14]. The Paleozoic-to-Cenozoic succession is formed mainly of siliciclastic and carbonate sedimentary rocks. The Mesozoic sequence includes thick and extensive successions of carbonate rocks that make up most of the aquifers in the region. This sequence is overlain by Cenozoic fluvial and lacustrine detrital and carbonate rocks [14]. A number of valleys include Pleistocene and Holocene conspicuous tufa buildups [15–19]. In many of these valleys, the present-day rivers continue to form tufa deposits via a long stretch downstream of the headwater springs [19–22].

The river sections studied by the authors of [19,21–23] in the four fluvial basins (Añamaza, Piedra, Mesa and Ebrón) have similar lengths, elevations and topographies, and they share the general Mediterranean continental climate conditions. The four rivers are mainly sourced from Jurassic carbonate rock aquifers (plus Upper Cretaceous carbonate rock aquifers in the case of the Piedra and Mesa Rivers), which supply water of the HCO_3–Ca (Piedra, Ebrón and Mesa) and SO_4–HCO_3–Ca (Añamaza) types. Most of the water is not only provided by the springs that are located in the uppermost reaches of each fluvial valley, but also by the variable contributions from the springs along the river paths, which are greater along the Ebrón River.

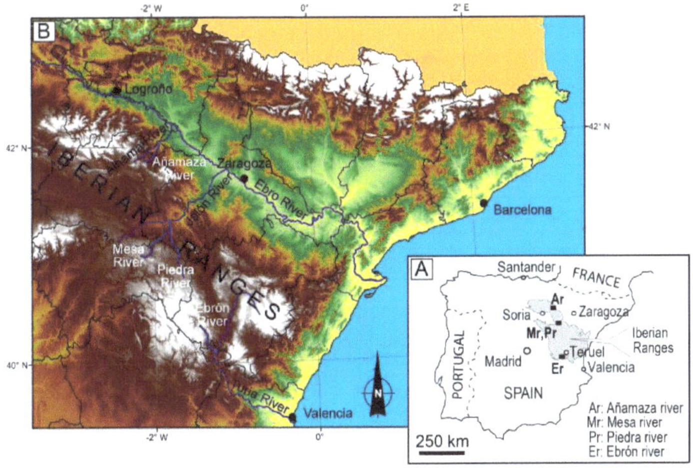

Figure 1. (**A**,**B**) Locations of the studied rivers in Iberian Ranges (NE Iberian Peninsula).

The Ebrón and Mesa Rivers have similar mean water discharge values (1.49 and 1.5 m^3/s, respectively), which are slightly higher than that of the Piedra River (1.26 m^3/s). The Añamaza River has a much smaller discharge value (0.21 m^3/s). The longitudinal profile of the Añamaza River has the highest gradient (mean slope of 1.9%), those of the Ebrón and Piedra Rivers have more moderate gradients (1.4% and 1.3%, respectively) and that of the Mesa River has the lowest gradient (1%).

3. Characteristics of Stromatolite Formation in the Studied Rivers

The authors of [19,21–23] performed sedimentary and hydrochemical studies on the calcite-depositing rivers: Añamaza, Ebrón, Mesa and Piedra Rivers (hereafter referred to as RA, RE, RM and RP, respectively), and they indicated that the deposition rates were strongly controlled by the mechanical CO_2 outgassing in the four cases, with minor contribution from the biological parameters, such as the photosynthetic CO_2 uptake. The highest deposition rates corresponded to the fast flow conditions along the gently to highly inclined river sites in which the stromatolite deposits were formed (mean measured deposition rates: RM: 4.0 mm/y; RE: 4.4 mm/y; RP: 13.7 mm/y; RA: 18.3 mm/y). The researchers measured from moderate to high deposition rates in the varied-sized waterfalls with continuous and turbulent flows, in which crudely laminated deposits of moss, macroscopic algae and minor stromatolites had formed (mean measured deposition rates: RM: 2.4 mm/y; in RE: 7.4 mm/y; RP: 8.2 mm/y; RA: 8.4 mm/y). The lower rates were those of the slow flowing water areas, in which the sediment consisted of loose lime mud, small phytoclasts and oncoids, as well as uneven and thin stromatolites. Other contexts, such as spray areas or caves, yielded much smaller deposition rates. In all cases, the deposition rates were much higher in the warm periods (i.e., spring and summer months) than in the cool periods (i.e., autumn and winter months). The six-month difference was explained by the differential effects of the temperature and temperature-related parameters on the calcite solubility in the water and on the biota activity, the latter either through plants or prokaryotes that act as substrates for calcite precipitation, or even through the photosynthesis process [24,25]. However, despite these similarities, there are large variations in the deposition rates between the four rivers due to differences in the discharges, water chemical compositions, riverbed slopes and water temperatures [21]. Among the studied facies, the most continuous records were for the stromatolite deposits, which allowed for the identification of the successive six-month period intervals that were monitored on site and in the laboratory (Figure 2). Therefore, we selected a number of stromatolite deposits formed on tablets for this study. These stromatolite deposits consist of thicker and denser composite laminae that alternate with thinner and more porous composite laminae. The former includes up to six simple laminae, and the latter includes up to four [26]. The laminae primarily consist of cyanobacterial calcite tubes that form palisades, fan-shaped calcite bodies and/or intertwined-calcite-tube mats (Figure 3). According to the RNA determinations from the same sites as the tablets, most of the studied samples coincided with *Phormidium incrustatum* [27,28]. In addition, diatoms, insect cavities and extracellular polymeric substances (EPSs) were variably present between the tubes (Figure 3).

Figure 2. (**A**) Field view of site where Tablet RP-16 was installed in the River Piedra [19]; (**B**) Plan view of Tablet RP-16 once removed from the river. Line indicates position for cutting; (**C**,**D**) Cross sections of Tablets RP-16 (P-13) and M-4, with sections perpendicular to flow direction. We indicate the thickness of each six-month interval measured at corresponding cutting sections.

We selected the stromatolite records that were suitable for this study as follows: For the RE and RA rivers, the highest deposition rates were measured in Tablets RE-8 and RA-6 (8.44 mm/y over 3.5 years, and 18.3 mm/y over 3 years, respectively). For RP, Tablet R-14 provided the longest and most continuous record (12 years, with a mean rate of 16.02 mm/y). For the RM, Tablet M-4 had a continuous record that was suitable for this study (3.5 years, with a mean rate of 5.45 mm/y).

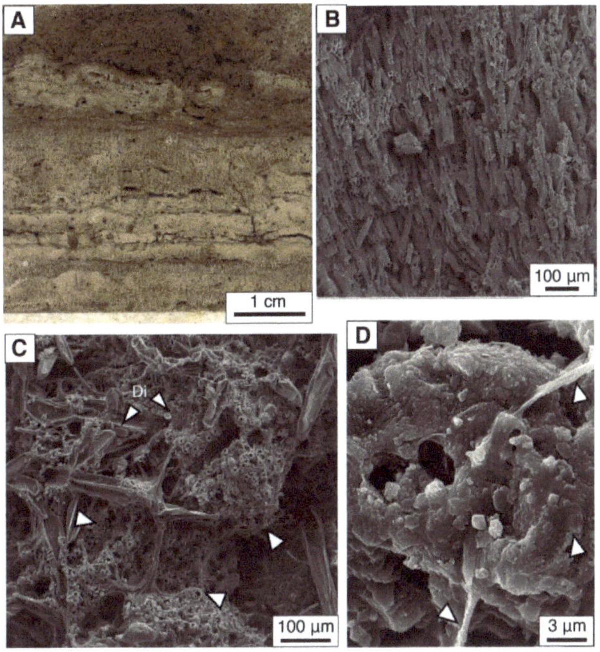

Figure 3. (**A**) Detail of lamination in cross-section of Tablet RP-16 [19] (six-month intervals as in Figure 2A); (**B**) SEM image of Tablet RP-14: palisades of calcite tubes from filamentous cyanobacteria; top is upward; (**C**,**D**) SEM images of Tablet M-4; (**C**) Calcite tubes (plan view) and EPS-containing diatoms. (**D**) Detail of calcite tubes and EPS (arrows).

4. Methods

4.1. Sediment Sampling and Analyses

The researchers obtained the sediment samples from tufa stromatolites developed on limestone tablets. The authors of [13,19,21–23,25] gathered the sediment records and the related water parameters in a multidisciplinary study performed between 1999 and 2012. They installed the tablets (25 × 15 × 2 cm) on the riverbeds at several places that represented different environmental conditions. They set the tablets at the end of the winter, and they removed them at the end of the summer to measure their thickness in the laboratory. After a week, they returned the same tablets back to the same places until the next semester. The authors of [25] explain the procedure in detail. Once they definitively removed the tablets, they cut them perpendicular to the depositional surfaces, and they identified the six-month intervals (hereafter, we refer to spring and summer as the warm period, and we refer to autumn and winter as the cool period) by plotting the corresponding measurements on the raw cuts (see [29] for details).

From these sections, they took from one to two sediment samples from each six-month interval with a punch and a microdrill (Navfram model N120 Micromotor 25.000 rpm with electronic speed regulator, AB SHOT TECNICS SL, Cervello, Barcelona, Spain). They ground the powdered matter and sieved it through a 50 µm size mesh, and they then separated it for different uses.

For the Ca and Mg analyses, they homogenised each stromatolite sample. They digested the subsamples of approximately 0.2 g using a mixture of concentrated Suprapur nitric acid (2.5 mL) and hydrochloric acid (7.5 mL) (all Merck), in closed Teflon crucibles (V = 35 cm^3) on a hotplate at a temperature of 170 °C, and then in open Teflon crucibles (V = 35 cm^3) at a temperature of 220 °C.

The researchers used a high-resolution inductively coupled plasma mass spectrometer (HR ICP-MS) (Element 2, Thermo Finnigan, Bremen, Germany) for the determination of the Ca and Mg concentrations. We present the results for all the six-month sediment intervals recorded on the tablets of the four rivers in Table S1 (Supplementary Material).

The stable isotope composition values of the sediment and water ($\delta^{13}C$ and $\delta^{18}O$ ‰VPDB) of the RA, RE and RP used in this paper are those published in [13]. The stable isotope composition values of the sediment and water of the RM presented in this contribution were obtained with the same technique described by [13].

The researchers determined the bulk mineralogical composition of the sediment by powder X-ray diffraction using a Phillips PW 1729 diffractometer (Crystallography and Mineralogy Division of the University of Zaragoza, Spain). According to the X-ray diffraction, all the stromatolite samples were composed of low-Mg calcite with minor amounts of detrital particles such as quartz, clay minerals and, occasionally, dolomite [19,21–23]). To assess the possible contribution of these detrital minerals to the chemical composition of the analysed samples, the researchers studied the clay mineral content using the RP as an example. For this purpose, they took five sediment samples (stromatolites) along the river in July 2011. They ground and sieved the samples, and they separated the < 2 μm sediment fraction by centrifugation, and they then analysed it for both the mineral (by means of oriented aggregates) and chemical (with the same methodology indicated above for the stromatolite samples) compositions. They performed a semi-quantitative determination of the clay minerals with the reference intensity ratio (RIR) method, and by using the RIR values of [30] taken from [31].

4.2. Water Sampling and Analysis

The researchers obtained the water samples for the chemical analysis at the sites in the four rivers that coincided with the tablet sites. They conducted biannual sampling (approximately in the middle of the warm and cool periods, i.e., at the end of June and in January, respectively) from June 2004 to December 2009 at the RP, from June 2004 to June 2006 at the RM, from December 2007 to December 2009 at the RA and from June 2007 to December 2009 at the RE.

They measured the water temperature (Tw) and pH on site using a portable pH meter (Jenway 4200; Bibby Scientific Limited, Stone, UK). They filtered the samples using a 0.45 μm Millipore cellulose filter, and they acidified them with ultra-pure HNO_3 to a pH of < 2 for the cation analyses. They determined the alkalinity, SO_4, Cl, Ca, Mg, Na and K with the same protocols and analytical methods (as is detailed, for instance, in [19]) for the four rivers. The charge imbalance percentage for the analytical data used in this study was always below 10% as calculated with the PHREEQC code (see below). We present the analytical results for the study sites on the four rivers in Table S2 (Supplementary Material).

4.3. Geochemical Modelling Calculations

The researchers performed the speciation–solubility calculations to obtain the values of the calcite saturation index (SIc), total dissolved inorganic carbon (TDIC) and partial pressure of CO_2 (pCO_2) values for the water samples with the PHREEQC code [32], and with the WATEQ4F thermodynamic database distributed with the code.

They calculated the inorganic precipitation rate for the calcite (mmol/cm^2/s) using the rate law in [33], which is commonly referred to as the PWP (Plummer, Wigley, Parkhurst) rate equation:

$$R = -\kappa_1 aH^+ - \kappa_2 aH_2CO_3{}^* - \kappa_3 aH_2O + \kappa_4 aCa^{2+} aHCO_3{}^- \quad (1)$$

where $H_2CO_3{}^* = H_2CO_3{}^0 + CO_{2(aq)}$, and $\kappa_1, \kappa_2, \kappa_3$ *and* κ_4 are the empirically determined rate constants. They used the temperature functions proposed by Plummer et al. (1978) for the κ_1, κ_2 and κ_3 rate constants [33]. The rate constant κ_4 depends on the temperature and pCO_2. The researchers used the equation proposed in [34] by fitting it to the empirical data of Plummer et al. (1978) [34].

The empirical rate (Equation (1)) was originally provided for calcite dissolution; however, it is also applicable to precipitation [33,35]. In the form presented in Equation (1), the negative values correspond to the dissolution rates and the positive values to the precipitation rates. Researchers frequently use this equation for tufa-depositing streams [2,19,21,23,36–42] because it provides the maximum rate of the inorganic precipitation in turbulent water [43]. The rate is usually larger than the actual precipitation rate; however, we can achieve a reasonable estimation by reducing the calculated PWP rate with Equation (1) by a factor of 10 [37,38,44]. The PWP values calculated in this study were transformed in this manner when they were assessed with the precipitation rates obtained in the literature by other methods.

We estimated the magnesium distribution coefficient between calcite and water from:

$$D_{Mg} = \frac{(Mg/Ca)_{calcite}}{(Mg/Ca)_{water}} \quad (2)$$

where $(Mg/Ca)_{calcite}$ denotes the molar ratio of the Mg and Ca concentrations in the stromatolites and $(Mg/Ca)_{water}$ denotes the molar ratio of the total dissolved Ca and Mg contents in the water. According to the results from the speciation–solubility calculations, there were no meaningful differences between the molar ratios calculated with the free Ca^{2+} and Mg^{2+} ion concentrations and those obtained with the total dissolved Ca and Mg contents; thus, we can neglect the effects of the Ca and Mg complexation in the estimation of the distribution coefficient.

4.4. Mg/Ca Thermometry

When using the Mg/Ca ratio in calcite as a water palaeothermometer, it is conventionally assumed that it is predominantly controlled by the Mg/Ca in the solution and a temperature-dependent partition coefficient. Although researchers have widely used Mg/Ca palaeothermometry in marine settings, to date, only a few researchers have utilized the Mg/Ca ratio for freshwater tufa deposits (e.g., [6], and the references therein). Moreover, the available experimental data for the Mg partitioning behaviour under karst/speleothem-specific conditions (e.g., for calcite precipitation from solutions of low ionic strength) are scarce, and to the best of our knowledge, only two experimental works are available in the literature: the Mg distribution coefficient D_{Mg}–temperature (D_{Mg}-T) relationships obtained by [45] and by [46].

The D_{Mg}-T relationships obtained by these authors have important differences. The equation in [46] always provides higher temperatures than that in [45], and in our case, it provided exceedingly higher temperatures with respect to the measured ones (see Figure S2, Supplementary Material). Furthermore, researchers have only used the D_{Mg}-T relationships in [45] to calculate the water temperatures from the Mg concentrations of tufa samples (e.g., [1,3,6]).

Therefore, in this work, we only used the D_{Mg}-T relationship from [45], as adjusted by [3]:

$$T\ (^{\circ}C) = \frac{D_{Mg} - 0.001}{0.0012} \quad (3)$$

where D_{Mg} is the Mg distribution coefficient (see Equation (2)), which is used to calculate the river Tw from the measured Mg/Ca ratios of the waters. Despite the adjustment-derived uncertainties of this equation, researchers have repeatedly used it to estimate the Tw in fluvial environments [1,3,6].

We compared the Mg-derived Tw values with the water temperatures measured in the sampling moment, as well as with the Tw values estimated from the $\delta^{18}O$ of both the stromatolite calcite and river water, by means of the formula in [47]. Except for the RM river, researchers have reported these temperatures in previous works (RP: [20]; RA: [23]; RE: [21]).

As the researchers measured most of the parameters on a six-month basis, we calculated the moving averages of two periods for the examined parameters, to remove the seasonality and to obtain the corresponding season-free evolutions.

5. Results

5.1. Hydrochemical Characteristics and Mg/Ca Ratios of the Waters

The hydrochemical characters of the waters at the studied sites are representative of the main compositional features of the four rivers. The water is of the HCO_3^-Ca type in the cases of the RE, RM and RP. In the case of the RA, the water is of the SO_4–HCO_3–Ca type (see Figure S1 Supplementary Material), which is the result of the intense interaction between the groundwater feeding the river and the bedrock evaporitic materials (gypsum/anhydrite) [23].

The Tw ranges measured in June and January at the studied sites (see Table S2, Supplementary Material) were similar in the four rivers: from 8 to 18 °C in the cases of the RA, RM and RP, and a bit narrower in the case of the RE (from 10.7 to 16 °C). The Tw values had a clear seasonal pattern (Figure 4). The differences between the mean temperatures for the warm and the cool periods were slightly higher in the RA and RP (6.3 and 5.2 °C, respectively) and lower in the RE and RM (2.6 and 2.7 °C, respectively) (Table S3, Supplementary Material).

The dissolved Mg concentrations were relatively similar in the four studied rivers (mean values between 0.8 and 1.04 mmol/L (Table 1)). The dissolved Ca contents had greater differences (mean values between 1.95 and 3.24 mmol/L (Table 1)). As a consequence, there were differences between the rivers for the Mg/Ca ratios of the waters (Mg/Ca_{water}) (Figure 5A and Table S2, Supplementary Material). The mean $(Mg/Ca)_{water}$ values exhibited an overall decreasing evolution from the RP to the RM, RE and RA, from 0.51 ± 0.08 in the RP to 0.27 ± 0.036 in the RA (Figure 5A, Table 1).

Table 1. Mean values (mean ± 1 σ) for dissolved Mg, Ca and Mg/Ca ratios in waters (Mg_{water}, Ca_{water} and Mg/Ca_{water}) and stromatolites ($Mg_{calcite}$, $Ca_{calcite}$ and $Mg/Ca_{calcite}$), and for the distribution coefficients in the studied sites at Piedra, Mesa, Añamaza and Ebrón Rivers. We also include data for the Kaisinger Creek [8] and Krka River [6] for comparison.

Chemical Components	Piedra River	Mesa River	Añamaza River	Ebrón River	Kaisinger Creek	Krka (2) River
Mg_{water} (mmol/L)	1.04 ± 0.19	0.87 ± 0.034	0.87 ± 0.07	0.80 ± 0.054	0.32 ± 0.03	0.94 ± 0.15
Ca_{water} (mmol/L)	2.04 ± 0.17	1.95 ± 0.22	3.24 ± 0.23	2.08 ± 0.17	3.16 ± 0.23	1.66 ± 0.1
Mg/Ca_{water} (molar)	0.51 ± 0.08	0.449 ± 0.045	0.27 ± 0.036	0.386 ± 0.033	0.101 ± 0.01	0.56 ± 0.08
$Ca_{calcite}$ (wt%)	37.0 ±2.56	35.2 ± 0.97	34.05 ± 2.08	34.8 ±2.53	37.1 ± 0.99	34.02 ±1.04
$Mg_{calcite}$ (g/kg)	2.46 ± 0.47	3.01 ± 0.38	1.73 ± 0.31	3.75 ± 0.8	0.58 ± 0.05	5.48 ± 0.71
$Mg/Ca_{calcite}$ (molar)	0.011 ± 0.002	0.014 ± 0.002	$8.38 \times 10^{-3} \pm 1.6 \times 10^{-3}$	0.018 ± 0.004	$2.58 \times 10^{-3} \pm 0.27 \times 10^{-3}$	0.027 ± 0.004
D_{Mg}	0.023 ± 0.008	0.032 ± 0.04	0.032 ± 0.012	0.046 ± 0.009	0.025 ± 0.003 (1)	0.044 ± 0.0068

(1) Averages from the tufas in three different sections of the Kaisinger Creek (from the data presented in [8]).
(2) Data taken from [6] in Table 1 and Table S3. For Ca, Mg and Mg/Ca values in waters, we present values from the sampling site W3. For Ca, Mg and Mg/Ca ratios in tufa in Table S3, samples from T1 to T10 are considered.

The $(Mg/Ca)_{water}$ values in the RE (RE-8) and RM (M-4) had a quasi-seasonal pattern, with higher values in the warm periods and lower values in the cool periods (Figure 4). This seasonal pattern was also present in the RP, from the beginning of the record until Cool 2007–2008; from this period onwards, the oscillations were wider than in the other rivers and did not present any regularity. Finally, the $(Mg/Ca)_{water}$ values in the RA (site RA-6) were rather constant over the periods, except for the last recorded period (Cool 2009–2010), which had a lower value (Figure 4).

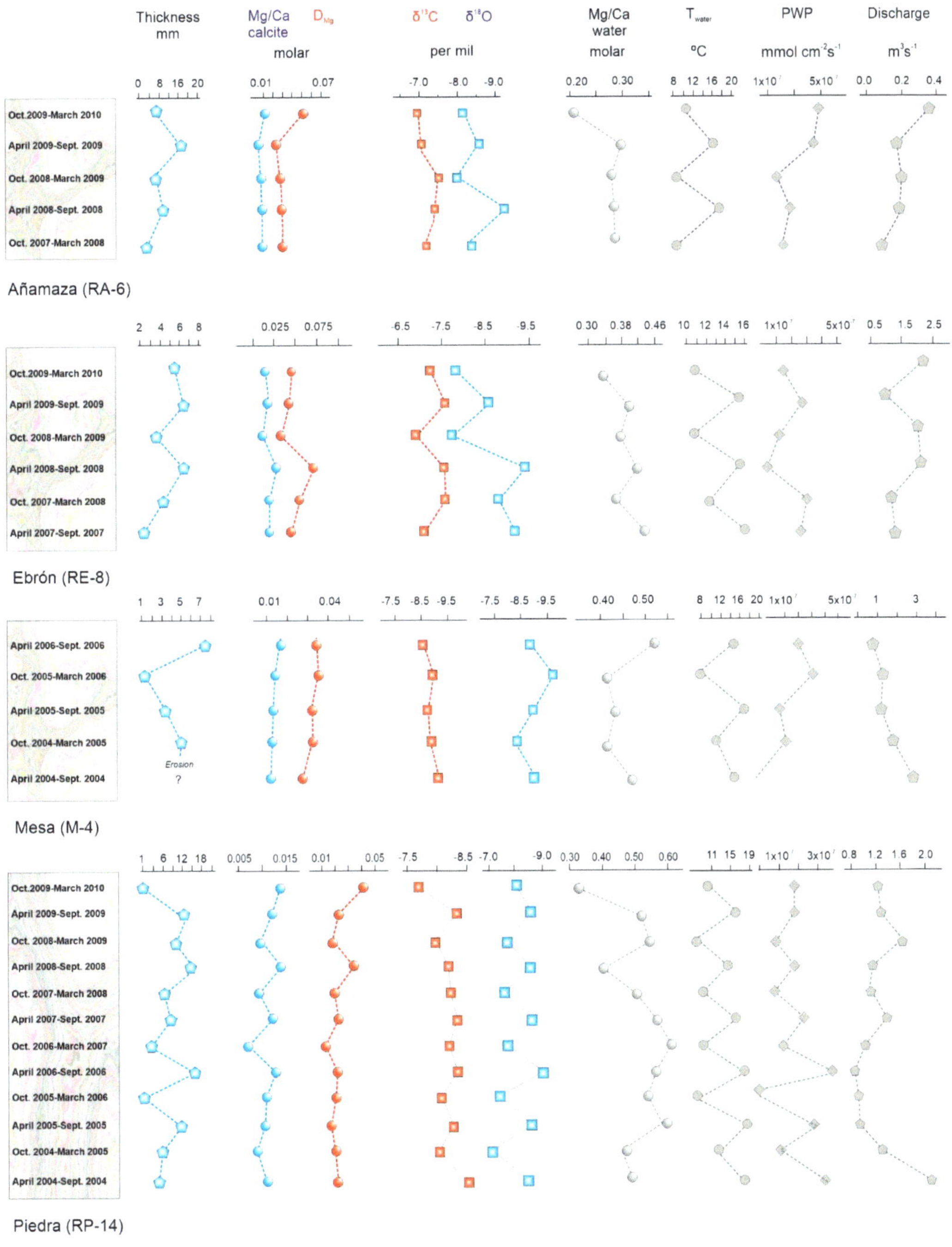

Figure 4. Temporal evolution of some parameters during examined sampling periods in water and stromatolites for the studied tablets. We present temperatures, Mg/Ca ratios, precipitation rates (as PWP values) and discharge values for waters, and six-month accumulated thicknesses, Mg/Ca ratios and $\delta^{18}O$ and $\delta^{13}C$ values for stromatolites. We indicate the units in the figure.

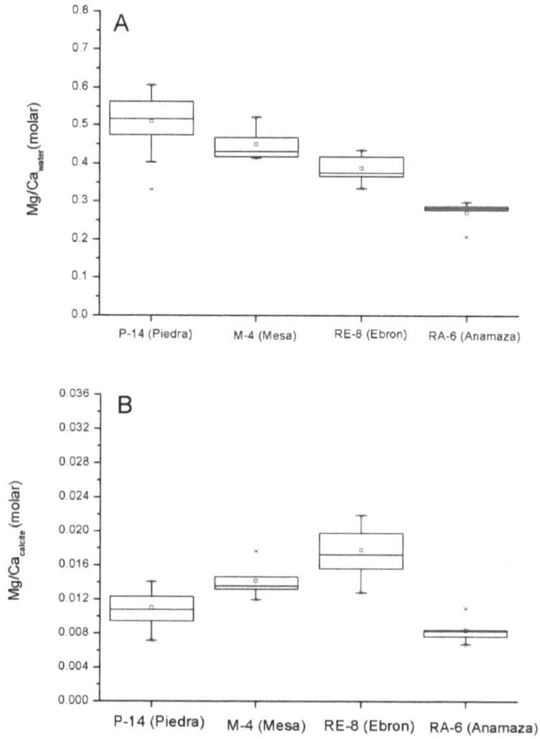

Figure 5. Box-and-whisker plots of statistical distributions of the Mg/Ca values in (**A**) waters and (**B**) stromatolites at studied sites in rivers. Statistical measures plotted are median (horizontal line inside the box), the 25th and 75th percentiles (bottom and top of the box respectively), mean (square), the 5th and 95th percentiles ("whiskers"), the 1st and 99th percentiles (crosses) and the maximum and minimum values (horizontal bars).

The waters were always oversaturated with respect to calcite, with the highest SIc values in the RA at the RA-6 site (mean value of 0.89 ± 0.23), and intermediate values in the RE at the RE-8 site (mean value of 0.73 ± 0.28), and in the RP at the RP-14 site (mean value of 0.67 ± 0.33). The lowest values were for the RM at the M-4 site (mean value of 0.57 ± 0.33). The calculated partial pressure of CO_2 (as log pCO_2) was always higher than the atmospheric values (mean values: RE: -2.79 ± 0.28; RA: -2.92 ± 0.25; RM: -2.62 ± 0.44; RP: -2.71 ± 0.35).

The mean values of the calcite precipitation rate ranged from $1.40 \times 10^{-8} \pm 1.13 \times 10^{-8}$ mmol/cm^2/s in the RM to $3.09 \times 10^{-7} \pm 1.44 \times 10^{-7}$ mmol/cm^2/s in the RA, with intermediate values in the RE and RP ($1.91 \times 10^{-8} \pm 1.02 \times 10^{-8}$ and $1.73 \times 10^{-8} \pm 1.16 \times 10^{-8}$ mmol/cm^2/s, respectively).

With some exceptions, the dissolved Ca, HCO_3 and TDIC concentrations at the studied sites had lower values in the warm periods due to the more intense calcite precipitation in these periods for most of the studied time intervals (which is in agreement with the overall trends previously detected in the studied rivers [19–23]). The PWP and SIc values are usually higher in the warm periods; however, in the case of Site M-4, the opposite trend was evident (Figure 4).

5.2. Geochemical Characteristics and Mg/Ca in Stromatolites

The calcium contents in the studied deposits ranged from 31.8 to 40 wt.% (Table S1, Supplementary Material), which corresponded to $CaCO_3$ percentages in the sediment from 80 to 100 wt.% (calculation based on Ca analytical data). Overall, these contents were consistent with those found in other modern (Krka River or Kaisenger Creek, Table 1) and older tufa deposits (e.g., [24,48–50]).

Although the analysed samples were primarily formed of calcite, there were some differences among the studied deposits, whereas the Ca contents in the deposits from the RA, RE, and RM were relatively similar, with mean values of 34–35 wt.% (85–87.5 wt.% as $CaCO_3$ (Table 1)), and the Ca concentrations in the stromatolite from the RP were higher, with a mean value of 37 wt.% (92.5 wt.% as $CaCO_3$).

The magnesium concentrations ranged from 1.28 to 5.15 g/kg, with the highest values in the RE (3.75 $\pm$ 0.8 g/kg) and the lowest in the RA (1.73 $\pm$ 0.31) (Table 1). The $MgCO_3$ contents (calculated assuming that Ca and Mg were the only elements in the carbonate fraction) ranged from 0.65 to 2.07 mol%, which indicated that the calcite in the stromatolites was low-Mg calcite.

The Mg/Ca ratio of the studied stromatolites (herein $(Mg/Ca)_{calcite}$) also exhibited differences between the studied rivers (Figure 5B, Table S1, Supplementary Material). The highest mean $(Mg/Ca)_{calcite}$ values were in the RE (0.018 $\pm$ 0.004), and the lowest were in the RA ($8.38 \times 10^{-3} \pm 1.6 \times 10^{-3}$) (Table 1). Overall, these values did not show a systematic evolution with respect to the corresponding $(Mg/Ca)_{water}$ (compare Figure 5A,B). The RP had the highest $(Mg/Ca)_{water}$ value and the second lowest $(Mg/Ca)_{calcite}$ value.

The six-month $(Mg/Ca)_{calcite}$ pattern over time was not the same for the four rivers (Figure 4). In the RE, and partially in the RP, there were six-month patterns with lower $(Mg/Ca)_{calcite}$ values in the cool periods and with higher values in the warm periods. In RP, this pattern was more distinct from the Warm 2006 onwards, except for the last period (Cool 2009–2010). The $(Mg/Ca)_{calcite}$ values in the RM did not show a cyclic pattern, and the values increased over time. The RA values had little variation, except for the last period (Cool 2009–2010). The overall $(Mg/Ca)_{calcite}$ trends of the RM and RP were parallel (Figure 4).

The calculated D_{Mg} values for the studied stromatolites in the four rivers ranged from 0.012 to 0.053, with the lowest values in the RP (mean 0.023 $\pm$ 0.008) and the highest in the RE (mean 0.046 $\pm$ 0.009) (Table 1). Overall, these values were in the range of the D_{Mg} determined in the experimental studies (from 0.01 to 0.06, as reviewed in [8,45,51,52]) and in other natural fluvial tufa systems (from 0.011 to 0.058, as reviewed in [8]).

Most of the studied rivers had seasonal D_{Mg} patterns over time, except for the RA (Figure 4). Most of the examined record of RE revealed a six-month pattern with higher D_{Mg} values in the warm periods, and from Cool 2005–2006 to Warm 2009 in the RP. In the case of the RM, the D_{Mg} had a faint seasonal pattern, but with higher values in the cool periods.

We present the $\delta^{18}O_{calcite}$ and $\delta^{13}C_{calcite}$ of the stromatolite records in Table S4 (Supplementary Material) and Figure 4. The $\delta^{18}O_{calcite}$ values in the four rivers were similar, although there were slightly higher in the RP. In the four rivers, the $\delta^{18}O$ patterns were evident in the alternating higher values in the cool periods and lower values in the warm periods, which is consistent with the temperature dependence of the oxygen isotope fractionation in calcite precipitation. Therefore, the stromatolite $\delta^{18}O$ values reflected the water temperature signature. The temperatures estimated from the $\delta^{18}O_{calcite}$ and $\delta^{18}O_{water}$ were close to the measured temperatures [13].

The $\delta^{13}C_{calcite}$ values were also similar in the four rivers, and the ranges of variation were narrower than in the $\delta^{18}O_{calcite}$ (Figure 4). The RM had the lowest values, and the RE and RA had the highest. In the RP, the $\delta^{13}C$ had an oscillating pattern that was roughly parallel to the $\delta^{18}O$. In the RE, the $\delta^{18}O$ and $\delta^{13}C$ patterns were similar. In contrast, in the RM and RA, the $\delta^{18}O$ and $\delta^{13}C$ did not display any significant relationship.

The clay composition of the five analysed stromatolites were similar: only illite and chlorite were present. The average proportion of both minerals was 65% for the illite and 35% for the chlorite, and the mean Mg content in the clay fraction was about 10,000 mg/kg. Therefore, the stromatolite Mg/Ca analysis could have included Mg from chlorite. However, considering a maximum clay mineral content of 5%, the chlorite contribution to the stromatolite sample would be 500 ppm, which is significantly low with respect to the entire Mg content (Table S1, Supplementary Material).

6. Discussion

The $(Mg/Ca)_{water}$ for the studied sites at the four rivers cover a relatively wide range of values (Table 1). If the $(Mg/Ca)_{calcite}$ ratio primarily depends mainly on the Mg/Ca in the solution and on a temperature-dependent partition coefficient (Equation (2)), then the theoretical $(Mg/Ca)_{calcite}$ values for the mean $(Mg/Ca)_{water}$ for the rivers should parallel the decreasing $(Mg/Ca)_{water}$ trends that can be observed in Figure 5A from the RP to RA rivers (given that the mean water temperatures were similar at the four sites, from 12.2 to 13.5 °C). However, this is far from reality, and there was even an increasing trend from the RP to the RE (Figure 5B). Thus, we need to consider other factors that participate in the control of the Mg partitioning in calcite when analysing the Tw derived from the Mg content in carbonates.

6.1. $(Mg/Ca)_{calcite}$ and Temperature

The temperatures derived from the stromatolite Mg/Ca content substantially differ from both the Tw calculated from the stromatolite $\delta^{18}O_{calcite}$ values and the measured Tw (Figure 4). Even in the RP, which had the best fitting between the Mg-derived Tw and measured Tw, the evolutions of the two temperatures were not parallel over time (Figure 6). Furthermore, in general, the Mg-derived Tw did not show the expected seasonality (e.g., higher values in the warm periods than in the cool periods), except partially in the RP. In the RE, the Mg-derived Tw had right oscillating behaviour in some periods. In the RA, there was no oscillation, and in the RM, the oscillation was reversed (Table S3, Supplementary Material, Figure 6).

The matching between the $(Mg/Ca)_{calcite}$ and measured Tw was of a diverse degree (Figure 4; Table 2), and with respect to the season-free evolutions, the $(Mg/Ca)_{calcite}$ and measured Tw exhibited a weaker similarity (Figure 6), even when there is a significant correlation between them (See Section A.1. in Appendix A for details). This inconsistency between the six-month pattern and the season-free evolution suggests that the correlation between the measured Tw and $(Mg/Ca)_{calcite}$ is caused by the common seasonal changes of both parameters, and not necessarily by the temperature dependence of the Mg partitioning. According to these results, it is clear that there is a complex interplay of diverse factors other than the temperature that influences the $(Mg/Ca)_{calcite}$ contents in the stromatolites and thus promotes different behaviours.

The abovementioned situation could explain the sparse results calculated with the D_{Mg}-T relation obtained by using the equation in [45]. The authors of [1] successfully used the D_{Mg}-T relationship to obtain the model temperatures from the Mg concentrations of tufa samples in Australia; however, there were discrepancies in the correlation between the tufa $(Mg/Ca)_{calcite}$ ratios and water temperature [1,7]. The use of the relationship described by the authors of [45] in other modern tufas has provided higher temperatures than the ones measured in rivers with temperature-controlled $(Mg/Ca)_{water}$ ratios (e.g., the Krka River in Croatia [3], and the Krka River in Slovenia [6]). The Mg/Ca paleothermometry in modern and ancient stream tufas is generally problematic due to the fact that the expected influence of the temperature on the $(Mg/Ca)_{calcite}$ can be obscured by diverse factors, such as the hydrogeochemical conditions, the changes in the $(Mg/Ca)_{water}$ and anthropogenic contamination [2,3,5,53], the influence of the microbial biofilm activity [4,7,8,54] and the presence of variable amounts of detrital limestone and dolomite in the tufa [12].

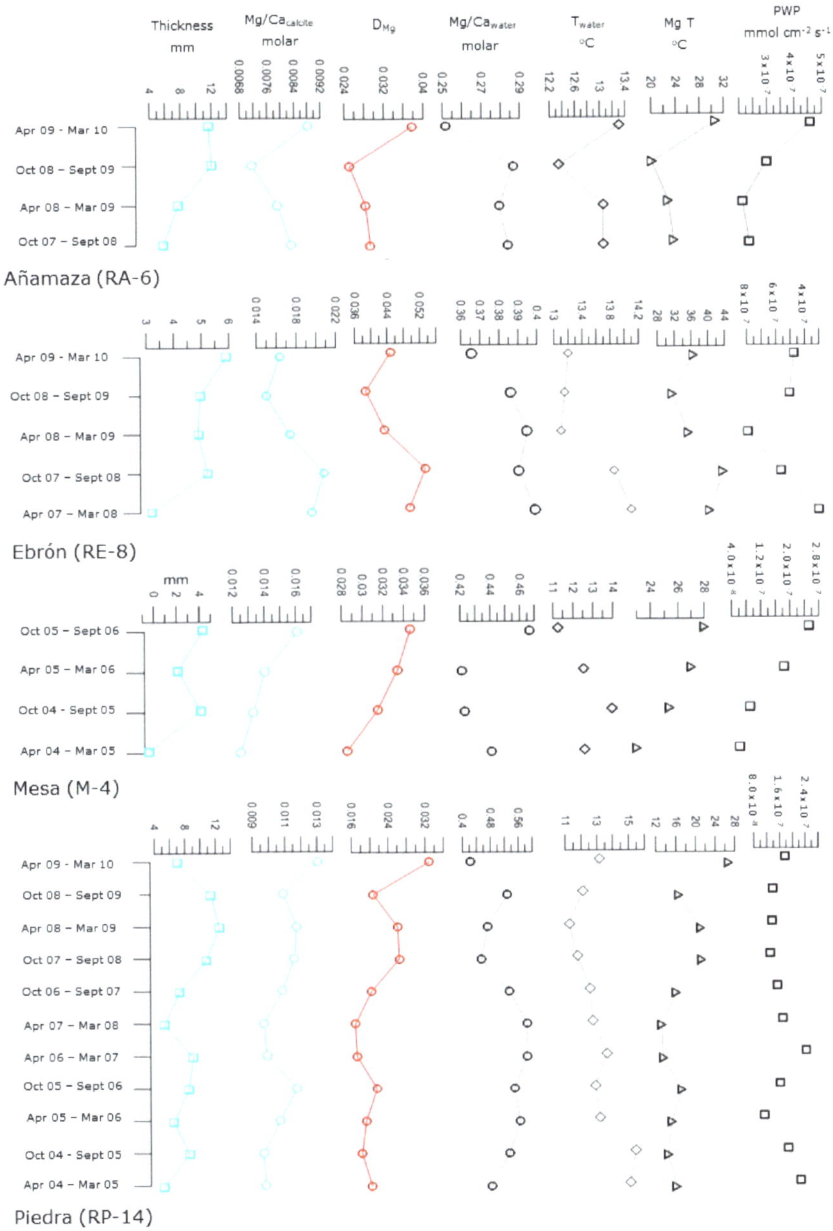

Figure 6. Temporal evolution of season-free interannual evolutions of some parameters during examined sampling periods in water and stromatolites for studied tablets. We present the six-month accumulated thicknesses, $Mg/Ca_{calcite}$ ratios, D_{Mg}, Mg/Ca_{water} ratios, measured water temperatures, Mg-derived water temperatures and stromatolite precipitation rates (as PWP values). We indicate the units in the figure.

Table 2. Correlations between some analytical parameters of studied tufa samples.

River (Number of Samples)	Mg/Ca_{water} vs. T	Mg/Ca_{tufa} vs. T	Mg/Ca_{tufa} vs. Mg/Ca_{water}	D_{Mg} vs. T	PWP vs. T	T_{calc}-$T_{measured}$ Mean
Piedra (n = 12)	+0.15 (p = 0.627)	+0.391 (p = 0.208)	−0.577 (p = 0.049)	+0.064 (p = 0.845)	+0.863 (p = 0.0003)	4.89
Piedra Cool 06-07 to Cool 09-10 (n = 5)	−0.03 (p = 0.945)	0.576 (p = 0.176)	−0.787 (p = 0.036)	0.213 (p = 0.646)	+0.842 (p = 0.017)	+7.57
Piedra Warm 04 to Warm 06 (n = 7)	+0.346 (p = 0.568)	+ 0.379 (p = 0.529)	0.587 (p = 0.298)	0.27 (p = 0.66)	+0.967 (p = 0.007)	+1.15
Añamaza (n = 5)	+0.326 (p = 0.59)	−0.338 (p = 0.577)	−0.939 (p = 0.018)	−0.312 (p = 0.608)	+0.39 (p = 0.516)	13.9
Mesa (n = 56)	+0.479 (p = 0.414)	−0.023 (p = 0.971)	+0.61 (p = 0.273)	−0.48 (p = 0.40)	−0.663 (p = 0.222)	12.5
Ebrón (n = 6)	+0.892 (p = 0.017)	+0.691 (p = 0.128)	+0.551 (p = 0.256)	+0.269 (p = 0.28)	+0.103 (p = 0.845)	24.0

6.2. *Other Factors That Influence* $(Mg/Ca)_{calcite}$

6.2.1. Mg/Ca of Water

The $(Mg/Ca)_{calcite}$ values not only depend on the temperature, but also on the $(Mg/Ca)_{water}$. Moreover, the empirical partitioning coefficient (D_{Mg}) depends on the $(Mg/Ca)_{water}$ value (see Equation (2)). Accordingly, variations in the $(Mg/Ca)_{water}$ may be involved in the systematic offset with respect to the measured temperatures. The authors of [2] found that the $(Mg/Ca)_{calcite}$ ultimately relies on rainfall and aquifer processes, which determine the $(Mg/Ca)_{water}$ content. The authors of [53] established that the seasonal variations in the $(Mg/Ca)_{calcite}$ do not depend on the temperature, but on the seasonal $(Mg/Ca)_{water}$ variations, which are induced by the preferential leaching of Mg from carbonates in dry periods [9]. Calcite precipitation can also modify the $(Mg/Ca)_{water}$ along streams by removing Ca from the water [55].

Thus, the $(Mg/Ca)_{water}$ variation may interfere with the temperature-dependent $(Mg/Ca)_{cacite}$ content. In general, the $(Mg/Ca)_{water}$ increases in warm and/or dry periods due to the greater Ca extraction by the calcite precipitation under these conditions, and in-aquifer effects produce the same result [53,55,56]. From a theoretical point of view, the increase in the $(Mg/Ca)_{water}$ in warm periods emphasizes the increase in the derived $(Mg/Ca)_{calcite}$ because the Mg content in calcite increases with increasing temperature. As the empirical D_{Mg} (Equation (2)) is the ratio between these coupled varying parameters, the simultaneous increases in both parameters cancel each other out, at least partially, which results in a smoothed D_{Mg}. This only occurred in the first interval of the RP (from Warm 2004 to Warm 2006), when oscillations in $(Mg/Ca)_{calcite}$ flatten the expected seasonal oscillations of the Mg-derived Tw, although these values are close to the measured Tw. In the rest of the cases, the Mg-derived Tw are higher than the measured Tw due to the anomalously high $(Mg/Ca)_{calcite}$. (See Section A.2. in Appendix A for details).

6.2.2. Detrital Minerals

Given the type of chemical attack used for the trace element analysis of the sediments, some detrital matter may have been dissolved and then included in the analysed aliquot, which may have caused the total Mg content to be higher than the total Mg in the calcite lattice. Moreover, the Al content of the analysed samples was significantly correlated to the $(Mg/Ca)_{calcite}$ in all the cases, except for the first time interval for the RP (RP from Warm 2004 to Warm 2006: r = 0.09; RP from Cool 2006–2007 to Cool 2009–2010: r = 0.62; RM: r = 0.88; RE: r = 0.96; RA: r = 0.87) (Table 3). Because of this, we cannot rule out the influence of Mg from a detrital source. The corresponding extra increase in the $(Mg/Ca)_{calcite}$ value may have been responsible for the anomalously higher-than-real Tw values observed in most of the cases. Moreover, the amount of detrital sediment may regularly oscillate and consequently overemphasize the $(Mg/Ca)_{calcite}$ values, and subsequently, the estimated Tw, while at least partially preserving the six-month oscillation, which may be the case for the RE, in which the $(Mg/Ca)_{calcite}$ exhibited a six-month oscillating pattern; however, the values were much higher than the expected ones from the measured $(Mg/Ca)_{water}$.

Table 3. Trends and correlations for some analytical parameters of studied tufa samples.

River	Mg/Ca_{tufa}	Al Contents in Tufas	Mg and Al Contents in Tufas	Correlation Coefficient (R, Pearson) for Al vs Mg/Ca (2) in Tufas		
				All Periods	Warm 04 to Warm 06	Cool 2006–07 to Cool 2009–10
Piedra	Almost seasonal (1)	Not seasonal	Al > Mg	0.598	0.055	0.756
Ebron	Almost seasonal (1)	Almost seasonal (1)	Al > Mg	+0.88		
Añamaza	Not seasonal	Seasonal (2)	Al > Mg	+0.965		
Mesa	Not seasonal	Almost seasonal (2)	Mg $\approx$ Al	+0.93		

(1) Higher values in the warm periods. (2) Higher values in the cool periods.

Assuming that the Mg contents in the stromatolites included 500 ppm of clay-derived Mg (see Methods and Results), we can eliminate the corresponding contribution to the Tw calculation by subtracting this amount from the total measured Mg contents. Then, by recalculating the Tw with the corrected Mg values, the mean D_{Mg} decreases and, as a consequence, so does the estimated Tw. Inversely, we can estimate the Mg content that corresponds to the measured Tw from the measured $(Mg/Ca)_{water}$ by using Equations (2) and (3), and we can then express the Mg excess (See Section A.3. in Appendix A for details).

The correction for the noncarbonate-derived Mg in the stromatolites was only able to cancel out the temperature offset in the RP, and likely in the RA, with small amounts of detrital Mg. In the first time interval of the RP (from Warm 2002 to Warm 2006), the detrital Mg fraction corresponded to probably less than the 5% of the clay minerals considered in the correction. In the second time interval of the RP (from Cool 2006–2007 to Cool 2009–2010) and in the RA, the detrital Mg fraction should have been higher, and it should have included not only clay minerals but also dolomite. In the other rivers (RE and RM), we have to invoke, most likely, additional factors to explain the high $(Mg/Ca)_{calcite}$.

6.2.3. Precipitation Rate

As stated above, in laboratory experiments, the $(Mg/Ca)_{calcite}$ ratio is predominantly controlled by the Mg/Ca in the solution and Tw, without the meaningful influence of other factors, such as the mineral precipitation rate (e.g., [6,7,57] and the references therein). However, according to the results obtained in natural settings, the $(Mg/Ca)_{calcite}$ values are dependent on additional factors, such as the mineral precipitation rate, crystal morphology and/or biological processes [7,8,57]. Furthermore, according to recent experimental results, the calcite precipitation rate may influence the $(Mg/Ca)_{calcite}$ and D_{Mg} under both biotic and abiotic conditions (e.g., [7,51]).

We can theoretically estimate the calcite precipitation rates (see Methods) as the PWP rates from the hydrochemical characteristics of the waters, or from the six-monthly thicknesses recorded on the tablets [13]. The deposition recorded on the tablets and PWP values are not fully equivalent because the latter corresponds to single sampling moments in the six-month periods, whereas the deposition rates from the tablets represent the six-month-period net deposition and could have been affected by erosive processes [21,23]. However, we use them in a complementary manner for the discussion.

In most of the studied cases, the thicknesses were higher in the warm periods than in the cool periods due to the influence of the temperature on the calcite precipitation, except for the RM [22]. This produced a six-month rhythmic pattern for the thicknesses that paralleled the $(Mg/Ca)_{calcite}$ in the cases in which the latter also exhibited a seasonal pattern (Figure 4). The PWP values, depending on the temperature and hydrochemical features, can also exhibit a six-month rhythmic pattern.

The calculated PWP values are in the upper range of the calculated precipitation rates in other tufa-depositing rivers (see the reviewed values in [8]). Overall, except for in the RM, there was a good correspondence between the PWP values and the thicknesses recorded on the tablets, with higher values in the warm periods (Figure 4) (See Section A.4. in Appendix A for details).

The best agreement between the measured thickness and PWP patterns occurred for the RP, for which both parameters exhibited six-month variations (Figure 4), especially for the first time interval (from Warm 2002 to Warm 2006). This differential behaviour of the RP with respect to the other rivers may be related to the lower partitioning coefficients (D_{Mg}) for this river. The inorganic precipitation rate influenced the Mg partitioning, at least in the RP, promoting a high $(Mg/Ca)_{calcite}$ and D_{Mg} values in the second time interval. According to recent experimental results, this issue may be related to the influence of the precipitation rate on the D_{Mg} values in abiotic conditions [51]. In this study we demonstrate, for the first time, the increase in the D_{Mg} values in calcite at 25 °C, with calcite precipitation rates from 5×10^{-10} to 2.5×10^{-8} mmol/cm^2/s; this range includes the values calculated for the studied rivers herein. Furthermore, according to the control experiments (abiotic experiments) performed by the authors of [7], there was a significant linear increment between the $(Mg/Ca)_{calcite}$ and precipitation rates (at values from around 5×10^{-9} to 2.2×10^{-7} mmol/cm^2/s). These results are also consistent with the theoretical fields [58] that predict an increase in the D_{Mg} values with the precipitation rate.

All these statements suggest a predominant inorganic precipitation rate control in the D_{Mg} values found in the RP, in which the abiotic parameters (e.g., temperature and CO_2 outgassing) importantly condition the tufa dynamics along the river [19].

6.2.4. Discharge

With respect to the discharge, the influence on the calcite composition is usually explained through changes in the $(Mg/Ca)_{water}$ [55], due to the residence time of the water in the aquifer: a longer residence time permits the dissolution of higher proportions of Mg from carbonate rocks [53], and it yields higher $(Mg/Ca)_{water}$ values. The studied rivers for this contribution are primarily fed by groundwater, for which the discharge oscillations are not high, except for the high discharge events linked to the punctual heavy rains (see mean discharge values for these rivers in "Location, geological context, climate and hydrology" section). The six-month accumulated discharge variation throughout the study time did not display any clear six-month oscillations, and it did not present any significant relationships with the $(Mg/Ca)_{water}$, except for some high discharge peaks (e.g., Ebrón: Cool 2009–2010, [21]; Añamaza: Cool 2008–2009; [23]) that coincided with decreases in the $(Mg/Ca)_{water}$ (Figure 4).

6.2.5. Sulphate Content

The onset of the calcite precipitation and the precipitation rate are inhibited by the presence of sulphate [59–63], and this effect could be significant in the rivers with a high sulphate-to-bicarbonate ratio (e.g., from 0.56 to 0.63 in the RA; from 0.18 to 0.33 in the RP and RE; from 0.09 to 0.15 in the RM); however, the authors of [45] did not take this into consideration in their experiment, in which they studied Mg partitioning during the precipitation from sulphate-free solutions.

In this study, the sulphate content in the water had a significant positive correlation with the Mg/Ca ratio of the water (R^2 = 0.48), and it had a negative correlation with the D_{Mg} (R^2 = 0.63) in the RP. In the RM, there was a negative correlation (R^2 = 0.838) between the sulphate concentration and Mg/Ca ratio, while the sulphate and D_{Mg} did not show any correlation (R^2 = 0.008). According to the data for the two sections investigated in the RA, there was no correlation between the dissolved SO_4 and $(Mg/Ca)_{water}$ (R^2 = 0.05) or D_{Mg} (R^2 = 0.015). In the other investigated streams, the sulphate did not seem to have any influence on any of the measured (or calculated) parameters. According to a study on the effect of dissolved sulphate on the deposition of tufa in the Trabaque River in Spain [63], the presence of sulphate can either limit or favour the precipitation of $CaCO_3$, depending on the occurrence of the incongruent dissolution of the bedrock dolomite, which subsequently either decreases or increases the D_{Mg}. However, the sulphate concentrations in our study were much lower than those discussed in [64] or [61], in which the authors found that the presence of sulphate decreased the calcite precipitation rate, which also affected the D_{Mg}.

Overall, the discussion on the effect of the sulphate ions on the Mg/Ca of the studied stromatolites requires more detailed hydrogeological study, which exceeds the scope of this paper.

6.2.6. Biogenic Influence on Stromatolite Mg/Ca

Biogenic processes, which arise from microbial metabolic activities and/or the presence of extracellular polymeric substances (EPSs), may also affect the D_{Mg} values through different mechanisms (e.g., the precipitation rate may be controlled by the biofilm growth rate [4,7] or by the presence of compositionally different EPSs [54]). Recently, the authors of [8] found that the spatial evolution of the D_{Mg} values along Kaisenger Creek (Germany) could be attributed to the changes in the relative proportions of the bioinfluenced and inorganically driven tufa formation processes, with a higher D_{Mg} when the bioinfluenced processes dominate.

Thus, we expected that the stromatolite formation would be enhanced, and higher D_{Mg} values would be favoured during the warm periods, whereas we expected lower D_{Mg} values in the cool periods, with reduced biological activity rates. If this is true, then the changes in the biological activity would move in the same direction as the observed seasonal effects of the inorganic precipitation rate and the decrease in the calcite solubility with higher temperatures.

6.3. *Differential Behaviours of the Studied Rivers*

6.3.1. Añamaza River

Neither the $(Mg/Ca)_{water}$ nor the $(Mg/Ca)_{calcite}$ exhibited seasonal evolution. Both had little variation, except in the last period (Cool 2009–2010), when they were inverse (Figure 4). The $(Mg/Ca)_{calcite}$ did not follow the temperature oscillations despite the $(Mg/Ca)_{water}$ stability. The sediment accumulation did not have any relationships with either the $(Mg/Ca)_{water}$ or $(Mg/Ca)_{calcite}$ (Figure 4). The Mg-derived Tw hardly showed seasonality (Table S3, Supplementary Material), and it was substantially higher than the measured Tw. Therefore, neither the temperature nor the $(Mg/Ca)_{water}$ controlled the $(Mg/Ca)_{calcite}$. The water discharge in this river is relatively constant throughout the year (minimum ecological discharge), which may explain the lack of seasonal changes in the $(Mg/Ca)_{water}$.

6.3.2. Ebrón River

The $(Mg/Ca)_{calcite}$ exhibited six-month oscillations that were parallel to the measured Tw, even in the season-free evolution (Figure 6). However, the Mg-derived Tw did not fully display the right seasonality, and the Tw values were systematically higher than the measured Tw, with differences of at least 17 °C (Table S3, Supplementary Material). Moreover, although the measured thickness data and $(Mg/Ca)_{water}$ had six-month oscillating patterns (Figure 4), their season-free evolutions were not parallel to the $(Mg/Ca)_{calcite}$. The PWP values did not show an oscillating pattern (Figure 6).

These results indicate that the $(Mg/Ca)_{calcite}$ is controlled by seasonal factors other than the temperature and $(Mg/Ca)_{water}$, which enhance the Mg content in the sediment. The six-month thickness evolution over time was opposite to the $(Mg/Ca)_{calcite}$, which researchers have interpreted as a sign of rapid calcite precipitation in other cases [7]. The higher-than-expected $(Mg/Ca)_{calcite}$ values may reflect the input of the detrital material in the sediment, and likely with six-month variations.

6.3.3. Mesa River

The $(Mg/Ca)_{calcite}$ did not display a rhythmic pattern; thus, it did not reflect either the $(Mg/Ca)_{water}$ oscillating pattern or the measured Tw oscillations (Figure 4). The estimated D_{Mg} and, consequently, the Mg-derived Tw, had higher values in the cool periods and lower values in the warm periods, with substantially higher values than the measured Tw (Table S3, Supplementary Material). Moreover, the Mg-derived Tw displayed an increasing season-free evolution over time that paralleled the $(Mg/Ca)_{calc}$ and PWP values, but not the measured Tw (Figure 6). Therefore, we can infer that the stromatolite $(Mg/Ca)_{calcite}$ does not only directly depend on the water temperature but is also controlled by nonseasonal factors that are linked to calcite precipitation, such as changes in the water discharge. Due to the significant correlation between the Al and $(Mg/Ca)_{calcite}$, we cannot rule out the influence of a Mg detrital source.

6.3.4. Piedra River

The $(Mg/Ca)_{calcite}$ follows a six-month pattern, which yielded the best agreement between the Mg-derived Tw and measured Tw (Figure 4), which occurred despite the fact that the $(Mg/Ca)_{water}$ was not stable and was even inverse with respect to the $(Mg/Ca)_{calcite}$ for many periods. In addition to the RA, this river had the lowest $(Mg/Ca)_{calcite}$ values, the lowest D_{Mg} and the more realistic Mg-derived Tw values.

According to the relationships between the $(Mg/Ca)_{calcite}$ and other parameters, we distinguished two different time intervals:

- In the first time interval (from Warm 2004 to Warm 2006), the $(Mg/Ca)_{water}$ values were higher in the warm periods and had lower values in the cool periods (Figure 4), which highlight the temperature-partitioning effect (higher temperatures allow more Mg to enter into the calcite structure). However, the $(Mg/Ca)_{calcite}$ oscillations were small, and thus, the empirical D_{Mg} was flattened (Figure 4), which determined that the Mg-derived Tw did not exhibit seasonal changes. Nevertheless, the corresponding Mg-derived Tw was the closest to the measured Tw as compared with the other rivers (Table S3, Supplementary Material). The PWP and sediment thickness values paralleled the Tw, with wide oscillations (Figure 4). Their season-free evolutions were also roughly parallel (Figure 6). Therefore, the $(Mg/Ca)_{calcite}$ maintains the general Tw signature, although the parallel evolution of the $(Mg/Ca)_{water}$ and $(Mg/Ca)_{calcite}$ can erase the six-month changes in the estimated Tw. We found no evidence of detrital Mg in the corresponding sediment.
- In the second time interval (from Cool 2006–2007 to Cool 2009–2010), the $(Mg/Ca)_{water}$ pattern did not match the $(Mg/Ca)_{calcite}$ variation, and it was even opposite to both the calculated and measured Tw. Therefore, the empirical D_{Mg} oscillations were amplified by the high $(Mg/Ca)_{water}$ values in the cool periods, when the $(Mg/Ca)_{calcite}$ was lower. The Tw values obtained from this D_{Mg}, and from the formula in [45], exhibited the expected six-month oscillations; however, these were systematically higher (Table S3, Supplementary Material) not only in the warm periods, when the estimated D_{Mg} was too high, but also in the cool periods, which means that the stromatolite $(Mg/Ca)_{calcite}$ was still higher than it should have been for the $(Mg/Ca)_{water}$ and water temperature at which it was formed. Despite the six-month oscillation of the Mg-derived Tw, the corresponding season-free evolution was not parallel to the measured Tw (Figure 6). Therefore, another control over Mg entry into the calcite lattice has overprinted the influence of temperature on the $(Mg/Ca)_{calcite}$. The significant correlation between the contents in Al and $(Mg/Ca)_{calcite}$ (Table 3) and the enhanced six-month oscillations of the $(Mg/Ca)_{calcite}$ (with respect to the first time interval) point to the influence of a Mg detrital source.

In the RP, the six-month thickness and PWP values mimic the $(Mg/Ca)_{calcite}$ pattern, except in the Cool 2009–2010 period, during which the $(Mg/Ca)_{calcite}$ and Mg-derived Tw were also anomalous (Figure 4). However, the PWP season-free evolution matched the measured Tw evolution (Figure 6), while the six-month thickness evolution was not well marked in any of the periods of this second time-interval (from Cool 2006–2007 to Cool 2009–2010). During this time interval, the sediment thickness increased and had high values, even for the cool periods. The influence of the growth rate on the Mg content in the sediment through the Mg entry into the calcite in natural systems may be responsible for the increase in the stromatolite $(Mg/Ca)_{calcite}$ [8], which would at least partially explain the anomalously high Mg-derived Tw. We cannot rule out the presence of a certain amount of Mg of detrital origin, as well as the biological influence.

In summary, the RP has the best adjustment between the Mg-derived and measured Tw. In addition, it also had the lowest estimated D_{Mg}, which was mostly below 0.025. According to the relationship between the Mg-derived and measured Tw, we distinguished two time intervals. In the first time interval (from Warm 2004 to Warm 2006), the Mg-derived Tw fit quite closely to the measured Tw, although neither of them exhibited seasonal oscillations. In the second time interval (from Cool 2006–2007 to Cool 2009–2010), the Mg-derived Tw displayed six-month changes; however, the values overpassed the measured Tw, and the season-free evolutions were very different. In the first interval, the $(Mg/Ca)_{calcite}$ reflected the average Tw, although it did not show a seasonal pattern linked to the $(Mg/Ca)_{water}$ six-month oscillations. In the second time interval, the different influences on the $(Mg/Ca)_{calcite}$ emphasize the six-month oscillations and magnify the Mg entry into the sediment. The presence of Mg from detrital particles (dolomite and chlorite) could have also contributed to the increase in the $(Mg/Ca)_{calcite}$.

6.4. Comparison with Other Systems

Until now, only a few researchers have investigated tufas and water chemistry in enough detail to evaluate the behaviour of the trace elements, such as Mg, as proxies of tufa palaeothermometry through the calculation of the D_{Mg} (e.g., [6,8]).

We present the mean D_{Mg} and Tw values obtained in other tufa-depositing rivers and streams from the scarce data in the literature [1,6,8,65] in Figure 7, which we compared with the mean values from the tufa deposits and rivers in this study. In addition, we included the values obtained from travertine deposits in low-temperature thermal waters [66] and speleothems [9,45,67,68] for comparison.

As we can see in Figure 7, the D_{Mg} values of the stromatolites in the RE were the highest out of the four compared examples, although they were only slightly higher than those found in the Krka River (Slovenia) by the authors of [6]. The partition coefficients in other tufa-depositing rivers (Hvanná, Krka, Gregory and Kainsinger) had an even wider range than of the rivers studied herein. Moreover, the D_{Mg} values of the compared speleothems had an even a wider range than those in tufas and travertines, although most of them (excluding the data from the cave in Jamaica studied in [67]) are usually lower.

The distribution of the plotted D_{Mg}-T points in Figure 7 again suggests that the temperature is not the only/main factor that conditions the partition coefficient values for the compared tufas, which explains the unrealistically high temperatures that were obtained from the available D_{Mg}-T experimental relationships (also presented in Figure 7).

However, the results obtained from the recent travertine deposits at the Embid and Alhama thermal springs almost perfectly fit with the D_{Mg}-T relationship in [45], as adjusted by the authors of [3]. The low-temperature thermal waters at Embid and Alhama are also dilute waters of the SO_4–HCO_3–Ca type, with constant temperatures (between 27.2 and 32.4 °C at the studied springs) and chemical compositions over time (e.g., [66]). The values of the travertine samples plotted in Figure 7 correspond to the deposits that precipitate near the spring orifice, and in most cases, they are of inorganic origin (or with minor biological participation and without the contribution of detrital sediment). Thus, it appears that the constant temperature and chemical composition (e.g., $(Mg/Ca)_{water}$ values (mostly around 0.7) [66]) of the waters during the precipitation process would favour the dominant role of the temperature over the Mg partitioning in the travertines, which additionally support the experimental D_{Mg}-T relationship in [45].

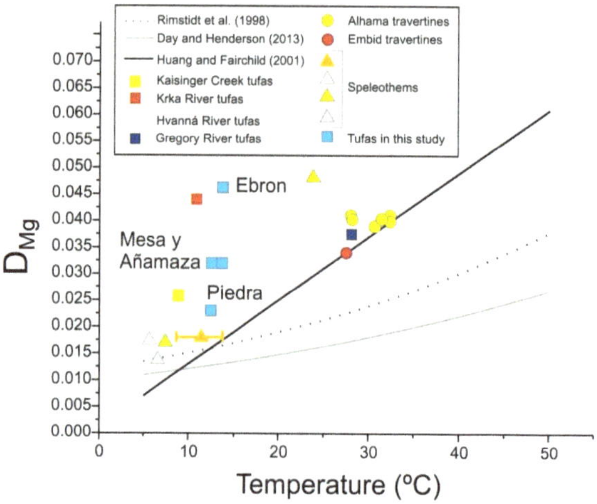

Figure 7. Mean D_{Mg} and water temperature values obtained for tufa deposits in studied rivers compared with available data for other tufa-depositing rivers and streams, for thermal spring deposits (travertines) and some cave speleothems. We present experimental relationships between water temperature and magnesium partition coefficient (D_{Mg}) determined by the authors of [45,46] under karst/speleothem-analogue conditions, as well as exponential best-fit curve to the data in [58], obtained by the authors of [46]. Data for the tufa deposits. Mean D_{Mg} and temperature values for Kainsinger Creek (Germany) were calculated from the data in [8] for sections T1, T2 and T3. Values for Krka River (Slovenia) were obtained from data presented in [6] at sampling points T1–T10 (see Table S3 in [6]) and excluding highest D_{Mg} values obtained at T11–T13 sampling points with the greatest dolomite contents. Values for Gregory River (Australia) obtained from data presented in [1] for tufa deposits at one sampling point: mean D_{Mg} value is that obtained in the review in [8], and mean water temperature value corresponds to the mean measured value (27.7 °C) for the studied period in [1]. Finally, the data for the calcite precipitated in Hvanná River (in the vicinity of the Eyjafjallajökull volcano, Iceland) were obtained from the range of D_{Mg} values presented in [65] and the measured temperatures at Sites 2–9 (see Tables 1 and 3 in [65]). Data for the travertine deposits. These data correspond to deposits sampled at the Alhama thermal spring system and at the Embid thermal spring (both in Zaragoza, Spain), presented in [66]. We only plotted samples consisting of pure calcite and precipitating near the orifices of the springs and from natural waters (e.g., not overheated) in the figure. We obtained the D_{Mg} values from the specific Mg/Ca values analysed in each deposit and the associated spring (water sample). Data for speleothems. Orange-filled triangle: mean D_{Mg} (0.0178 ± 0.0005) and cave temperature values presented in [68] from recent carbonate crystals precipitated on watch glasses and glass plates in seven caves in Germany, Morocco and Romania; we also indicate a range of cave temperatures. Empty triangle: mean D_{Mg} and drip water temperature presented in [9] for annually laminated stalagmites in Alpine cave (Obir, Austria). Yellow-filled triangles: mean D_{Mg} and temperature values for seepage waters and associated calcites in speleothems from caves on Vancouver Island and in Jamaica studied in [67]. Grey-filled triangle: mean D_{Mg} values and cave air temperature (assumed to be constant) presented in [45] for modern speleothems in Grotta di Ernesto cave (Italy).

The travertine points in Figure 7 are also close to the Gregory River mean D_{Mg} and temperature values provided in [1]. This study is the only successful "thermometrical" application of the Mg partitioning in fluvial tufas from the literature. As in other tufa deposits, other factors besides the temperature may influence Mg partitioning (e.g., the biological influence through the effects of EPSs); however, the high deposition temperature in the Gregory River (with a mean value of 27.7 °C for the studied period, which was the highest among all of the compared rivers (Figure 7)) apparently dominates the partitioning process. Whether this high-temperature value is enough to dominate the process or whether the effects of other influencing processes must decrease at the same time should be further studied.

7. Conclusions

We analysed the Mg/Ca ratios of the recent stromatolite records of four rivers on the Iberian Peninsula and the different linked parameters, and we compared them to similar systems. We derived several conclusions from this work:

- We tested the equation in [45] in four different fluvial stromatolite records, and we found limited application. Except for the work of [1], this study is one of the few cases (i.e., Piedra River) in which the Mg-derived water temperatures match the measured ones, in the fluvial environment.
- The degree of the consistency between the Mg-derived water temperatures and measured water temperatures was variable, depending on the studied cases, which indicates that factors other than the temperature influence the $(Mg/Ca)_{calcite}$. This result is contrary to the high degree of agreement between the $\delta^{18}O$-derived and measured temperatures for the same samples.
- The Mg contents of detrital minerals were responsible for the offset of the Mg-derived water temperatures towards higher values than the measured water temperatures in at least one of the studied rivers (Piedra River), in which the correction of the detrital Mg content yielded water temperatures that matched the measured ones.
- In the three other studied cases, we need to consider the interference of several other factors, either seasonal or not, to explain the lack of agreement between the estimated and measured temperatures, such as the changes in the calcite precipitation rates and water discharges. The seasonal behaviour of the $Mg_{calcite}$, and its circumstantial correlation with the temperature, may be due to the influence of other seasonal parameters.
- The presence of noncarbonate Mg (and Ca) minerals, and the occurrence of nonequilibrium conditions in natural systems (e.g., variable flow rates, turbulent flows, the presence of biofilms or plant substrates), substantially limit the use of the $(Mg/Ca)_{cacite}$ as a geochemical thermometer in continental sedimentary environments, except for well-defined systems (e.g., laminar flow, small seasonal variations, and in some hydrothermal systems).
- For the first time, we demonstrate the seasonal variation in the $(Mg/Ca)_{calcite}$ and D_{Mg} in fluvial carbonates. Moreover, the seasonal variations in these two parameters are not necessarily associated with the temperature or temperature-dependent parameters, which researchers have recorded in the Piedra River and partially in the Ebrón River stromatolite records.

Supplementary Materials: The following supporting information can be downloaded at: https://www.mdpi.com/article/10.3390/min13010057/s1, Figure S1: Piper diagrams for the waters of (A) Piedra, (B) Mesa, (C) Ebrón and (D) Añamaza Rivers; Figure S2: Box-and-whisker plots of the statistical distribution of Mg/Ca values in water samples biannually taken at studied sites in Piedra, Mesa, Ebron and Añamaza Rivers (P-14, M-4, RE-8 and RA-6, respectively); Figure S3: Experimental relationships between water temperature and magnesium partition coeficient DMg determined in [45,46] under karst/speleothem-analogue conditions. Table S1: Concentrations of Ca and Mg (in mg/kg) in the tufa samples recorded on Tablets P-14 (Piedra River), M-4 (Mesa River), RA-6 (Añamaza River) and RE-8 (Ebrón River) over time; Table S2: Hydrochemical characteristics of monitored sites in Piedra, Mesa, Añamaza and Ebrón Rivers. Concentrations in ppm; Table S3: Measured water (T_w), calculated T_{Mg} (using D_{Mg}–T relationship in [45]) and calculated T_{18O} (from $\delta^{18}O$ values in calcite and water, using the equation of [47]) temperatures for studied sites at Piedra, Mesa, Añamaza and Ebrón Rivers; Table S4. Values of δ^{18} O and δ^{13} C (in ‰ VPDB) of tufa calcite recorded over time on Tablets P-14 (Piedra River), M-4 (Mesa River), RA-6 (Añamaza River) and RE-8 (Ebrón River).

Author Contributions: C.A., L.F.A., M.C.O. and C.S. performed the field work, laboratory sampling and the interpretation of the sediment and water results. N.C. performed the trace element analyses of the sediment and water. S.L. participated in the analysis of the stable isotope compositions of the sediment samples. All authors have read and agreed to the published version of the manuscript.

Funding: This research was funded by the Spanish Ministry through grants REN2002-3575/CLI, CGL2006-05063/BTE, CGL2009-09216/BTE and PID2019-106440GB-C22 (MCIN/AEI/ 10.13039/ 501100011033) and by the Slovenian Research Agency, research project J1-2478 and research program P1-0143.

Data Availability Statement: The data are reported either in the main text or in the Supplementary Materials.

Acknowledgments: The authors would like to acknowledge the use of the Servicio General de Apoyo a la Investigación-SAI, Universidad de Zaragoza, Spain, and the Servicios Científico-Técnicos (CCIT-UB Serveis), University of Barcelona, Spain. We are grateful to the management and staff of the Monasterio de Piedra Natural Park for facilitating our fieldwork. We would like to kindly acknowledge Marta Vázquez Urbez for her scientific help and the two anonymous reviewers for their comments on the manuscript. This paper is dedicated to the memory of our colleague and friend, Carlos Sancho, who passed away during the preparation of this manuscript.

Conflicts of Interest: The authors declare no conflict of interest.

Appendix A. Influence of the Parameters Controlling $(Mg/Ca)_{calcite}$ on the Four Rivers

Appendix A.1. $(Mg/Ca)_{calcite}$ and Measured Tw

The RE values had the highest direct correlation between the $(Mg/Ca)_{calcite}$ and measured Tw (r = 0.69, N = 6), and both patterns were roughly parallel. In the RM, there was no correlation (r = −0.02, N = 5), and in the RA, there was a hardly significant correlation (r = −0.31, N = 5). In the RP, the correlation was r = 0.39 (N = 12); however, it was poorer over the first interval (from Warm 2004 to Warm 2006) (r = 0.38, N = 5) and higher over the second interval (from Cool 2006–2007 to Cool 2009–2010) (r = 0.57, N = 7). The significant positive correlations do not necessarily imply the dependence of the $(Mg/Ca)_{calcite}$ on the temperature; however, they may depend on the six-month oscillations of both the $(Mg/Ca)_{calcite}$ and Tw parameters. In the RE, which had the highest differences between the $(Mg/Ca)_{calcite}$-based, estimated Tw and the measured Tw (Figure 4), the highest correlation was between the Tw and $(Mg/Ca)_{calcite}$, and it had the clearest $(Mg/Ca)_{calcite}$ pattern. On the contrary, the RP only had a roughly rhythmic $(Mg/Ca)_{calcite}$ pattern, and it exhibited the lowest differences between the $(Mg/Ca)_{calcite}$-derived and measured Tw. In the RA and RM, there were no six-month variations in the $(Mg/Ca)_{calcite}$, and there was no significant correlation between the $(Mg/Ca)_{calcite}$ and measured Tw.

With respect to the season-free evolutions, in the RE and RA, the $(Mg/Ca)_{calcite}$ and measured Tw were only vaguely parallel. (Figure 6). In the RM, the evolutions of the $(Mg/Ca)_{calcite}$ and Tw were opposite. Even in the RP, which had the best fitting between the Mg-derived Tw and measured Tw, the season-free evolutions were not parallel.

Appendix A.2. Mg/Ca of Water

In the RE, the $(Mg/Ca)_{water}$ and $(Mg/Ca)_{calcite}$ had seasonal variations, and although the correlation was not strong (r = 0.55, N = 6), both had similar season-free evolutions (Figure 6). Moreover, the corresponding D_{Mg} parallels the $(Mg/Ca)_{calcite}$, especially in terms of the season-free evolution (Figure 6). Therefore, the Mg-derived Tw, through the D_{Mg}, exhibited a broad six-month oscillation, with some exceptions. Despite this positive outcome, the estimated Tw values (minimum value = 27.7 °C, maximum value = 44.4 °C) were not plausible, due to the high $(Mg/Ca)_{calcite}$ values with respect to the $(Mg/Ca)_{water}$ values (Table 1), which is why the D_{Mg} predominantly reflects the $(Mg/Ca)_{calcite}$ pattern.

According to these results, additional factors other than the temperature and $(Mg/Ca)_{water}$ influence the $(Mg/Ca)_{calcite}$, and these also operate on a seasonal basis. Unlike in the

RP, in the RE, the incorporation of Mg into calcite was favoured above the theoretical partitioning values.

In the RA, in which the $(Mg/Ca)_{water}$ was quite constant over time, the $(Mg/Ca)_{calcite}$ was also constant, except in the last period (Cool 2009–2010), when both parameters and their inter-annual evolutions were opposite (Figure 4). The ratio of these parameters results is a quasistable D_{Mg}, with a sharp increase in the last period and the consequent temperature spikes over 43 °C (Table S3, Supplementary Materials). The D_{Mg} (and the Mg-derived Tw) mimics the $(Mg/Ca)_{calcite}$ variations; therefore, the Mg-derived Tw values did not exhibit six-month oscillations, and they were also higher than the real ones, which means that the $(Mg/Ca)_{calcite}$ was higher than it should have been with respect to the $(Mg/Ca)_{water}$.

In these cases, the temperature effect on Mg partitioning in calcite was overprinted by the influence of parameters other than the $(Mg/Ca)_{water}$ on the $(Mg/Ca)_{calcite}$ content. Unlike the first time interval of the RP (from Warm 2004 to Warm 2006), the Mg-derived Tw were higher than the measured Tw, evidencing higher $(Mg/Ca)_{calcite}$ values than were expected from the $(Mg/Ca)_{water}$ content. In the RA, there was no relationship between the $(Mg/Ca)_{calcite}$ and Tw measures.

In the RM, the $(Mg/Ca)_{water}$ had a dizzy seasonal pattern; however, the oscillations were small, except in the last period (Warm 2006) (Figure 4). Like in the RA, there was no apparent influence of the $(Mg/Ca)_{water}$ variations on the $(Mg/Ca)_{calcite}$ of the RM. Thus, the $(Mg/Ca)_{calcite}$ irregularly varied with respect to the $(Mg/Ca)_{water}$ (Figure 4), and its season-free evolutions were apparently not related to each other (Figure 6). The result was an oscillating D_{Mg} with an increasing season-free evolution, which mimicked the $(Mg/Ca)_{calcite}$ season-free evolution (Figure 6). Consequently, the Mg-derived Tw had a reversed oscillating pattern, with higher values in the cool periods (Table S3, Supplementary Materials), and a range that was narrower than that of the measured Tw.

In the RP, in the first interval (from Warm 2004 to Warm 2006), when both the $(Mg/Ca)_{water}$ and $(Mg/Ca)_{calcite}$ displayed parallel six-month oscillations ($r = 0.59$ ($N = 5$)), the six-month changes were almost fully compensated for the D_{Mg} calculation. As a consequence, the Mg-derived Tw (Equation (3)) did not show six-month variations, although the calculated Tw values fit quite closely with the average measured Tw values. On the contrary, in the interval spanning from Cool 2006-2007 to Cool 2009-2010, which had one of the highest correlations between the $(Mg/Ca)_{calcite}$ and measured Tw ($r = 0.57$, $N = 7$), the $(Mg/Ca)_{water}$ was negatively correlated to the $(Mg/Ca)_{calcite}$ ($r = -79$, $N = 7$), and their season-free evolutions were almost opposite (Figure 6). Therefore, the high $(Mg/Ca)_{water}$ in the cool periods should, to some extent, compensate for the temperature Mg-partitioning effects and produce a relatively stable $(Mg/Ca)_{calcite}$ pattern. On the contrary, in this time interval, the six-month oscillations of the $(Mg/Ca)_{calcite}$ were wider than in the first interval (from Warm 2002 to Warm 2006) (Figure 4), and this amplitude was transferred to the empirical D_{Mg}. The highest D_{Mg} values corresponded to Warm 2008 and Warm 2009 (Figure 4), in which the minimum $(Mg/Ca)_{water}$ values coincided with the maximum $(Mg/Ca)_{cacite}$ values. Therefore, although the Mg-derived Tw values had a realistic, six-month oscillating pattern, the estimated Tw values were substantially higher than the measured Tw, and the season-free evolution of the Mg-derived Tw was not parallel to the measured Tw (Figure 6). The different behaviours during the two time intervals in the same river suggest a quick change in the controls on the $(Mg/Ca)_{calcite}$, which in the second time interval (from Cool 2006–2007 to Cool 2009–2010), overprint the variations in the $(Mg/Ca)_{water}$.

Appendix A.3. Detrital Minerals

If the Tw is recalculated after subtracting 500 ppm from the Mg values, the mean D_{Mg} and the corresponding estimated Tw decreases in the four rivers:

- In the RP, the mean D_{Mg} decreases from 0.023 to 0.018, and the calculated Mg-derived Tw is 3.7 °C lower. This value cancels out the mean difference between the estimated and measured Tw in the first time interval (from Warm 2004 to Warm 2006) (3.02 °C,

Table 2); in the second time interval (from Cool 2006–2007 to Cool 2009–2010), the difference of 7.6 °C is substantially reduced with this correction.

- In the RE, the mean D_{Mg} decreases from 0.046 to 0.040, and the calculated Tw is 5 °C lower. These temperatures are still far from the measured ones, which are 27 °C lower as an average. Therefore, a larger amount of detrital matter would be required to fit that value.
- In the RA, the mean D_{Mg} decreases from 0.032 to 0.023, and the calculated Tw is 7.4 °C lower. Again, this value is rather behind 13.9 °C, which is the mean difference between the estimated and measured Tw.
- In the RM, the mean D_{Mg} decreases from 0.032 to 0.026, and the calculated Tw is 4.2 °C lower than the previously estimated value. This correction does not compensate for the 12.7 °C offset between the measured and calculated Tw.

If the Mg content corresponding to the measured Tw is estimated from the measured $(Mg/Ca)_{water}$, by using Equations (2) and (3), it can be shown that the first time interval for the RP (from Warm 2002 to Warm 2006) had the smallest mean Mg excess (568 ppm), which was quite close to the Mg that corresponded to the detrital fraction; the second interval (from Cool 2006–2007 to Cool 2009–2010) had a higher Mg excess (1003 ppm), which was similar to that of the RA (959 ppm). This Mg excess was much higher in the RM (1608 ppm) and, especially, in the RE (2494 ppm). The latter values are difficult to explain based on the Mg content from the detrital fraction, even though there was a larger proportion of detrital Mg (from both clay minerals and dolomite) in the sample.

Appendix A.4. Precipitation Rate

In the RA, both measured thicknesses and calculated PWP values displayed the expected rhythmic pattern (higher in the warm periods), except for the PWP values in the last period (Cool 2009–2010) (Figure 4). Therefore, there was a good correspondence between the calculated PWP values (higher in the warm periods) and measured thicknesses. However, their season-free evolutions were not parallel to each other. The $(Mg/Ca)_{calcite}$ and D_{Mg} values only had minor variations over time, for which they did not mimic either of the two growth rates. The $(Mg/Ca)_{calcite}$ season-free evolution was closer to the PWP value than to the sediment thickness, which was opposite to the $(Mg/Ca)_{calcite}$ (Figure 6).

In the RE, the measured thickness had the expected six-month rhythmic pattern, only it was reversed in the first period, and the PWP values only displayed this pattern in the last three periods (Figure 4). Thus, they had neither parallel patterns, nor parallel season-free evolutions (Figure 6). The $(Mg/Ca)_{calcite}$ had six-month variations, and it matched the thicknesses measured on the tablets (except for Warm 2007) and the PWP values in the last three periods. However, the corresponding season-free evolutions of the $(Mg/Ca)_{calcite}$ and thicknesses were not parallel; the measured thicknesses and the season-free evolution were opposite to the $(Mg/Ca)_{calcite}$. Researchers have described this reverse relationship between the $(Mg/Ca)_{calcite}$ and both types of deposition rates for other tufa sediments, in which the kinetic effects of the rapid calcite precipitation yield to a Mg-impoverished calcite [7].

In the RM, the rhythmic pattern of the PWP values was reversed compared with the other studied rivers, (i.e., higher values for the cool periods and lower values for the warm periods; Figure 4). Thus, the PWP patterns did not match the measured thickness rates, although both displayed increasing season-free evolutions (Figure 6). The PWP values paralleled the reversed pattern of the D_{Mg}. Contrarily, the $(Mg/Ca)_{calcite}$ values did not mimic the PWP pattern; however, their corresponding season-free evolutions followed parallel trends (Figure 6), as did the measured thickness rates, despite the fact that we detected erosional processes in this record.

The best agreement between the measured thickness and PWP patterns occurred for the RP, for which both parameters exhibited six-month variations (Figure 4). However, their season-free evolutions were not fully parallel (Figure 6): in the second time interval (from Cool 2006–2007 to Cool 2009–2010), they were reversed. The $(Mg/Ca)_{calcite}$ was broadly coupled with the sediment thickness and PWP values in this second time interval, in which it exhibited a six-month oscillation. In this time interval, the season-free evolution of the $(Mg/Ca)_{calcite}$ roughly paralleled the sediment thickness, but not the PWP values. This time interval coincided with an increase in the sediment accumulation, which was higher than those in the RE and RM and similar to that of the RA.

References

1. Ihlenfeld, C.; Norman, M.D.; Gagan, M.K.; Drysdale, R.N.; Maas, R.; Webb, J. Climatic significance of seasonal trace element and stable isotope variations in a modern freshwater tufa. *Geochim. Cosmochim. Acta* **2003**, *67*, 2341–2357. [CrossRef]
2. Kawai, T.; Kano, A.; Matsuoka, J.; Ihara, T. Seasonal variation in water chemistry and depositional processes in a tufa-bearing stream in SW-Japan, based on 5 years of monthly observations. *Chem. Geol.* **2006**, *232*, 33–53. [CrossRef]
3. Lojen, S.; Trkov, A.; Scancar, J.; Vázquez-Navarro, J.A.; Cukrov, N. Continuous 60-year stable isotopic and earth-alkali element records in a modern laminated tufa (Jaruga, river Krka, Croatia): Implications for climate reconstruction. *Chem. Geol.* **2009**, *258*, 242–250. [CrossRef]
4. Rogerson, M.; Pedley, H.M.; Wadhawan, J.D.; Middleton, R. New insights into biological influence on the geochemistry of freshwater carbonate deposits. *Geochim. Cosmochim. Acta* **2008**, *72*, 4976–4987. [CrossRef]
5. Brasier, A.T.; Andrews, J.E.; Marca-Bell, A.D.; Dennis, P.F. Depositional continuity of seasonally laminated tufas: Implications for $\delta^{18}O$ based palaeotemperatures. *Glob. Planet. Change* **2010**, *71*, 160–167. [CrossRef]
6. Zavadlav, S.; Rožič, B.; Dolenec, M.; Lojen, S. Stable isotopic and elemental characteristics of recent tufa from a karstic Krka River (south-east Slovenia): Useful environmental proxies? *Sedimentology* **2017**, *64*, 808–831. [CrossRef]
7. Saunders, P.; Rogerson, M.; Wadhawan, J.D.; Greenway, G.; Pedley, H.M. Mg/Ca ratios in freshwater microbialcarbonates: Thermodynamic, kinetic and vital effects. *Geochim. Cosmochim. Acta* **2014**, *147*, 107–118. [CrossRef]
8. Ritter, S.M.; Isenbeck-Schröter, M.; Schröder-Ritzrau, A.; Scholz, C.; Rheinberger, S.; Höfle, B.; Frank, N. Trace element partitioning in fluvial tufa reveals variable portions of biologically influenced calcite precipitation. *Geochim. Cosmochim. Acta* **2018**, *225*, 176–191. [CrossRef]
9. Fairchild, I.J.; Borsato, A.; Tooth, A.F.; Frisia, S.; Hawkesworth, C.J.; Huang, Y.-M.; McDermott, F.; Spiro, B. Controls on trace element (Sr-Mg) compositions of carbonate cave waters: Implications for speleothem climatic records. *Chem. Geol.* **2000**, *166*, 255–269. [CrossRef]
10. Pogge von Strandmann, P.; Burton, K.; Snæbjörnsdóttir, S.O.; Sigfusson, B.; Aradottir, E.; Gunnarsson, I.; Alfredsson, H.; Mesfin, K.; Oelkers, E.; Gislason, S. Rapid CO_2 mineralisation into calcite at the CarbFix storage site quantified using calcium isotopes. *Nat. Commun.* **2019**, *10*, 1983. [CrossRef]
11. Zhao, M.; Zheng, Y.F.; Zhao, Y. Seeking a geochemical identifier for authigenic carbonate. *Nat. Commun.* **2016**, *7*, 10885. [CrossRef] [PubMed]
12. Rovan, L.; Zuliani, T.; Horvat, B.; Kanduč, T.; Vreča, P.; Jamil, Q.; Čermelj, B.; Bura-Nakić, E.; Cukrov, N.; Štrok, M.; et al. Uranium isotopes as a possible tracer of terrestrial authigenic carbonate. *Sci. Total Environ.* **2021**, *797*, 194103. [CrossRef]
13. Osácar, M.C.; Arenas, C.; Auqué, L.; Sancho, C.; Pardo, G.; Vázquez-Urbez, M. Discerning the interactions between environmental parameters reflected in $\delta^{13}C$ and $\delta^{18}O$ of recent fluvial tufas: Lessons from a Mediterranean climate región. *Sediment. Geol.* **2016**, *345*, 126–144. [CrossRef]
14. Vera, J.A. *Geología de España*; SGE-IGME: Madrid, Spain, 2004; p. 884.
15. Lozano, M.V.; Sancho, C.; Arenas, C.; Vázquez-Urbez, M.; Ortiz, J.E.; Torres, T.; Pardo, G.; Osácar, M.C.; Auqué, L. Analisis preliminar de las tobas cuaternarias del Río Ebrón (Castielfabib, Valencia, Cordillera Ibérica). *Geogaceta* **2012**, *51*, 51–54.
16. Vázquez-Úrbez, M.; Arenas, C.; Pardo, G. A sedimentary facies model for stepped, fluvial tufa systems in the Iberian Range (Spain): The Quaternary Piedra and Mesa valleys. *Sedimentology* **2012**, *59*, 502–526. [CrossRef]
17. Rico, M.T.; Sancho-Marcén, C.; Arenas-Abad, M.C.; Vázquez-Urbez, M.; Valero-Garcés, B.L. El sistema de barreras tobáceas Holocenas de las Parras de Martín (Cordillera Ibérica, Teruel). *Cuad. Investig. Geogr.* **2013**, *39*, 141–158. [CrossRef]
18. Luzón, M.A.; Pérez, A.; Borrego, A.G.; Mayayo, M.J.; Soria, A.R. Interrelated continental sedimentary environments in the central Iberian Range (Spain): Facies characterization and main palaeoenvironmental changes during the Holocene. *Sediment. Geol.* **2011**, *239*, 87–103. [CrossRef]
19. Arenas, C.; Vázquez-Urbez, M.; Auqué, L.; Sancho, C.; Osácar, M.C.; Pardo, G. Intrinsic and extrinsic controls of spatial and temporal variations in modern fluvial tufa sedimentation: A thirteen-year record from a semi-arid environment. *Sedimentology* **2014**, *61*, 90–132. [CrossRef]
20. Osácar, C.; Arenas, C.; Vázquez-Urbez, M.; Sancho, C.; Auqué, L.F.; Pardo, G. Environmental factors controlling the $\delta^{13}C$ and $\delta^{18}O$ variations of recent fluvial tufas: A 12-year record from the Monasterio de Piedra Natural Park (NE Iberian Peninsula). *J. Sediment. Res.* **2013**, *83*, 309–322. [CrossRef]
21. Arenas, C.; Auqué, L.; Osácar, M.C.; Sancho, C.; Lozano, M.V.; Vázquez-Urbez, M.; Pardo, G. Current tufa sedimentation in a high discharge river: A comparison with other synchronous tufa records in the Iberian Range (Spain). *Sediment. Geol.* **2015**, *325*, 132–157. [CrossRef]

22. Auqué, L.; Arenas, C.; Osácar, M.C.; and Pardo, G.; Sancho, C.; Vázquez-Urbez, M. Tufa sedimentation in changing hydrological conditions: The River Mesa (Spain). *Geol. Acta* **2013**, *11*, 85–102.
23. Auqué, L.; Arenas, C.; Osácar, C.; Pardo, G.; Sancho, C.; Vázquez-Urbez, M. Current tufa sedimentation in a changing-slope valley: The River Añamaza (Iberian Range, NE Spain). *Sediment. Geol.* **2014**, *303*, 26–48. [CrossRef]
24. Pentecost, A. *Travertine*; Springer: Berlin/Heidelberg, Germany; New York, NY, USA,, 2005; 445p.
25. Vázquez-Urbez, M.; Arenas, C.; Sancho, C.; Osácar, C.; Auqué, L.; Pardo, G. Factors controlling present-day tufa dynamics in the Monasterio de Piedra Natural Park (Iberian Range, Spain): Depositional environmental settings, sedimentation rates and hydrochemistry. *Int. J. Earth Sci.* **2010**, *99*, 1027–1049. [CrossRef]
26. Arenas, C.; Jones, B. Temporal and environmental significance of microbial lamination: Insights from Recent fluvial stromatolites in the River Piedra, Spain. *Sedimentology* **2017**, *64*, 1597–1629. [CrossRef]
27. Beraldi-Campesi, H.; Arenas-Abad, C.; Garcia-Pichel, F.; Arellano-Aguilar, O.; Auque, L.; Vazquez-Urbez, M.; Sancho, C.; Osácar, C.; Ruiz-Velasco, S. Benthic bacterial diversity from freshwater tufas of the Iberian Range (Spain). *FEMS Microbiol. Ecol.* **2012**, *80*, 363–379. [CrossRef] [PubMed]
28. Berrendero, E.; Arenas, C.; Matero, P.; Jones, B. Cyanobacterial diversity and related sedimentary facies as a function of water flow conditions: Example from the Monasterio de Piedra Natural Park (Spain). *Sediment. Geol.* **2016**, *337*, 12–28. [CrossRef]
29. Arenas, C.; Osácar, C.; Sancho, C.; Vázquez-Urbez, M.; Auqué, L.; Pardo, G. Seasonal record from recent fluvial tufa deposits (Monasterio de Piedra, NE Spain): Sedimentological and stable isotope data. *Geol. Soc. Lond. Spec. Publ.* **2010**, *336*, 119–142. [CrossRef]
30. Moore, D.M.; Reynolds, R.C., Jr. *X-ray Diffraction and the Identification and Analysis of Clay Minerals*, 2nd ed.; Oxford University Press: Oxford, UK, 1997; pp. xvii + 378.
31. Hillier, S. Quantitative Analysis of Clay and Other Minerals in Sandstones by X-ray Powder Diffraction (XRPD). In *Clay Minerals Cements in SANDSTONES. Int. Assoc. Sedimento*; Worden, R.H., Morad, S., Eds.; Blackwell Science Ltd.: Hoboken, NJ, USA, 2003; Volume 34, pp. 213–251.
32. Parkhurst, D.L.; Appelo, C.A.J. Description of Input and Examples for PHREEQC Version 3. A Computer Program for Speciation, Batch Reaction, One Dimensional Transport, and Inverse Geochemical Calculations. In *Techniques and Methods, Book 6, Chap. A43*; U.S. Geological Survey, Ed.; U.S. Geological Survey: Denver, CO, USA, 2013; 497p. Available online: https://www.usgs.gov/software/phreeqc-version-3 (accessed on 28 December 2022).
33. Plummer, L.N.; Wigley, T.M.L.; Parkhurst, D.L. The kinetics of calcite dissolution in CO_2–water system at 5° to 60 C and 0.0 to 1.0 atm CO_2. *Am. J. Sci.* **1978**, *278*, 179–216. [CrossRef]
34. Kaufmann, G.; Dreybrodt, W. Calcite dissolution kinetics in the system $CaCO_3$–H_2O–CO_2 at high undersaturation. *Geochim. Cosmochim. Ac.* **2007**, *71*, 1398–1410. [CrossRef]
35. Reddy, M.M.; Plummer, L.N.; Busenberg, E. Crystal growth of calcite from calcium bicarbonate solutions at constant P_{CO2} and 25 °C: A test of a calcite dissolution model. *Geochim. Cosmochim. Acta* **1981**, *45*, 1281–1289. [CrossRef]
36. Lorah, M.M.; Herman, J.S. The chemical evolution of a travertine depositing stream: Geochemical processes and mass transfer reactions. *Water Resour. Res.* **1988**, *24*, 1541–1552. [CrossRef]
37. Dreybrodt, W.; Buhmann, D.; Michaelis, J.; Usdowski, E. Geochemically controlled calcite precipitation by CO_2 outgassing: Field measurements of precipitation rates in comparison to theoretical predictions. *Chem. Geol.* **1992**, *97*, 285–294. [CrossRef]
38. Liu, Z.; Svensson, U.; Dreybrodt, W.; Yuan, D.; Buhmann, D. Hydrodynamic control of inorganic calcite precipitation in Huanglong ravine, China: Field measurements and theoretical prediction of deposition rates. *Geochim. Cosmochim. Acta* **1995**, *59*, 3087–3097.
39. Kano, A.; Matsuoka, J.; Kojo, T.; Fujii, H. Origin of annual laminations in tufa deposits, southwest Japan. *Palaeogeogr. Palaeocl. Palaeoecol.* **2003**, *191*, 243–262. [CrossRef]
40. Kano, A.; Hagiwara, R.; Kawai, T.; Hori, M.; Matsuoka, J. Climatic conditions and hydrological change recorded in a high-resolution stable-isotope profile of a recent laminated tufa on a subtropical island, southern Japan. *J. Sediment. Res.* **2007**, *77*, 59–67. [CrossRef]
41. Kawai, T.; Kano, A.; Hori, M. Geochemical and hydrological controls on biannual lamination of tufa deposits. *Sediment. Geol.* **2009**, *213*, 41–50. [CrossRef]
42. Shiraishi, F.; Reimer, A.; Bisset, A.; de Beer, D.; Arp, G. Microbial effects on biofilm calcification, ambient water chemistry and stable isotope records in a highly supersaturated setting (Westerhöfer Bach, Germany). *Palaeogeogr. Palaeocl. Palaeoecol.* **2008**, *262*, 91–106. [CrossRef]
43. Dreybrodt, W.; Buhmann, D. A mass transfer model for dissolution and precipitation of calcite from solutions in turbulent motion. *Chem. Geol.* **1991**, *90*, 107–122. [CrossRef]
44. Lu, G.; Zheng, C.; Donahoe, R.J.; Lyons, B.W. Controlling processes in a $CaCO_3$ precipitating stream in Huanglong Natural Scenic District, Sichuan, China. *J. Hydrol.* **2000**, *230*, 34–54. [CrossRef]
45. Huang, Y.M.; Fairchild, I.J. Partitioning of Sr^{2+} and Mg^{2+} into calcite under karst-analogue experimental conditions. *Geochim. Cosmochim. Acta* **2001**, *65*, 47–62. [CrossRef]
46. Day, C.C.; Henderson, G.M. Controls on trace-element partitioning in cave-analogue calcite. *Geochim. Cosmochim. Acta* **2013**, *120*, 612–627. [CrossRef]

47. O'Brien, G.R.; Kaufman, D.S.; Sharp, W.D.; Atudorei, V.; Parnell, R.A.; Crossell, L.J. Oxygen isotope composition of annually banded modern and mid-Holocene travertine and evidence of paleomonsoon floods, Grand Canyon, Arizona, USA. *Quat. Res.* **2006**, *65*, 366–379. [CrossRef]
48. Özkul, M.; Gökgöz, A.; Horvatinčić, N. Depositional properties and geochemistry of Holocene perched springline tufa deposits and associated spring waters: A case study from the Denizli Province, Western Turkey. In *Tufas and Speleothems: Unravelling the Microbial and Physical Controls*; Pedley, H.M., Rogerson, M., Eds.; Geological Society: London, UK, 2010; Volume 336, pp. 245–262.
49. Dabkowski, J.; Limondin-Lozouet, N.; Andrews, J.; Marca-Bell, A.; Antoine, A. Climatic and environmental variations during the last interglacial recorded in a Northern France tufa (Caours, Somme basin). Comparisons with regional records. *Quaternaire* **2016**, *27*, 249–261. [CrossRef]
50. Dabkowski, J.; Frodlová, J.; Hájek, M.; Hájková, P.; Petr, L.; Fiorillo, D.; Dudová, L.; Horsák, M. A complete Holocene climate and environment record for the Western Carpathians (Slovakia) derived from a tufa deposit. *Holocene* **2018**, *29*, 493–504. [CrossRef]
51. Mavromatis, V.; Gautier, Q.; Bosc, O.; Schott, J. Kinetics of Mg partition and Mg stable isotope fractionation during its incorporation in calcite. *Geochim. Cosmochim. Acta* **2013**, *114*, 188–203. [CrossRef]
52. Wassenburg, J.A.; Riechelman, S.; Schoder-Ritzrau, A.; Riechelman, D.F.C.; Richter, D.K.; Immenhauser, A.; Terente, M.; Constantin, S.; Hachenberg, A.; Hansen, M.; et al. Calcite Mg and Sr partition coefficients in cave environments: Implications for interpreting prior calcite precipitation in speleothems. *Geochim. Cosmochim. Acta* **2020**, *269*, 581–596. [CrossRef]
53. Garnett, E.R.; Andrews, J.E.; Preece, R.C.; Dennis, P.F. Climatic change recorded by stable isotopes and trace elements in a British Holocene tufa. *J. Quat. Sci.* **2004**, *19*, 251–262. [CrossRef]
54. Rogerson, M.; Pedley, H.M.; Greenway, G.M.; Wadhawan, J.D. Interaction of temperature, salinity and extracellular polymeric substances controls trace element incorporation into tufa calcite. *Depos. Rec.* **2022**, *8*, 210–219. [CrossRef]
55. Sironić, A.; Barešić, J.; Horvatinčić, N.; Brozinčević, A.; Vurnek, M.; Kapelj, S. Changes in the geochemical parameters of karst lakes over the past three decades. The case of Plitvice Lakes, Croatia. *Appl. Geochem.* **2017**, *78*, 12–22. [CrossRef]
56. Karmann, I.; Cruz, F.W., Jr.; Viana, O., Jr.; Burns, S.J. Climate influence on geochemistry parameters of waters from Santana–Pérolas cave system, Brazil. *Chem. Geol.* **2007**, *244*, 232–247. [CrossRef]
57. Fairchild, I.J.; Treble, P.C. Trace elements in speleothems as recorders of environmental change. *Quat. Sci. Rev.* **2009**, *28*, 449–468. [CrossRef]
58. Rimstidt, J.D.; Balog, A.; Webb, J. Distribution of trace elements between carbonate minerals and aqueous solutions. *Geochim. Cosmochim. Acta* **1998**, *62*, 1851–1863. [CrossRef]
59. Walter, L.M. Relative efficiency of carbonate dissolution and precipitation during diagenesis: A progress report on the role of solution chemistry. In *Roles of Organic Matter in Sediment Diagenesis*; Gautier, D.L., Ed.; Society of Economic Paleontologists and Mineralogists Special Publication: Tulsa, OK, USA, 1986; Volume 38, pp. 1–11.
60. Sheikholeslami, R.; M. Ng. Calcium Sulfate Precipitation in the Presence of Nondominant Calcium Carbonate: Thermodynamics and Kinetics. *Ind. Eng. Chem. Res.* **2001**, *40*, 3570–3578. [CrossRef]
61. Flaathen, T.K.; Oelkers, E.H.; Gislason, S.R.; Aagaard, P. The effect of dissolved sulphate on calcite precipitation kinetics and consequences for subsurface CO_2 storage. *Enrgy. Proced.* **2011**, *4*, 5037–5043. [CrossRef]
62. Mejri, W.; Korchef, A.; Tlili, M.; Ben Amor, M. Effects of temperature on precipitation kinetics and microstructure of calcium carbonate in the presence of magnesium and sulphate ions. *Desalin. Water Treat.* **2014**, *52*, 25–27. [CrossRef]
63. Cuesta-Mayorga, I.; Astilleros, J.M. Fernández-Díaz, L Precipitation of $CaCO_3$ Polymorphs from Aqueous Solutions: The Role of pH and Sulphate Groups. *Minerals* **2019**, *9*, 178. [CrossRef]
64. Domínguez-Villar, D.; Vázquez-Navarro, J.A.; Krklec, K. The role of gypsum and/or dolomite dissolution in tufa precipitation: Lessons from the hydrochemistry of a carbonate-sulphate karst system. *Earth Surf. Proc. Land.* **2017**, *42*, 245–258. [CrossRef]
65. Olsson, J.; Stipp, S.L.S.; Makovicky, E.; Gislason, S.R. Metal scavenging by calcium carbonate at the Eyjafjallajökull volcano: A carbon capture and storage analogue. *Chem. Geol.* **2014**, *384*, 135–148. [CrossRef]
66. Asta, M.P.; Auqué, L.F.; Sanz, F.J.; Gimeno, M.J.; Acero, P.; Blasco, M.; García-Alix, A.; Gómez, J.; Delgado-Huertas, A.; Mandado, J. Travertines associated with the Alhama-Jaraba thermal Waters (NE, Spain): Genesis and geochemistry. *Sediment. Geol.* **2017**, *347*, 100–116. [CrossRef]
67. Gascoyne, M. Trace-element partition coefficients in the calcite-water system and their paleoclimatic significance in cave studies. *J. Hydrol.* **1983**, *61*, 213–222. [CrossRef]
68. Riechelmann, D.F.C.; Deininger, M.; Scholz, D.; Riechelmann, S. Schröder-Ritzrau, A.; Spötl, C.; Richter, D.K.; Mangini, A.; Immenhauser, A. Disequilibrium carbon and oxygen isotope fractionation in recent cave calcite: Comparison of cave precipitates and model data. *Geochim. Cosmochim. Acta* **2013**, *103*, 232–244. [CrossRef]

minerals

Article

Stable Carbon and Oxygen Isotopic Features of Banded Travertines from the Xiagei Fissure Ridge System (Shangri-La, China)

Yaxian You [1,2], Huaguo Wen [1,2,3,4,*], Lianchao Luo [5], Zhipeng Lu [1,2] and Liang Li [6]

1. State Key Laboratory of Oil and Gas Reservoir Geology, Chengdu University of Technology, Chengdu 610059, China
2. Institute of Sedimentary Geology, Chengdu University of Technology, Chengdu 610059, China
3. CNPC Key Laboratory of Carbonate Reservoir, Chengdu University of Technology, Chengdu 610059, China
4. Key Laboratory of Deep-Time Geographical Environment Reconstruction and Application, Ministry of Natural Resources, Chengdu 610059, China
5. Department of Earth Sciences, University of Florence, 50121 Florence, Italy
6. College of Earth Science, Chengdu University of Technology, Chengdu 610059, China

* Correspondence: wenhuaguo08@cdut.edu.cn

Abstract: Banded travertines are important parts of fissure ridge systems, but studies on geochemical characterization of banded travertines are limited. This study investigated the lithofacies and stable carbon and oxygen isotopic features of banded travertines from Xiagei (southwestern China) to examine their formation mechanisms. Petrographic analyses of the banded travertines revealed two lithotypes: thick-laminated palisade crystalline crust and thin-laminated composite crystalline crust. $\delta^{13}C$ and $\delta^{18}O$ of the Xiagei banded travertines range from 2.82‰ to 4.50‰ V-PDB, and from −25.86‰ to −20.90‰ V-PDB. Parent CO_2 evaluation shows that the Xiagei banded travertines mainly received CO_2 from the decarbonation of marine carbonates, but the contributions of magmatic CO_2 and the dissolution of marine carbonates are also unneglectable. Significantly, the magmatic-derived CO_2 might indicate that the delamination of the lithosphere along with the asthenosphere upwelling could be taking place in the eastern Tibetan plateau. Paleotemperature calculation shows that the Xiagei travertines were precipitated from moderate- to high-temperature hot springs (44.3 to 86.8 °C). Interestingly, the thick-laminated palisade crystalline crust and thin-laminated composite crystalline crust display calculated paleotemperature between 66.6 and 86.8 °C and between 56.6 and 77.7 °C, respectively, reflecting the great role of water temperature in controlling the lithofacies of banded travertines. A comparison between the banded travertines at Xiagei and other areas also shows temperature is a non-negligible factor controlling banded travertine precipitation. However, this does not mean that water temperature is the decisive controlling factor and more studies on banded travertines are still indispensable to disclose the potential factors controlling the factors/processes affecting banded travertine lithofacies. This study provides a good example for understanding the relationship between lithofacies and stable isotopic geochemical characteristics of travertine deposits.

Keywords: banded travertine; fissure ridge; lithofacies; stable carbon and oxygen isotopes; Shangri-La

Citation: You, Y.; Wen, H.; Luo, L.; Lu, Z.; Li, L. Stable Carbon and Oxygen Isotopic Features of Banded Travertines from the Xiagei Fissure Ridge System (Shangri-La, China). *Minerals* **2023**, *13*, 76. https://doi.org/10.3390/min13010076

Academic Editors: Francesca Giustini and Mauro Brilli

Received: 19 November 2022
Revised: 27 December 2022
Accepted: 28 December 2022
Published: 3 January 2023

1. Introduction

Fissure ridges are common elongated travertine build-ups with open or (partly) sealed central fissures and inclined flanks dipping away from the central fissures in geothermal areas [1–10]. Travertine fissure ridges are closely related with tectonic activity, such as faulting and earthquakes [2–7,10–16]; and, as one of carbonate sediments, their geochemistry bears abundant tectonic-environment information [17]. Travertine deposits in fissure ridge systems can be divided into two types: bedded travertines and banded travertines (also named banded veins, carbonate/calcite veins, etc.) [3–6,11,16]. The former are epigean/surface carbonate deposits after the emergence of thermal water from orifices/vents and are deposits dipping gently away from nearly vertical, irregular central fissures, while the latter are less porous hypogean/underground subvertical to vertical carbonate deposits (i.e., formed during the upwelling of thermal fluid along the conduit) and are often small in scale (commonly millimeter-scale to decimeter-scale in width although meter-scale or even more large banded travertines were also reported) [1,4–6,11,12,18–20].

Previous studies on banded travertines in fissure ridge systems mainly focus on their distribution, geometry, mineralogy, texture, and tectonic implications [1,3,4,6,9,11,12,15,16,20–22]. There were also some studies on the geochemical and lithofacies characteristics of banded travertines [18,23–26], but rare studies focused on the relationship between geochemical signatures and lithofacies of banded travertines. Theoretically, the chemical composition of fluids in hypogean conduits might change during the formation of banded travertines and such variation might change both the geochemical signatures and lithofacies of banded travertines.

In this study, careful lithofacies and stable carbon and oxygen isotope analyses of banded travertines at Xiagei were performed to determine their genesis and the relationship between their lithofacies and stable isotope carbon and oxygen isotope compositions. Hydrochemical analyses of modern hot springs at Xiagei and mineralogical and petrological investigations of the Xiagei travertines were also conducted to aid in the interpretation of the Xiagei banded travertines. The findings may help us better understand the formation mechanism of banded travertines.

2. Geological Background

The Shangri-La area, Yunnan Province, China is located near the Jinshajiang fault zone, one of the most active tectonic zones in the Tibetan plateau (Figure 1A). Deposits exposed near the Shangri-La city are mainly composed of Pleistocene to Holocene lacustrine and alluvial-pluvial sediments (e.g., sandy conglomerate, gravel and clay) and Late Triassic sedimentary rocks (including both siliciclastic and carbonate rocks), but Late Paleozoic carbonate rocks and Triassic intrusive igneous rocks are also partly exposed in Shangri-La (Figure 1B) [27–32]. This area was continuously impacted by the subduction of the Paleo- and Meso-Tethyan oceans and continental collisions following their closures during the Mesozoic and the India-Asian collision during the Cenozoic [33–35]. There are lots of N-S trending faults and associated secondary faults [30]. These large-scale faults and fractures provided favorable conduits for the circulation and upwelling of geothermal fluids in Shangri-La and generated a large number of hot springs and associated spring deposits [29–31]. Five key fault zones controlling the hydrothermal activities have been reported in the Shangri La area: N-W striking Zhongdian reverse fault zone (F1), NWW-SEE-striking Yanggu fault zone (F2), N-S striking Gezanhe-Are reverse fault zone (F3), N-NE-striking Kangshi normal fault zone (F4), near E-W trending Tianshengqiao reverse fault zone (F5) (Figure 1B) [30].

The Xiagei area (99°51′20″ E, 27°47′00″ N), is situated to the northeast of the Shangri-La city and lies on the Tianshengqiao fault zone (Figure 1B). Some NW-SE trending and NE-SW trending travertine fissure ridges were developed at Xiagei (Figure 2A,B) and ranged from 25 to 140 m long and up to 120 m wide. The longest fissure ridge at Xiagei (i.e., Fr1 in Figure 2) is a NW-SE trending system and laterally extends around 140 m long (Figure 2A–C). In addition to travertine fissure ridges, some meter-scale travertine mounds/cones, waterfalls, and terraced slopes were also observed near the fissure ridges. Furthermore, a few hot springs occurred in the Xiagei fissure ridge system although their volumetric flow rates were often very low. Active hot springs at Xiagei discharged mid-temperature (41.4 to 63.1 °C) HCO_3-Ca and HCO_3-Ca-Na type waters with slightly acidic to near neutral pH (5.57 to 7.67) and high TDS (1249 to 1468 mg/L) [32]. Isotopic values of the Xiagei hot spring waters were reported to be between −15‰ and −16.2‰ V-SMOW for $\delta^{18}O_{water}$ and between −128.3‰ and −133.4‰ V-SMOW for δD_{water}, which are close to the meteoric water line and are indicative of a meteoric origin [32].

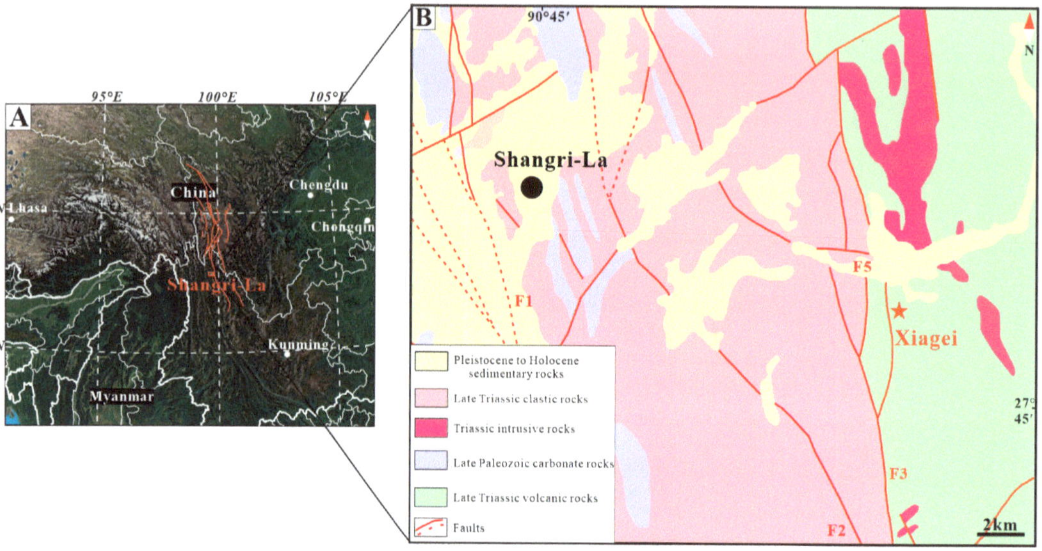

Figure 1. (**A**) Location of the Shangri-La area, northwest of Yunnan Province (China); Red lines: Jinshajiang fault zone. (**B**) Geologic map of Xiagei and surrounding region (modified from Zheng [32]). F2: NWW-SEE-striking Yanggu fault zone, F3: N-S striking Gezanhe-Are reverse fault zone, F5: near E-W trending Tianshengqiao reverse fault zone.

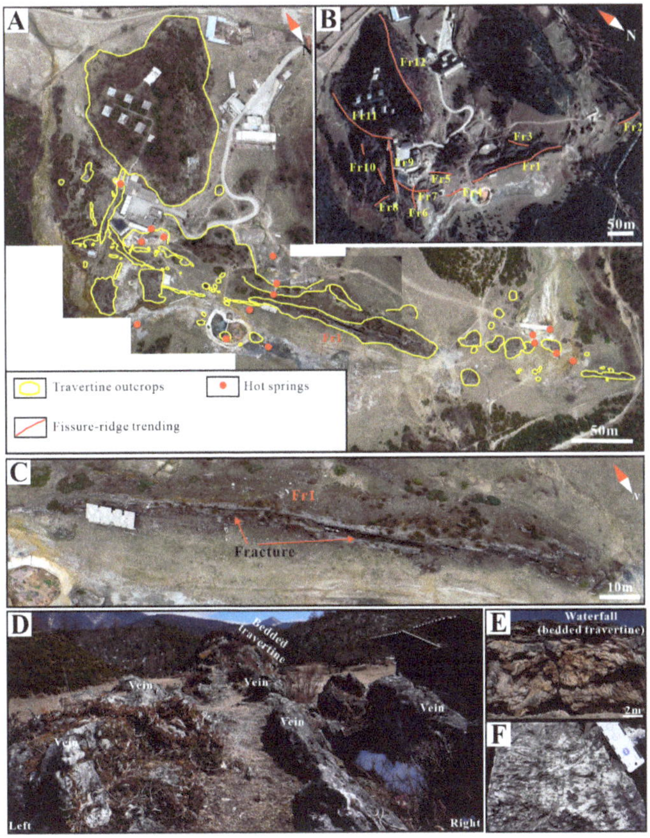

Figure 2. Photographs illustrating the travertine deposits at Xiagei. (**A**) UAV images splicing, indicating the travertine deposit outcrops and the distribution of hot springs. (**B**) An aerial photograph from Google Maps, highlighting the main fissure ridges. (**C**) Images of the studied fissure ridge travertines. (**D**) A general view of the fissure-ridge-type travertines at Xiagei with banded and bedded travertines. (**E**) Examples of travertine waterfalls with thick-bedded travertines. (**F**) Details of thick-laminated palisade crystalline crust.

3. Methods

Banded travertine samples were collected from the top area of Fr1 fissure ridge at Xiagei, where banded travertines were well exposed (Figure 2D). Some fossil bedded travertine samples and modern travertine samples were also collected at Xiagei. To analyze their petrological, mineralogical, and textural characteristics, nineteen thin sections were made from eighteen fossil travertine samples and were examined under an Eclipse LV100POL optical polarizing microscope (Nikon, Tokyo, Japan) at Chengdu University of Technology.

Stable carbon and oxygen isotope analyses were performed on 80 banded travertine samples, 12 bedded travertine samples, and three modern travertine samples at Yangtze University. Samples for $\delta^{13}C$ and $\delta^{18}O$ analyses were first dried and ground into powder. A Delta V Advantage isotope ratio mass spectrometer (Thermo Fisher Scientific, Waltham, MA, USA) was employed to measure the $\delta^{13}C$ and $\delta^{18}O$ values and the final results were calibrated to V-PDB (Vienna Pee Dee Belemnite) scale. Analytical errors of $\delta^{13}C$ and $\delta^{18}O$ are $\leq$0.2‰ and $\leq$0.3‰, respectively.

Water samples of three hot springs were also collected and analyzed to aid in the interpretation of banded travertines. Main physicochemical parameters of the water samples, including water temperature (T_{water}), pH, and total dissolved solids (TDS), were measured in the field using a portable water quality meter (Eutech CyberScan COND 610 meter, Eutech Instruments, San Francisco, CA, USA) and a portable pH meter (Lichen, Shanghai, China). Water samples for lab analyses were filtered using a 0.45 μm filtration membrane and were stored in 50 ml polyethylene plastic bottles. Concentrations of CO_3^{2-} and HCO_3^- of water samples were determined by titration [36] soon after the sampling. Concentrations of other anions (F^-, Br^-, Cl^- and SO_4^{2-}) and common cations (Na^+, K^+, Ca^{2+} and Mg^{2+}), and dissolved SiO_2 were measured with an ion chromatography system (Dionex ICS-1100, Thermo Fisher Scientific, Waltham, MA, USA; 883 Basic IC plus, Metrohm, Herisau, Switzerland) at Analytical Laboratory of Beijing Research Institute of Uranium Geology (China). PHREEQC (ver. 3) with the WATEQ4F database [37] was utilized to calculate the saturation index of calcite ($SI_{calcite}$) and partial pressure of (dissolved) CO_2 (log P_{CO_2}). Stable hydrogen (δD_{water}) and oxygen ($\delta^{18}O_{water}$) isotopes of the three collected water samples were also measured using an isotope ratio mass spectrometer (Delta V Advantage, Thermo Fisher Scientific, Waltham, MA, USA) at Analytical Laboratory of Beijing Research Institute of Uranium Geology (China). $\delta^{18}O_{water}$ and δD_{water} values were calibrated based on the Vienna Standard Mean Ocean Water(V-SMOW) scale.

4. Results

4.1. Fissure Ridge Travertines at Xiagei

The Xiagei hydrothermal system is characterized by the occurrence of multiple active springs and widespread (modern and fossil) travertines (Figure 2). Travertines cropping out in the study area are ca. 0.08 km^2 and contain twelve fissure ridges. The spatial distribution of the Xiagei fissure ridges is mapped in Figure 2A,B. It is striking that the Xiagei fissure ridges developed with two entirely different trends (NE-trending and NW-trending) and displayed varying lengths (25 to 140 m) and widths (1 to 120 m). Interestingly, some NE-trending fissure ridges were found to cut through some NW-trending fissure ridges, implying that there were at least two periods of fault activity at Xiagei.

A common characteristic of the Xiagei fissure ridges is that their central fissures are more or less filled with banded travertines. However, not all the banded travertines were well exposed and easily accessible, and only the longest fissure ridge (i.e., Fr1 in Figure 2A–C) cropped out thick and relatively complete banded travertines. Fr1 fissure ridge is a NW-trending system with a length of around 140 m, a maximum width of near 20 m, and a maximum height of about 20 m (Figure 2A–C). Several small travertine mounds and a 35-meter-long fissure ridge developed to the northwestern of the Fr1 fissure ridge and exhibited the same trend to the Fr1 fissure ridge (Figure 2A). The central fissure of Fr1 fissure ridge was not fully sealed by banded travertines. Soils, vegetation, and open fissures were also observed within the central fissure of Fr1 (Figure 2C).

4.2. Spring Water

Chemical composition and $\delta^{18}O_{water}$ and δD_{water} of hot spring waters at Xiagei were reported in Table 1. The results show that the studied hot springs are characterized by slightly acidic to near neutral pH values and water temperatures between 31.7 °C and 58 °C. HCO_3^- (953 to 998 mg/L), Na^+ (231 to 249 mg/L), and Ca^{2+} (120 to 151 mg/L) are the main components of the spring waters, while some SO_4^{2-} (36.9 to 133 mg/L), Cl^- (26.5 to 31.3 mg/L), K^+ (16.4 to 20.5 mg/L), SiO_2 (20.4 to 83.2 mg/L), and minor F^- (2.95 to 9.84 mg/L) and Mg^{2+} (8.82 to 9.89 mg/L) are also present. $\delta^{18}O_{water}$ and δD_{water} values of the three studied springs are between −15.3‰ and −14.5‰ V-SMOW (average = −14.9‰ V-SMOW), and between −133.9‰ and −128.7‰ (average = −133.3‰ V-SMOW), respectively (Table 1). The here obtained ion concentrations, $\delta^{18}O_{water}$, and δD_{water} are very close to the data reported earlier by Zheng [32]. Calculated $SI_{calcite}$ and log P_{CO_2} in hot spring waters from Xiagei range from 0.14 to 0.69 and from −0.54 to −0.96, respectively (Table 1).

Table 1. Water chemistry, δ^2H_{water} and $\delta^{18}O_{water}$ values of hot springs at Xiagei.

Location	T (°C)	pH_{field}	Alkalinity (mg/L)	HCO_3^- (mg/L)	F^- (mg/L)	Cl^- (mg/L)	SO_4^{2-} (mg/L)	Na^+ (mg/L)	K^+ (mg/L)	Mg^{2+} (mg/L)	Ca^{2+} (mg/L)	SiO_2 (mg/L)	δ^2H_{water} (‰, V−SMOW)	$\delta^{18}O_{water}$ (‰, V−SMOW)	SI_C	log pCO_2
S1	32	6.98	842	998	3.0	27.5	65.1	231	16.4	8.8	148	20.4	−133.9	−14.9	0.64	−0.96
S2	58	6.81	825	993	3.2	31.3	36.9	249	20.5	9.9	120	25.4	−137.3	−15.3	0.69	−0.61
S3	31.7	6.52		953	9.8	26.5	133	234	17.3	9.6	151	83.2	−128.7	−14.5	0.14	−0.54
YX2-1 [a]	63	6.70	762	928.7	5.9	24.5	33	250	14.9	29.2	65.9	85.4	−130	−16.2	0.36	−0.49
YX2-3 [a]	50.5	5.57	649	790.8	5.3	21.3	29.8	219	27.6	25.5	60.1	68.2	−130.2	−15.1	−1	0.48
YX2-5 [a]	41.4	7.66	622	759.1	5.8	26.1	35.1	255	29.9	16.9	35.9	80.8	−128.3	−15	0.72	−1.7
YX2-6 [a]	47.3	7.67	745	908	4.3	26.6	35.7	234	19.7	15.6	89.8	85.4	−131.5	−15.2	1.25	−1.6
YX2-7 [a]	63.1	7.56	781	951.9	5.9	26.1	35.1	251	15	34.0	63.9	83.1	−133.4	−15.6	1.18	−1.35

(blank = no specific data); [a] reference from Zheng [32].

4.3. Lithofacies of Banded Travertines

According to the travertine lithofacies classification of Gandin and Capezzuoli [38], the Xiagei banded travertines belong to abiotic crystalline crust. The Xiagei banded travertines is characteristic by striking lamination, but laminae of the Xiagei banded travertines show varying thickness from tens of micrometers to several centimeters. Additionally, the main components of the Xiagei banded travertines are diverse and include dendritic crystals, fan crystals, platy crystals, banded palisade crystals and granular crystals. Thus, the Xiagei banded travertines were here further divided into thick-laminated palisade crystalline crust and thin-laminated composite crystalline crust.

The thick-laminated palisade crystalline crust is not very common in the Xiagei banded travertines and stands out by the alternation of thick layers over one centimeter in thickness (Figure 3). Each thick layer is composed of abundant large palisade crystals often ranging from 1 to 3 cm long and over 0.4 mm wide and may contain several growth lines (Figure 3A,C). The growth lines were visible in both hand specimens and thin sections, but they were not laterally continuous. Palisade crystal growth was not terminated at most of the growth lines (e.g., Figure 3A).

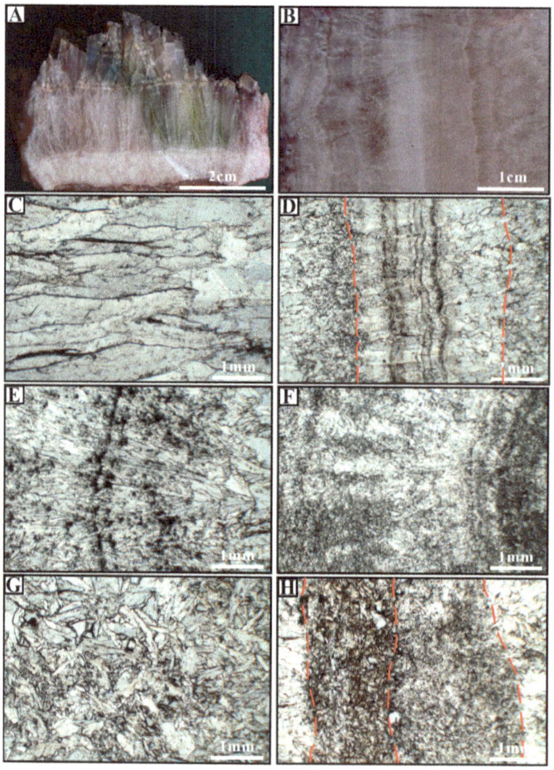

Figure 3. Hand specimens (**A**,**B**) and microphotographs (**C**–**H**) of the main petrographic characteristics of the studied banded travertines. (**A**) Cross-section of thick-laminated palisade crystalline crust. (**B**) Polished side of thin-laminated composite crystalline crust. (**C**) cm-size elongated palisade crystals. (**D**) Short-columinar calcite crystals. (**E**) Feather-like dendritic crystals. (**F**) Dendritic crystals with blurry edges and numerous microcrystals filling intercrystalline pores. (**G**) Willow leaf bladed-like calcite crystals, and intercrystalline microcrystals. (**H**) Bands of dark granular crystals and calcite micrite.

Compared with the thick-laminated crystalline crust, the thin-laminated composite crystalline crust is characterized by the alternation of thin layers (less than 1 cm in thickness). Composition of the thin layers is very complex. Short-columnar crystals, dendritic crystals, platy crystals, fan crystals, granular crystals and calcite micrite were all found as possible main components of the thin layers. In many cases, the thin layers in the thick-laminated crystalline crust only consist of one component, such as the short palisade crystal layer in Figure 3D and the dendritic crystal layer in Figure 3E. However, in some cases, thin layers of the thick-laminated crystalline crust may contain several different components, such as the co-occurrence of dendritic crystals and micrite in Figure 3F.

Short-columnar crystals are the most common components in the thin-laminated composite crystalline crust from Xiagei and is dominantly composed of columnar-like calcite crystal often ranging from 3 to 8 mm in length, significantly shorter than the palisade crystals forming the thick-laminated palisade crystalline crust. Calcite micrite was observed among the short-palisade crystals but was not common in all samples. Dendrite crystals were found in many samples and were characterized by elongated trunks with plentiful branches. In terms of crystal morphologies, the dendrite crystals from Xiagei can be divided into two types: dendritic crystals (3 to 5 mm in length) coexist with micrites (Figure 3F) and typical feather-like calcite crystals (up to 4mm in length; Figure 3E). Platy crystals (Figure 3G) were present in some calcite veins at Xiagei and consisted of willow leaf bladed-like calcite with lengths commonly lower than 1 mm and thicknesses below 0.3 mm. The platy crystals co-appeared with some micrite (Figure 3G) and/or fan crystals. A few samples from the thin-laminated composite crystalline crust contained fan crystals and granular crystals. The fan crystals generally have high length-width ratio ($\approx$1.5) and have varying sizes (up to 3.5 mm long and 2 mm wide). Granular crystals (Figure 3H) here refer to subhedral to anhedral calcite crystals. The granular crystals in the Xiagei thin-laminated composite crystalline crust display varying sizes from 0.15 mm to tens of microns in diameter.

4.4. Stable Isotopes

Stable oxygen and carbon isotope compositions of the Xiagei travertines were listed in Table 2 and illustrated in Figures 4 and 5. The studied banded travertines show narrow variations of δ^{13}C from 2.82‰ to 4.50‰ V-PDB (average = 3.52‰ V-PDB) and δ^{18}O from −25.86‰ to −20.90‰ V-PDB (average = −23.66‰ V-PDB). The thick-laminated palisade banded travertines display different δ^{13}C and δ^{18}O ranges from the thin-laminated composite banded travertines. The former show δ^{13}C between 2.82‰ and 3.74‰ V-PDB and δ^{18}O between −25.86‰ and −22.64‰ V-PDB, while the later display more positive δ^{13}C (3.37‰ and 4.50‰ V-PDB) and δ^{18}O (−24.45‰ and −20.90‰ V-PDB). A good positive correlation (R^2 = 0.8, n= 80) was also found between δ^{13}C and δ^{18}O of the banded travertines.

Table 2. δ^{13}C, δ^{18}O, calculated paleo-temperature (T_{cal}), $\delta^{13}C_{parent\,CO_2}$ of the Xiagei fossil and recently formed travertines. The depositional paleo-temperature (°C) of the travertines was calculated following the equation from Kele et al. [39] and using $\delta^{18}O_{water}$ = −15.2‰; the $\delta^{13}C_{parent\,CO_2}$ was calculated following the equation from Chacko et al. [40] and using the calculated paleo-temperatures.

Sample ID	Travertine Classification	δ^{13}C (‰, V-PDB)	δ^{18}O (‰, V-PDB)	δ^{18}O (‰, V-SMOW)	T_{cal} (°C)	$\delta^{13}C_{parent\text{-}CO_2}$ (‰, V-PDB)
1	TP	3.63	−22.98	7.23	68.6	−2.68
2		3.74	−22.97	7.24	68.6	−2.57
3		3.56	−22.64	7.58	66.6	−2.93
4		3.03	−23.82	6.37	73.7	−2.85
5		3.14	−24.15	6.02	75.8	−2.55
6		3.15	−23.93	6.25	74.4	−2.67

Table 2. *Cont.*

Sample ID	Travertine Classification	$\delta^{13}C$ (‰, V-PDB)	$\delta^{18}O$ (‰, V-PDB)	$\delta^{18}O$ (‰, V-SMOW)	T_{cal} (°C)	$\delta^{13}C_{parent\text{-}CO_2}$ (‰, V-PDB)
7		3.05	−25.52	4.61	84.5	−1.96
8		2.94	−25.68	4.45	85.5	−1.98
9		2.93	−25.72	4.4	85.9	−1.98
10		3.10	−25.21	4.93	82.5	−2.07
11		3.09	−25.68	4.45	85.5	−1.83
12		3.29	−25.41	4.73	83.7	−1.77
13		3.01	−24.98	5.17	81	−2.27
14		2.97	−25.37	4.77	83.5	−2.11
15		3.08	−25.58	4.55	84.9	−1.89
16		2.90	−25.40	4.73	83.7	−2.16
17		3.01	−25.55	4.58	84.7	−1.98
18		3.00	−25.86	4.26	86.8	−1.83
19		3.21	−25.47	4.66	84.2	−1.82
20		3.09	−25.61	4.52	85.1	−1.86
21		3.22	−25.39	4.74	83.7	−1.84
22		2.83	−25.46	4.67	84.1	−2.21
23		2.82	−25.51	4.62	84.4	−2.19
24		3.10	−25.21	4.93	82.5	−2.07
25		2.91	−25.63	4.5	85.2	−2.04
26	TC	3.54	−22.95	7.26	68.4	−2.78
27		3.54	−23.01	7.2	68.8	−2.76
28		3.37	−23.98	6.2	74.7	−2.42
29		3.80	−23.21	6.99	70	−2.38
30		3.63	−23.48	6.71	71.7	−2.42
31		3.62	−23.27	6.93	70.4	−2.53
32		3.60	−23.16	7.04	69.7	−2.62
33		3.88	−22.86	7.35	67.9	−2.50
34		3.54	−23.26	6.94	70.3	−2.63
35		3.66	−23.76	6.43	73.3	−2.25
36		3.74	−23.82	6.37	73.7	−2.13
37		3.49	−23.39	6.81	71.1	−2.61
38		3.58	−23.59	6.6	72.3	−2.42
39		3.81	−23.13	7.08	69.5	−2.43
40		3.93	−22.78	7.44	67.4	−2.48
41		3.43	−24.02	6.16	74.9	−2.34
42		3.48	−23.93	6.25	74.4	−2.34
43		3.70	−23.41	6.78	71.3	−2.38
44		3.70	−23.81	6.37	73.7	−2.17
45		3.57	−23.48	6.71	71.7	−2.48
46		3.40	−23.62	6.57	72.5	−2.57

Table 2. *Cont.*

Sample ID	Travertine Classification	$\delta^{13}C$ (‰, V-PDB)	$\delta^{18}O$ (‰, V-PDB)	$\delta^{18}O$ (‰, V-SMOW)	T_{cal} (°C)	$\delta^{13}C_{parent\text{-}CO_2}$ (‰, V-PDB)
47		4.19	−21.50	8.75	60	−2.90
48		3.73	−22.29	7.94	64.5	−2.95
49		3.49	−22.78	7.44	67.4	−2.92
50		3.60	−23.00	7.21	68.7	−2.71
51		3.54	−23.39	6.81	71.1	−2.56
52		3.85	−22.80	7.41	67.6	−2.55
53		4.11	−22.03	8.21	63	−2.70
54		3.45	−23.27	6.93	70.4	−2.70
55		3.41	−23.48	6.72	71.6	−2.64
56		3.69	−23.65	6.54	72.6	−2.29
57		3.66	−23.45	6.74	71.5	−2.41
58		3.67	−23.34	6.86	70.8	−2.44
59		3.67	−23.53	6.67	71.9	−2.36
60		3.80	−23.13	7.08	69.5	−2.44
61		4.04	−22.74	7.47	67.2	−2.39
62		3.51	−23.19	7.01	69.9	−2.70
63		3.80	−22.98	7.23	68.6	−2.51
64		3.84	−22.98	7.23	68.6	−2.47
65		3.69	−23.21	6.99	70	−2.50
66		3.56	−23.39	6.8	71.1	−2.54
67		3.48	−23.16	7.04	69.7	−2.74
68		3.69	−23.18	7.02	69.8	−2.51
69		3.85	−22.12	8.12	63.6	−2.91
70		3.64	−22.62	7.6	66.5	−2.86
71		3.76	−23.00	7.21	68.7	−2.55
72		3.53	−23.04	7.16	69	−2.74
73		3.75	−22.61	7.62	66.4	−2.75
74		3.88	−21.95	8.29	62.6	−2.97
75		4.24	−21.46	8.8	59.7	−2.88
76		3.80	−22.68	7.54	66.8	−2.67
77		3.73	−22.72	7.5	67	−2.72
78		3.75	−20.90	9.38	56.6	−3.65
79		4.50	−22.39	7.84	65.1	−2.13
80		3.66	−24.45	5.71	77.7	−1.88

Table 2. *Cont.*

Sample ID	Travertine Classification	$\delta^{13}C$ (‰, V-PDB)	$\delta^{18}O$ (‰, V-PDB)	$\delta^{18}O$ (‰, V-SMOW)	T_{cal} (°C)	$\delta^{13}C_{parent\text{-}CO_2}$ (‰, V-PDB)
81	BT	4.49	−21.79	8.45	61.7	−2.45
82		4.49	−21.96	8.29	62.6	−2.36
83		4.71	−20.90	9.37	56.7	−2.69
84		4.84	−20.97	9.3	57	−2.52
85		5.16	−20.23	10.06	53	−2.59
86		4.88	−20.93	9.34	56.8	−2.52
87		4.68	−21.29	8.98	58.7	−2.53
88		4.98	−20.91	9.36	56.7	−2.42
89		5.31	−20.68	9.6	55.4	−2.20
90		4.40	−22.00	8.24	62.9	−2.43
91		4.38	−21.91	8.33	62.4	−2.49
92		5.51	−20.41	9.88	53.9	−2.16
93	RFT	5.98	−21.28	8.98	58.7	−1.24
94		5.66	−21.86	8.39	62	−1.25
95		6.81	−18.58	11.76	44.3	−1.80

TP = thick-laminated crystalline crust; TC = thin-laminated composite crystalline crust; BT = bedded travertines; RFT = recently formed travertines.

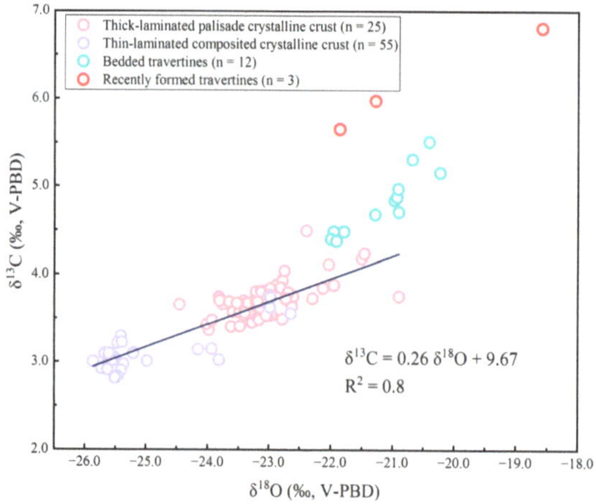

Figure 4. $\delta^{13}C$ and $\delta^{18}O$ of thick-laminated palisade crystalline crust, thin-laminated composite crystalline crust, bedded travertines and recently formed travertines at Xiagei. n = the number of samples.

Bedded travertines from Xiagei show narrower $\delta^{13}C$ and $\delta^{18}O$ ranges compared to the fossil banded travertines (Table 2, Figure 5). They have $\delta^{13}C$ values from 4.40‰ to 5.51‰ V-PDB (average = 4.82‰ V-PDB) and $\delta^{18}O$ values from −22.00‰ to −20.23‰ V-PDB (average = −21.17‰ V-PDB). $\delta^{13}C$ values of recently formed travertines from Xiagei are from 5.66‰ to 6.81‰ V-PDB (average = 6.15‰ V-PDB), slightly higher that $\delta^{13}C$ of the banded travertines. Comparatively, $\delta^{13}O$ of the recently formed travertines from Xiagei is more negative than that of the banded travertines and varies from −21.86‰ to −18.58‰ V-PDB (average = −20.57‰ V-PDB).

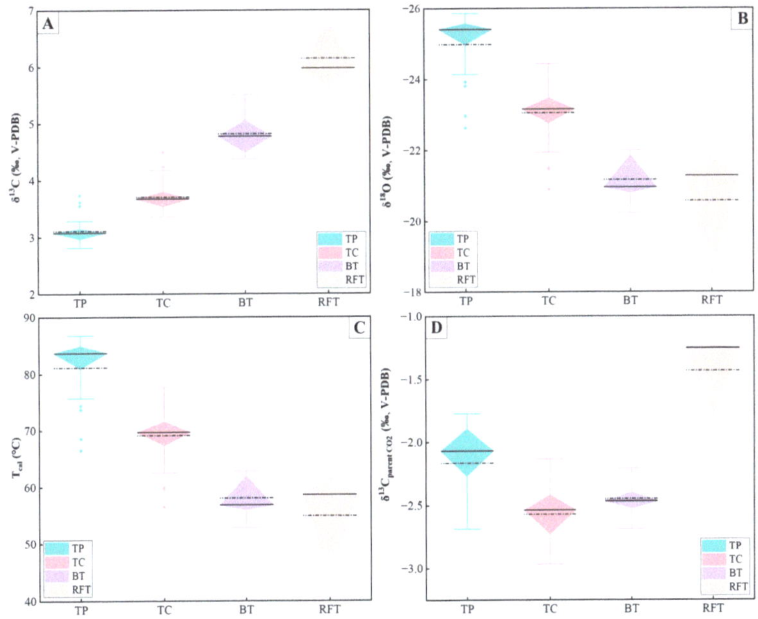

Figure 5. Diamond box plots showing the relationships between lithotypes and δ^{13}C (**A**), between lithotypes and δ^{18}O (**B**), between lithotypes and T_{cal} (**C**), and between lithotypes and $\delta^{13}C_{parent\ CO_2}$ (**D**). thick-laminated crystalline crust (TP), thin-laminated composite crystalline crust (TC), bedded travertines (BT) and recently formed travertines (RFT). Horizontal dark lines referred to the median values, while dot-dash lines referred to the mean values.

5. Interpretation and Discussion

5.1. Paleo-Temperatures for Banded Calcite Precipitation

Paleotemperature evaluation is of great importance in studies on fossil travertine systems because it can disclose the depositional conditions of travertines. The $CaCO_3$-H_2O oxygen isotopic geothermometer is considered to be a powerful tool for assessing the temperature of carbonate precipitation since the oxygen isotope fractionation will take place between the calcite/aragonite and water during naturally precipitated carbonates [5,6,39,41–43]. The calcite–water oxygen isotope fractionation equation developed by Kele et al. [39] (i.e., Equation (1)) is commonly utilized to predict the water temperature of fossil travertines because this equation decreases the influences of degassing and evaporation on the calculated results:

$$1000 \ln\alpha_{(travertine\text{-}water)} = 20{,}000/T_{cal} - 36 \quad (1)$$

where T_{cal} = calculated paleotemperature in Kelvin and $\alpha_{(travertine\text{-}water)}$ is expressed as:

$$\alpha_{(travertine\text{-}water)} = (\delta^{18}O_{travertine} + 1)/(\delta^{18}O_{water} + 1) \quad (2)$$

where $\delta^{18}O_{travertine}$ and $\delta^{18}O_{water}$ refer to the δ^{18}O values of travertines and spring waters, respectively.

In this study, the average $\delta^{18}O_{water}$ value of active hot springs at Xiagei ($\delta^{18}O_{water}$ = −15.2‰ V-SMOW, Table 1) was used in the paleotemperature calculation.

The calculated results (Table 2, Figure 5) show that banded travertines, bedded travertines, recently formed travertines from Xiagei exhibit T_{cal} from 56.6 to 86.8 °C, from 53.0 to 62.9 °C, and from 44.3 to 62.0 °C, respectively. This indicates that the studied travertines were formed in moderate- to high-temperature hot spring environments. Water temperatures of active hot springs at Xiagei vary from 32.0 to 63.1 °C (Table 1). However, according to our field observations, hot springs with high volumetric flow rates at Xiagei (e.g., S2; Table 1) often have vent temperatures of near 60.0 °C, slightly higher than T_{cal} of the recently formed travertines and close to T_{cal} of the bedded travertines (Table 2, Figure 5). This reflects that there were no significant vent temperature changes after the deposition of the studied bedded travertines.

The calculated paleotemperatures of the studied banded travertines generally range from 68 °C to 85 °C, higher than the vent temperatures of main active hot springs at Xiagei (ca. 60 °C; Table 2, Figure 5). This may indicate an unneglectable water temperature decrease between the places forming banded travertines and spring vents (i.e., vent temperature was lower than water temperature in the conduit). However, it must be noted that the vent temperatures in this study were measured in the margin areas of meter-scale vent pools. Thus, our measured temperatures might be slightly lower than the temperatures of spring waters outflowing from orifices just now, because the pool waters may have experienced important heat exchange with cool air (i.e., cooling). Such sample location-related temperature difference is common and unavoidable, especially for the large vent pools. Peng and Jones [44], for example, found that the spring water in the marginal areas of Eryuan pool (China) is 20 °C lower than the spring water near the Eryuan pool central vent. Therefore, simply extrapolating that there was a significant temperature decline between the places forming banded travertines and spring vents at Xiagei is still questionable. However, it is clear that fluid temperature changed a lot during the formation of the studied banded travertines, as shown by their great T_{cal} variation (56.6 °C to 86.8 °C).

5.2. Parent CO_2 Origin of Travertines

Parent CO_2 origins of spring-related carbonate deposits decide whether the deposits are tufas (also named meteogene travertines) or travertines (also called thermogene travertines) [45–47]. In this study, the equation (Equation (3)) from Chacko et al. [40] and the evaluated paleotemperatures (T_{cal}) were utilized to calculate the stable carbon isotope composition of parent CO_2 of the Xiagei Travertines ($\delta^{13}C_{parent\ CO_2}$).

$$1000 \ln\alpha_{(parent\ CO_2\text{-}travertine)} = -0.10028 + 5.4173x - 2.5076x^2 + 0.47193x^3 - 0.0027046x^5 - 0.000059409x^6 \quad (3)$$

where $x = 10^6/T_{cal}^2$ and $\alpha_{(parent\ CO_2\text{-}travertine)}$ is:

$$\alpha_{(parent\ CO_2\text{-}CaCO_3)} = (\delta^{13}C_{parent\ CO_2} + 1)/(\delta^{13}C_{travertine} + 1) \quad (4)$$

The calculated results were listed in Table 2 and illustrated in Figures 5D and 6. Banded travertines at Xiagei have $\delta^{13}C_{parent\ CO_2}$ from −3.65‰ to −1.77‰ V-PDB (average = −2.44‰ V-PDB). In more detail, thick-laminated palisade banded travertines from Xiagei show $\delta^{13}C_{parent\ CO_2}$ between −2.93‰ to −1.77‰ V-PDB (average = −2.16‰ V-PDB), whereas thin-laminated composite banded travertines from Xiagei show a wider $\delta^{13}C_{parent\ CO_2}$ range (−3.65‰ to −1.88‰ V-PDB, average = −2.57‰ V-PDB). $\delta^{13}C_{parent\ CO_2}$ of bedded travertines from Xiagei falls into the banded travertine range and varies from −2.69‰ to −2.16‰ V-PDB (average = −2.45‰ V-PDB). In contrast, $\delta^{13}C_{parent\ CO_2}$ of recently formed travertines at Xiagei is very different from $\delta^{13}C_{parent\ CO_2}$ of the Xiagei banded travertines and bedded travertines and is between −1.80‰ and −1.24‰ V-PDB (average = −1.43‰ V-PDB).

Parent CO_2 sources of travertines are complex, but they mainly include magmatic CO_2, soil-derived CO_2, and CO_2 related to (marine) carbonate decarbonation [45]. It is reported that soil-derived CO_2 has $\delta^{13}C_{CO_2}$ values lower than −10‰ V-PDB (C3 plant soils = −30‰ to −16‰ V-PDB, C4 plant soils = −18‰ to −10‰ V-PDB; [48,49]. Magmatic CO_2 has $\delta^{13}C$ values from −7‰ to −5‰ V-PDB [50], while CO_2 related to (marine) carbonates decarbonation often has more positive $\delta^{13}C$ values (>1‰ V-PDB) [43,51]. The calculated $\delta^{13}C_{parent\ CO_2}$ ranges of the Xiagei fossil banded travertines, bedded travertines, and recently formed travertines are all between the (marine) carbonate-related CO_2 zone and the magmatic CO_2 zone (Figure 6), instead of falling into any of the three typical CO_2 source areas. This largely suggests a possible mixing from (marine) carbonate decarbonation-related CO_2 and magmatic CO_2 and confirms that the Xiagei travertines are 'real' (thermogene) travertines, instead of tufas. Such inference (i.e., combined contribution of (marine) carbonate decarbonation-related CO_2 and magmatic CO_2) is consistent with early studies on hot spring gases at Xiagei and surrounding areas [52,53].

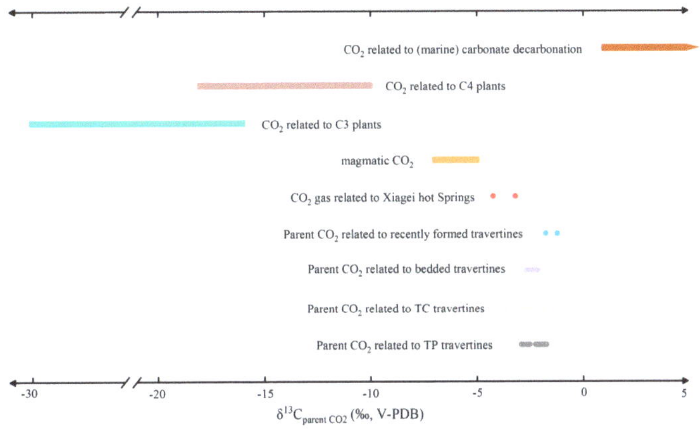

Figure 6. $\delta^{13}C_{parent\ CO_2}$ of the travertines from Xiagei, Shangri-La, China, according to the equation established by Chacko et al. [40]. In comparison with the $\delta^{13}C$ values from CO_2 gas related to Xiagei hot springs [53], magmatic CO_2 [50], CO_2 related to C3 and C4 plants [48,49], and marine carbonate decarbonation [43,51]. TP: thick-laminated crystalline crust; TC: thin-laminated composite crystalline crust.

Notably, the magmatism is lacking now in the Xiagei hydrothermal system (near the Jinshajiang fault zone) in the eastern Tibetan plateau. Thus, the detected magmatic source CO_2 for travertine samples might indicate a hypogene origin. As shown in Figure 1A, the Jinshajiang fault zone is important for the formation of Xiagei travertines as active large-scale strike-slipping, apart-pull basins and metamorphic core complexes are present [54,55], which provides ideal tectonic-environments for Paleogene mantle-derived high-K magmatism [54]. In addition, the Tibetan plateau has been a site for lithospheric delamination since about 40 Ma and is marked by intense and intensive intraplate, mantle-derived high-K magmatism [33,54]. Therefore, the magmatic-derived CO_2 for the Xiagei travertines probably indicates that the lithospheric mantle beneath the Jinshajiang fault zone is/was sinking, which led to the asthenosphere upwelling and subsurface accumulation of magmatic-derived CO_2 that was transported through the lithosphere via these large-scale faults.

Both (marine) carbonate decarbonation-related CO_2 and magmatic CO_2 are 'external' carbon sources. The Xiagei travertines were here also considered to receive some 'internal' carbon from marine carbonate rocks. As shown in Figure 1 and Table 1, marine carbonate rocks (mainly Late Paleozoic) are distributed in the study area and active hot springs have high Ca^{2+} concentrations (e.g., Springs S1 to S3 show Ca^{2+} concentrations from 120 to 151 mg/L) and low SO_4^{2-} concentrations (average = 50.5 mg/L). This largely reflects that marine carbonate dissolution also play a role in supplying carbon to the Xiagei travertines. Assuming that (1) HCO_3^- in spring waters at Xiagei was the mixing product of external carbon (i.e., magmatic CO_2 and carbonate decarbonation-related CO_2: $CO_2 + 2H_2O \rightarrow HCO_3^- + H^+$) and internal carbon (i.e., dissolution of Ca-carbonate rocks: $CaCO_3 + CO_2$

$\rightarrow 2HCO_3^- + Ca^{2+}$) and (2) Ca^{2+} in spring waters from Xiagei originated mainly from Ca-carbonate rock dissolution, the following equation can be obtained:

$$C_{total} = C_{external} + C_{internal} = C_{external} + C_{Ca} \quad (5)$$

where C_{total}, $C_{external}$, $C_{internal}$, and C_{Ca} represent the molar concentrations of HCO_3^-, external carbon, internal carbon, and Ca^{2+}, respectively.

Using the water data in Table 1, C_{total} and C_{Ca} in hot spring waters from Xiagei can be calculated. The calculated C_{total} and C_{Ca} are between 12.4 and 16.4 mM and between 0.90 and 3.78 mM, respectively, and the calculated C_{Ca}/C_{total} ratios vary from 10.2% to 24.2%, indicating the unneglectable role of Ca-carbonate dissolution in supplying carbon to the Xiagei travertines. However, it must be emphasized that the above calculations are not exact, because other processes (such as passive CO_2 degassing and carbonate precipitation) which are common in travertine systems [56–59] and can change HCO_3^- and Ca^{2+} concentrations in spring waters were not taken into consideration in our calculations.

5.3. Lithofacies of Banded Travertines Controlled by Water Temperature

There were two lithofacies of banded travertines at Xiagei: thick-laminated crystalline crust and thin-laminated composite crystalline crust. These two lithofacies display different features, especially in crystal sizes and morphologies. Calcite crystals in the thick-laminated crystalline crust are large and palisade-like, whereas calcite crystals in the thin-laminated composite crystalline crust are diverse but small in size. Recent studies have shown that the fluctuations of hydrodynamic conditions and hydrochemical composition of thermal fluids and microbial mediation can lead to the formation of calcite crystals of various sizes at thermal springs [28,60,61]. Jones and Renaut [62] and Jones [60] adopted a general term, 'driving force', to describe the growth condition of $CaCO_3$ precipitation in hot springs: the larger driving force, the more irregular $CaCO_3$ crystals. However, directly measuring 'driving force' is impossible for researchers, because it may be affected by many parameters, such as CO_2 degassing, water temperature, saturation index of $CaCO_3$, flow velocity, pH, and microbial effect [10,63–65]. These indicate that banded travertine precipitation is a complex process controlled by multiple factors/processes.

With respect to the two lithofacies of banded travertines at Xiagei, one possible factor controlling their crystal shapes and sizes might be water temperature. As shown in Table 2 and Figure 5D, the thick-laminated crystalline crust at Xiagei mainly displays T_{cal} over 80 °C, which is significantly higher than most of the T_{cal} values of the thin-laminated composite crust (generally below 72 °C). Theoretically, temperature increase would lower $CaCO_3$ and CO_2 solubility (when water temperature is below 155 °C) [66,67], promoting $CaCO_3$ precipitation. As a result, the thick-laminated crystalline crust was interpreted to be formed with a precipitation rate faster than the thin-laminated composite crystalline crust. Such inference agrees with the crystal size and morphology of these two lithofacies: rapid $CaCO_3$ precipitation formed large palisade crystals of the thick-laminated palisade crystalline crust, while relatively slow $CaCO_3$ precipitation generated small crystals of the thin-laminated composite crystalline crust.

However, some of the thick-laminated palisade crystalline crust samples (e.g., Samples 1, 2 and 3; Figures 4 and 5B,C; Table 2) show similar $\delta^{18}O$ and T_{cal} to the thin-laminated composite crystalline crust. Therefore, water temperature is an important but not decisive factor controlling the calcite crystal size and shape in the Xiagei banded travertines. Indeed, recently, Luo et al. [68] argued that $CaCO_3$ precipitation at the hypogean solid-water interface (i.e., banded travertine precipitation environments) was (at least) controlled by original fluid chemical composition, temperature, pressure, and fluid pathways. Thus, any change in fluid chemical composition (e.g., Ca^{2+} and HCO_3^- concentrations), CO_2 partial pressure in surrounding environments, and conduit morphology may significantly alter the precipitation rate of $CaCO_3$ at the hypogean solid–water interface. Unfortunately, these changes were difficult to evaluate only based on carbon and oxygen isotope compositions of banded travertines.

The thin-laminated composite crystalline crust is characterized by the appearance of diverse calcite crystals with different shapes and sizes (dendritic crystals, fan crystals, platy crystals, granular crystals, and micrite), whereas the thick-laminated palisade crystalline crust is compositionally simple. This may indicate a steady formation environment of the thick-laminated palisade crystalline crust and reflect that the formation condition of the thin-laminated composite crystalline crust underwent striking and rapid changes. Indeed, dendritic crystals, fan crystals, platy crystals, granular crystals, and micrite are often interpreted to be formed in different environments [38,60,69]. For example, fan crystals tend to be generated in turbulent water conditions [38,70–72], while platy crystals are often produced at near boiling conditions (>80 °C) [62,73,74].

5.4. Comparison with Banded Travertines from Other Areas

In Table 3 and Figure 7, the Xiagei banded travertines were compare with banded travertines (also sometimes named travertine veins) from other areas around the word [13,18,24,75,76] to examine their microscopic and stable carbon and oxygen isotopic differences. All banded travertines are found to be mainly composed of calcite and /or aragonite crystalline crusts (Table 3), but their microscopic features, such as crystal sizes and types are diverse, indicating the complex precipitation conditions/processes of banded travertines. For example, diverse calcite crystals have been described in the thin-laminated crystalline crust at Xiagei, probably representing changeable travertine formation environments. Note that elongated palisade calcite and/or aragonite crystals are commonly described in banded veins and are hardly seen in non-vein sediments of travertines. It may indicate that this texture can represent a typical lithofacies related to banded travertines. In addition, the banded travertines with diverse crystal morphologies, such as those from Semproniano (Italy), Kamara (Turkey), Gölemezli (Turkey), and Xiagei (China), are usually characterized by wide paleo-temperature ranges (Table 3). This might indicate that temperature plays an important role in controlling the microscopic features of banded travertines.

All the summarized banded travertines, excluding the Semproniano village giant veins [75], are characterized by good positive correlations between $\delta^{13}C$ and $\delta^{18}O$ values (Figure 7), probably indicating these banded travertines precipitated under non-equilibrium conditions (in more detail, fast CO_2 degassing). Using $\delta^{13}C$ values to gain the fluid CO_2 origins of spring related carbonate deposits is commonly used in earlier studies [13,18,39,75]. The parent CO_2 from all banded travertine was found to be largely derived from mixing sources (Table 3). However, the result shows that the $\delta^{13}C$ ranges of the banded travertine in different places are diverse. Especially, $\delta^{13}C$ of the Semproniano giant vein falls into a wider range, whereas $\delta^{13}C$ of travertines from Xiagei and other locations (Figure 7) exhibits relatively narrow isotopic variation. Berardi et al. [75] proposed that the long-lived giant vein underwent multiple tectonic events, spanning at least 650 and 85 ka. Thus, tectonic activities may play a non-negligible role in changing the $\delta^{13}C$ composition of the travertine vein. For instance, frequent tectonic activities may cause rapid $\delta^{13}C$ variations, while strong tectonic activities, like big earthquakes, may even modify thermal fluid circulation pathways and lead to significant $\delta^{13}C$ changes of banded travertines. According to Kele et al. [43], $\delta^{18}O$ values of travertines can be easily affected by fluid origins, evaporation rates and fluid temperature changes. In most travertine systems, the $\delta^{18}O$ values of travertines can be used to reflect the water temperature and $\delta^{18}O$ of the parent water. Taking into account the calculated paleo-temperatures, all the banded travertines in Table 3 are precipitated from moderate- to high-temperature hot springs, which further indicates that the fissure ridge is related to hot springs rather than cold springs. Overall, the comparison shows that $\delta^{13}C$ and $\delta^{18}O$ records in banded travertines might be useful for the assessment of tectonic activities and paleo-fluids.

Table 3. Comparison of banded veins from Xiagei and those from other places. Except Pamukkale and Reşadiye, the calculated temperature of other locations and Xaigei using the equation from Kele et al. [39]. Blank = no data.

Location	Classification	Characteristics	Lithofacies	Components	T (°C)	Fluid Origins
Semproniano Village, Italy [75]	Banded travertine	A giant structure characterized by a minimum thickness of 50 m	Crystalline crust	Complex, such as elongate V-like-shapes crystals, can reach to 2 cm.	34 ± 2 to 71 ± 7	Water is meteoric origin. CO_2 originated from limestone decarbonation with CO_2 of igneous origin
Semproniano Village, Italy [75]	Calcite vein	Veins cutting through carbonate rocks, reach to 5 cm thick			49 ± 4 to 56 ± 5	
I Vignacci, Italy [75]	Banded travertine	Composed of subvertical NW–SE-striking bands, located at an altitude of about 430 m a.s.l.			43 ± 3	Water is meteoric origin. CO_2 originated from limestone decarbonation with CO_2 of igneous origin
Kamara, Turkey [13]	Banded travertine	Bands up to 1 m thick		(a) The palisade/columnar structure made of bladed, rhombohedral/prismatic elongated crystals; (b) The botryoidal structure composed of fan-like splays of acicular/needle-like crystals, mm-to cm thick.	55 to 80	Water is meteoric origin. Parent CO_2 from magmatic, decarbonation of marine carbonate rocks and organic-sedimentary
Kamara, Turkey [13]	Calcite vein	These calcite veins isolate cm- to dm-thick volumes of hosting sediments.		Mm-thick, fibrous onyx-like calcite/aragonite crystals.	44 to 57	
Gölemezli, Turkey [18]	Banded travertine	Onyx-like, up to 12 m in width.		(a)Type A band, palisade/columnar, acicular/needle-like crystals, up to cm-long; (b) Type B band, micro-crystalline calcite, up to mm-thick.	53 to 76	Water is meteoric origin; Parent CO_2 from crustal- and mantle-derived CO_2.
Gölemezli, Turkey [18]	Calcite vein	Localized brecciation, formed in late deformational process, thickness ranges from 15 to 50 cm.		Few millimeter calcite crystals	60 to 65	
Pamukkale, Turkey [24,43]	Banded travertine	The band thickness varies from a few centimeters to 10 cm, mineral type: calcite and aragonite.		Radial, needle-shaped elongated crystals with length of about 0.5 to 1 mm	<60	Water is meteoric origin; Parent CO_2 from crustal- and mantle-derived CO_2.
Reşadiye, Turkey [24,77]	Banded travertine	Except typical bands, the travertines are detritus-free, porous, moderately crystalline, mineral type: calcite		Radial, needle-shaped elongated crystals with length of about 1 to 2 mm	<60	Water is meteoric origin; Parent CO_2 from crustal- and mantle-derived CO_2.
Ballık, Turkey [78]	Calcite vein	Cross-cutting the micritic travertine host rock, millimeter-thick to centimeter-thick		Micrite and fine-crystalline (<50 μm) calcite crystals, elongated coarse-crystalline sparite		
Utah, USA [76]	Carbonate vein	(a)Thin veins with characterized by isolated millimeter-thick calcium carbonate veins and 3D dense network veins; (b) Thick carbonate veins, up to 1 m thick; mineral type: calcite and aragonite.		Elongate V-like-shapes crystals, can reach to 1 mm.		
Xiagei, China this study	Banded travertine	Exposed in the center of fissure ridge, approximately 1.5 m.		Thick-laminated palisade crystalline crust: palisade crystals (from 1 to 3 cm long), thin-laminated composite crystalline crust: short-columnar crystals, dendritic crystals, platy crystals, fan crystals, granular crystals, and calcite micrite (less than 1 cm).	68 to 85	Water is meteoric origin; Parent CO_2 from (marine) carbonate decarbonation-related CO_2 and magmatic CO_2.

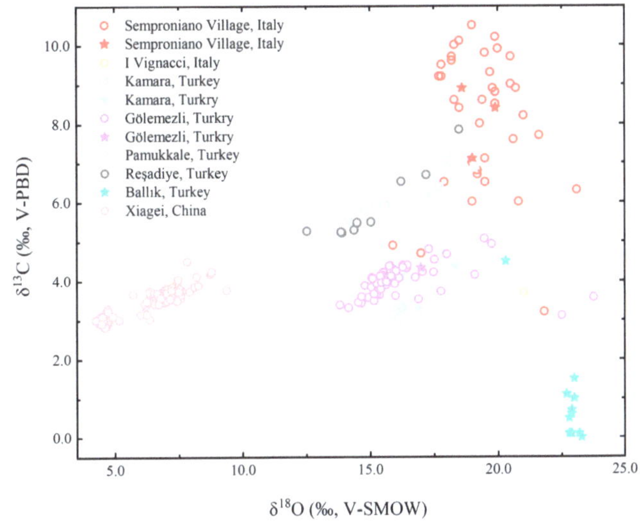

Figure 7. $\delta^{13}C$ and $\delta^{18}O$ values of banded calcite veins from Xiagei and those from other examples.

6. Conclusions

This study investigated the lithofacies and stable carbon and oxygen isotope geochemistry of banded travertines from fissure ridge-type travertine (Xiagei, China). Two lithofacies were recognized in the Xiagei banded travertines: thick-laminated palisade crystalline crust and thin-laminated composite crystalline crust. Parent CO_2 of the Xiagei banded travertines mainly originated from limestone decarbonation of marine carbonates, but they also received the contribution from magmatic CO_2. The magmatic-derived CO_2 may also demonstrate the sinking of the lithospheric mantle beneath the Jinshajiang fault zone. Additionally, dissolution of marine carbonate rock is considered to be important in providing carbon and calcium to the paleo-spring waters. Paleotemperatures precipitating the Xiagei banded travertines are from 56.6 °C to 86.8 °C, indicating moderate- to high-temperature spring environments. However, different lithofacies of banded travertines from Xiagei show distinct calculated paleotemperatures, suggesting temperature is a non-negligible control factor on the lithofacies of banded travertines. This study highlights the importance of depositional conditions (especially water temperature) on banded travertine lithofacies and reflects the significance of careful stable carbon and oxygen isotope analyses in the determination of paleo-fluid evaluation. The comparison between the banded travertines at Xiagei and other areas shows that microscopic features of banded travertines in different places is diverse, and banded travertines formed by calcite/aragonite with diverse morphologies are usually characterized by large paleo-temperature variations, further indicating that temperature plays a crucial role in controlling the microscopic characteristic of banded travertines. The comparison also implies that the stable carbon and oxygen isotopic characterization of banded travertines is likely an effective tool for assessing tectonic activities and paleo-fluids.

Author Contributions: Y.Y., L.L. (Lianchao Luo) and H.W. conceived this contribution. Y.Y., Z.L. and L.L. (Liang Li) conducted the field investigation. Y.Y., Z.L. and L.L. (Liang Li) performed the sample handling and data analysis. Y.Y. wrote the original draft of the paper. L.L. (Lianchao Luo) reviewed the original draft of the paper. All authors have read and agreed to the published version of the manuscript.

Funding: This research was financially supported by the National Natural Science Foundation of China (grant Nos. 41972116 and 41572097).

Data Availability Statement: The data that support the findings of this study are available in the text, figures and tables of this manuscript.

Acknowledgments: We are grateful to Lei Du for his help in field works. We also thank four anonymous reviewers for their detailed and constructive comments.

Conflicts of Interest: The authors declare no conflict of interest from their affiliation or funding.

References

1. Altunel, E.; Karabacak, V. Determination of horizontal extension from fissure-ridge travertines: A case study from the Denizli Basin, southwestern Turkey. *Geodin. Acta* **2005**, *18*, 333–342. [CrossRef]
2. Brogi, A.; Capezzuoli, E. Travertine deposition and faulting: The fault-related travertine fissure-ridge at Terme S. Giovanni, Rapolano Terme (Italy). *Int. J. Earth Sci.* **2008**, *98*, 931–947. [CrossRef]
3. Brogi, A.; Capezzuoli, E.; Alçiçek, M.C.; Gandin, A. Evolution of a fault-controlled fissure-ridge type travertine deposit in the western Anatolia extensional province: The Çukurbağ fissure-ridge (Pamukkale, Turkey). *J. Geol. Soc.* **2014**, *171*, 425–441. [CrossRef]
4. Brogi, A.; Capezzuoli, E.; Karabacak, V.; Alcicek, M.C.; Luo, L. Fissure ridges: A reappraisal of faulting and travertine deposition (travitonics). *Geosciences* **2021**, *11*, 278. [CrossRef]
5. De Filippis, L.; Anzalone, E.; Billi, A.; Faccenna, C.; Poncia, P.P.; Sella, P. The origin and growth of a recently-active fissure ridge travertine over a seismic fault, Tivoli, Italy. *Geomorphology* **2013**, *195*, 13–26. [CrossRef]
6. De Filippis, L.; Faccenna, C.; Billi, A.; Anzalone, E.; Brilli, M.; Soligo, M.; Tuccimei, P. Plateau versus fissure ridge travertines from Quaternary geothermal springs of Italy and Turkey: Interactions and feedbacks between fluid discharge, paleoclimate, and tectonics. *Earth-Sci. Rev.* **2013**, *123*, 35–52. [CrossRef]
7. Guo, L.; Riding, R. Rapid facies changes in Holocene fissure ridge hot spring travertines, Rapolano Terme, Italy. *Sedimentology* **1999**, *46*, 1145–1158. [CrossRef]
8. Karabacak, V.; Uysal, I.T.; Ünal-İmer, E.; Mutlu, H.; Zhao, J. U-Th age evidence from carbonate veins for episodic crustal deformation of Central Anatolian Volcanic Province. *Quat. Sci. Rev.* **2017**, *177*, 158–172. [CrossRef]
9. Mancini, A.; Capezzuoli, E.; Erthal, M.; Swennen, R. Hierarchical approach to define travertine depositional systems: 3D conceptual morphological model and possible applications. *Mar. Pet. Geol.* **2019**, *103*, 549–563. [CrossRef]
10. Della Porta, G. Carbonate build-ups in lacustrine, hydrothermal and fluvial settings: Comparing depositional geometry, fabric types and geochemical signature. In: Bosence, D.W.J., Gibbons, K.A., Le Heron, D.P., Morgan, W.A., Pritchard, T., Vining, B.A. Microbial Carbonates in Space and Time: Implications for Global Exploration and Production. *Geol. Soc. Lond. Spec. Publ.* **2015**, *418*, 17–68.
11. Altunel, E.; Hancock, P.L. Structural attributes of travertine-filled extensional fissures in the Pamukkale Plateau, Western Turkey. *Int. Geol. Rev.* **1996**, *38*, 768–777. [CrossRef]
12. Brogi, A.; Alçiçek, M.C.; Liotta, D.; Capezzuoli, E.; Zucchi, M.; Matera, P.F. Step-over fault zones controlling geothermal fluid-flow and travertine formation (Denizli Basin, Turkey). *Geothermics* **2021**, *89*, 101941. [CrossRef]
13. Brogi, A.; Alçiçek, M.C.; Yalçıner, C.Ç.; Capezzuoli, E.; Liotta, D.; Meccheri, M.; Rimondi, V.; Ruggieri, G.; Gandin, A.; Boschi, C.; et al. Hydrothermal fluids circulation and travertine deposition in an active tectonic setting: Insights from the Kamara geothermal area (western Anatolia, Turkey). *Tectonophysics* **2016**, *680*, 211–232. [CrossRef]
14. Brogi, A.; Capezzuoli, E. Earthquake impact on fissure-ridge type travertine deposition. *Geol. Mag.* **2014**, *151*, 1135–1143. [CrossRef]
15. De Filippis, L.; Billi, A. Morphotectonics of fissure ridge travertines from geothermal areas of Mammoth Hot Springs (Wyoming) and Bridgeport (California). *Tectonophysics* **2012**, *548–549*, 34–48. [CrossRef]
16. De Filippis, L.; Faccenna, C.; Billi, A.; Anzalone, E.; Brilli, M.; Ozkul, M.; Soligo, M.; Tuccimei, P.; Villa, I.M. Growth of fissure ridge travertines from geothermal springs of Denizli Basin, western Turkey. *Geol. Soc. Am. Bull.* **2012**, *124*, 1629–1645. [CrossRef]
17. Zhang, K.-J.; Li, Q.-H.; Yan, L.-L.; Zeng, L.; Lu, L.; Zhang, Y.-X.; Hui, J.; Jin, X.; Tang, X.-C. Geochemistry of limestones deposited in various plate tectonic settings. *Earth-Sci. Rev.* **2017**, *167*, 27–46. [CrossRef]
18. Capezzuoli, E.; Ruggieri, G.; Rimondi, V.; Brogi, A.; Liotta, D.; Alçiçek, M.C.; Alçiçek, H.; Bülbül, A.; Gandin, A.; Meccheri, M.; et al. Calcite veining and feeding conduits in a hydrothermal system: Insights from a natural section across the Pleistocene Gölemezli travertine depositional system (western Anatolia, Turkey). *Sediment. Geol.* **2018**, *364*, 180–203. [CrossRef]
19. Hancock, P.L.; Chalmers, R.M.L.; Altunel, E.; Çakir, Z. Travitonics: Using travertines in active fault studies. *J. Struct. Geol.* **1999**, *21*, 903–916. [CrossRef]
20. Sağlam Selçuk, A.; Erturaç, M.K.; Üner, S.; Özsayın, E.; Pons-Branchu, E. Evolution of Çamlık fissure-ridge travertines in the Başkale basin (Van, Eastern Anatolia). *Geodin. Acta* **2016**, *29*, 1–19. [CrossRef]
21. Haluk Selim, H.; Yanik, G. Development of the Cambazlı (Turgutlu/MANISA) fissure-ridge-type travertine and relationship with active tectonics, Gediz Graben, Turkey. *Quat. Int.* **2009**, *199*, 157–163. [CrossRef]
22. Curewitz, D.; Karson, J.A. Structural settings of hydrothermal outflow: Fracture permeability maintained by fault propagation and interaction. *J. Volcanol. Geotherm. Res.* **1997**, *79*, 149–168. [CrossRef]

23. Matera, P.F.; Ventruti, G.; Zucchi, M.; Brogi, A.; Capezzuoli, E.; Liotta, D.; Yu, T.-L.; Shen, C.-C.; Huntington, K.W.; Rinyu, L.; et al. Geothermal fluid variation recorded by banded Ca-carbonate veins in a fault-related, fissure ridge-type travertine depositional system (Iano, southern Tuscany, Italy). *Geofluids* **2021**, *2021*, 1–28. [CrossRef]
24. Rizzo, A.L.; Uysal, I.T.; Mutlu, H.; Ünal-İmer, E.; Dirik, K.; Yüce, G.; Caracausi, A.; Italiano, F.; Misseri, M.; Temel, A.; et al. Geochemistry of Fluid Inclusions in Travertines From Western and Northern Turkey: Inferences on the Role of Active Faults in Fluids Circulation. *Geochem. Geophys. Geosyst.* **2019**, *20*, 5473–5498. [CrossRef]
25. Uysal, I.; Feng, Y.; Zhao, J.; Altunel, E.; Weatherley, D.; Karabacak, V.; Cengiz, O.; Golding, S.; Lawrence, M.; Collerson, K. U-series dating and geochemical tracing of late Quaternary travertine in co-seismic fissures. *Earth Planet. Sci. Lett.* **2007**, *257*, 450–462. [CrossRef]
26. Uysal, I.T.; Feng, Y.-x.; Zhao, J.-x.; Isik, V.; Nuriel, P.; Golding, S.D. Hydrothermal CO_2 degassing in seismically active zones during the late Quaternary. *Chem. Geol.* **2009**, *265*, 442–454. [CrossRef]
27. Chen, Q. Studies on Geochemistry of the Permian Basalts in Shangri-La County YunNan Province China. Master's Thesis, University of Geosciences, Beijing, China, 2017.
28. Jones, B.; Peng, X. Intrinsic versus extrinsic controls on the development of calcite dendrite bushes, Shuzhishi Spring, Rehai geothermal area, Tengchong, Yunnan Province, China. *Sediment. Geol.* **2012**, *249–250*, 45–62. [CrossRef]
29. Li, L.; Wang, Y. Application of geophysical method to geothermal resource prospecting in Xianggelila Basin, Yunnan. *Carsologica Sin.* **2004**, *23*, 187–195.
30. Wang, Y. *Research Report on Geothermal Resources Exploration in Shangri La Basin, Diqing Tibetan Autonomous Prefecture, Yunnan Province*; Yunnan Exploration Development Bureauof Geology Mineral Resources: Yunnan, China, 2003.
31. Wang, Y.; Yang, S.Y. The geothermal feature and exploring prospect of the Xianggelila Basin in Narthwestern Yunnnan, China. *Bull. Mineral. Petrol. Geochem.* **2003**, *33*, 265–269, 272.
32. Zheng, Y.H. *Characteristics of the Tianshengqiao and Xiagei Hots Prings in Shangrila County of Yunnan and Analysis of the Formation of the Travertine*; China University of Geosciences: Beijing, China, 2015.
33. Zhang, K.-J.; Zhang, Y.-X.; Tang, X.-C.; Xia, B. Late Mesozoic tectonic evolution and growth of the Tibetan plateau prior to the Indo-Asian collision. *Earth-Sci. Rev.* **2012**, *114*, 236–249. [CrossRef]
34. Zhang, K.-J.; Zhang, Y.-X.; Li, B.; Zhong, L.-F. Nd isotopes of siliciclastic rocks from Tibet, western China: Constraints on provenance and pre-Cenozoic tectonic evolution. *Earth Planet. Sci. Lett.* **2007**, *256*, 604–616. [CrossRef]
35. Zhang, K.-J.; Xia, B.; Liang, X. Mesozoic–Paleogene sedimentary facies and paleogeography of Tibet, western China: Tectonic implications. *Geol. J.* **2002**, *37*, 217–246. [CrossRef]
36. Jackson, M.L. *Soil Chemical Analysis*; Prentice Hall: Englewood Cliffs, NJ, USA, 1958.
37. Parkhurst, D.L.; Appelo, C.A.J. Description of input and examples for PHREEQC version 3—A computer program for speciation, batch-reaction, one-dimensional transport, and inverse geochemical calculations. *US Geol. Surv. Tech. Methods* **2013**, *6*, 497.
38. Gandin, A.; Capezzuoli, E. Travertine: Distinctive depositional fabrics of carbonates from thermal spring systems. *Sedimentology* **2014**, *61*, 264–290. [CrossRef]
39. Kele, S.; Breitenbach, S.F.M.; Capezzuoli, E.; Meckler, A.N.; Ziegler, M.; Millan, I.M.; Kluge, T.; Deák, J.; Hanselmann, K.; John, C.M.; et al. Temperature dependence of oxygen- and clumped isotope fractionation in carbonates: A study of travertines and tufas in the 6–95 °C temperature range. *Geochim. Et Cosmochim. Acta* **2015**, *168*, 172–192. [CrossRef]
40. Chacko, T.; Cole, D.R.; Horita, J. Equilibrium Oxygen, Hydrogen and Carbon Isotope Fractionation Factors Applicable to Geologic Systems. *Rev. Mineral. Geochem.* **2001**, *43*, 1–81. [CrossRef]
41. Chafetz, H.S.; Guidry, S.A. Deposition and diagenesis of Mammoth Hot Springs travertine, Yellowstone National Park, Wyoming, U.S.A. *Can. J. Earth Sci.* **2003**, *40*, 1515–1529. [CrossRef]
42. Karabacak, V.; Uysal, I.T.; Mutlu, H.; Ünal-İmer, E.; Dirik, R.K.; Feng, Y.x.; Akıska, S.; Aydoğdu, İ.; Zhao, J.x. Are U-Th dates correlated with historical records of earthquakes? Constraints from coseismic carbonate veins within the North Anatolian Fault Zone. *Tectonics* **2019**, *38*, 2431–2448. [CrossRef]
43. Kele, S.; Özkul, M.; Fórizs, I.; Gökgöz, A.; Baykara, M.O.; Alçiçek, M.C.; Németh, T. Stable isotope geochemical study of Pamukkale travertines: New evidences of low-temperature non-equilibrium calcite-water fractionation. *Sediment. Geol.* **2011**, *238*, 191–212. [CrossRef]
44. Peng, X.; Jones, B. Patterns of biomediated $CaCO_3$ crystal bushes in hot spring deposits. *Sediment. Geol.* **2013**, *294*, 105–117. [CrossRef]
45. Pentecost, A. *Travertine*; Springer: Berlin/Heidelberg, Germany, 2005.
46. Pentecost, A.; Viles, H. A Review and Reassessment of Travertine Classification. *Géographie Phys. Et Quat.* **1994**, *48*, 305–314. [CrossRef]
47. Teboul, P.A.; Durlet, C.; Gaucher, E.C.; Virgone, A.; Girard, J.P.; Curie, J.; Lopez, B.; Camoin, G.F. Origins of elements building travertine and tufa: New perspectives provided by isotopic and geochemical tracers. *Sediment. Geol.* **2016**, *334*, 97–114. [CrossRef]
48. Claes, H.; Soete, J.; Van Noten, K.; El Desouky, H.; Marques Erthal, M.; Vanhaecke, F.; Özkul, M.; Swennen, R.; Reijmer, J. Sedimentology, three-dimensional geobody reconstruction and carbon dioxide origin of Pleistocene travertine deposits in the Ballık area (south-west Turkey). *Sedimentology* **2015**, *62*, 1408–1445. [CrossRef]
49. Vogel, J.C. Variability of Carbon Isotope Fractionation during Photosynthesis. In *Stable Isotopes and Plant Carbon-Water Relations*; Ehleringer, J.R., Hall, A.E., Farquhar, G.D., Eds.; Academic Press: San Diego, CA, USA, 1993; pp. 29–46.

50. Hoefs, J. *Stable Isotope Geochemistry*; Springer: Berlin/Heidelberg, Germany, 2009; Volume 201.
51. Rollinson, H. *Using Geochemical Data*; Cambridge University Press: Cambridge, UK, 1993; Volume 1.
52. Liu, Z.; Yuan, D. Features of geochemical variations in typical epikarst systems of China and thier environment significance. *Geol. Rev.* **2000**, *46*, 324–327.
53. Sheng, L.C. The Study of Deep Source CO_2 Degasification and Carbon Cycle in the Southwest of China. Ph.D. Thesis, Southwest University (China), Chongqing, China, 2007.
54. Xu, Y.; Zhu, J.; Hu, R.; Bi, X.; Yu, H.; Xu, L.; Liu, B.; Huang, M.; Sheng, X. Heterogeneous lithospheric mantle beneath the southeastern Tibetan Plateau: Evidence from Cenozoic high-Mg potassic volcanic rocks in the Jinshajiang–Ailaoshan Cenozoic magmatic belt. *J. Asian Earth Sci.* **2019**, *180*, 103849. [CrossRef]
55. Tapponnier, P.; Lacassin, R.; Leloup, P.H.; Schärer, U.; Dalai, Z.; Haiwei, W.; Xiaohan, L.; Shaocheng, J.; Lianshang, Z.; Jiayou, Z. The Ailao Shan/Red River metamorphic belt: Tertiary left-lateral shear between Indochina and South China. *Nature* **1990**, *343*, 431–437. [CrossRef]
56. Mancini, A.; Frondini, F.; Capezzuoli, E.; Galvez Mejia, E.; Lezzi, G.; Matarazzi, D.; Brogi, A.; Swennen, R. Porosity, bulk density and $CaCO_3$ content of travertines. A new dataset from Rapolano, Canino and Tivoli travertines (Italy). *Data Brief* **2019**, *25*, 104158. [CrossRef]
57. Mancini, A.; Frondini, F.; Capezzuoli, E.; Mejia, E.G.; Lezzi, G.; Matarazzi, D.; Brogi, A.; Swennen, R. Evaluating the geogenic CO_2 flux from geothermal areas by analysing quaternary travertine masses. New data from western central Italy and review of previous CO_2 flux data. *Quat. Sci. Rev.* **2019**, *215*, 132–143. [CrossRef]
58. Mancini, A.; Capezzuoli, E.; Brogi, A.; Swennen, R.; Ricci, L.; Frondini, F. Geogenie CO_2 flux calculations from the Late Pleistocene Tivoli travertines (Acque Albule Basin, Tivoli, Central Italy). *Ital. J. Geosci.* **2020**, *139*, 374–382. [CrossRef]
59. Uysal, I.T.; Ünal-İmer, E.; Shulmeister, J.; Zhao, J.-X.; Karabacak, V.; Feng, Y.-X.; Bolhar, R. Linking CO_2 degassing in active fault zones to long-term changes in water balance and surface water circulation, an example from SW Turkey. *Quat. Sci. Rev.* **2019**, *214*, 164–177. [CrossRef]
60. Jones, B. Review of aragonite and calcite crystal morphogenesis in thermal spring systems. *Sediment. Geol.* **2017**, *354*, 9–23. [CrossRef]
61. Mors, R.A.; Gomez, F.J.; Astini, R.A.; Mlewski, E.C.; Gérard, E. Physico-chemical and biological controls in a travertine system in the high Andes of northwestern Argentina. *Sediment. Geol.* **2022**, *439*, 106214. [CrossRef]
62. Jones, B.; Renaut, R.W. Origin of platy calcite crystals in hot-spring deposits in the Kenya Rift valley. *J. Sediment. Res.* **1995**, *68*, 913–927. [CrossRef]
63. Minissale, A.; Kerrick, D.M.; Magro, G.; Murrell, M.T.; Paladini, M.; Rihs, S.; Sturchio, N.C.; Tassi, F.; Vaselli, O. Geochemistry of Quaternary travertines in the region north of Rome (Italy): Structural, hydrologic and paleoclimatic implications. *Earth Planet. Sci. Lett.* **2002**, *203*, 709–728. [CrossRef]
64. Minissale, A. Origin, transport and discharge of CO_2 in central Italy. *Earth-Sci. Rev.* **2004**, *66*, 89–141. [CrossRef]
65. Della Porta, G.; Croci, A.; Marini, M.; Kele, S. Depositional architecture, facies character and geochemical signature of the Tivoli travertines (Pleistocene, Acque Albule Basin, Central Italy). *Riv. Ital. Di Paleontol. E Stratigr.* **2017**, *123*, 487–540.
66. Duan, Z.; Sun, R. An improved model calculating CO_2 solubility in pure water and aqueous NaCl solutions from 273 to 533 K and from 0 to 2000 bar. *Chem. Geol.* **2003**, *193*, 257–271. [CrossRef]
67. Spycher, N.; Pruess, K. A Phase-Partitioning Model for CO_2–Brine Mixtures at Elevated Temperatures and Pressures: Application to CO_2-Enhanced Geothermal Systems. *Transp. Porous Media* **2010**, *82*, 173–196. [CrossRef]
68. Luo, L.; Capezzuoli, E.; Rogerson, M.; Vaselli, O.; Wen, H.; Lu, Z. Precipitation of carbonate minerals in travertine-depositing hot springs: Driving forces, microenvironments, and mechanisms. *Sediment. Geol.* **2022**, *438*, 106207. [CrossRef]
69. Luo, L.; Wen, H.; Li, Y.; You, Y.; Luo, X. Mineralogical, crystal morphological, and isotopic characteristics of smooth slope travertine deposits at Reshuitang, Tengchong, China. *Sediment. Geol.* **2019**, *381*, 29–45. [CrossRef]
70. Guo, L.I.; Riding, R. Aragonite laminae in hot water travertine crusts, Rapolano Terme, Italy. *Sedimentology* **1992**, *39*, 1067–1079. [CrossRef]
71. Liu, Z.; Li, H.; You, C.; Wan, N.; Sun, H. Thickness and stable isotopic characteristics of modern seasonal climate-controlled sub-annual travertine laminas in a travertine-depositing stream at Baishuitai, SW China: Implications for paleoclimate reconstruction. *Environ. Geol.* **2006**, *51*, 257–265. [CrossRef]
72. Okumura, T.; Takashima, C.; Shiraishi, F.; Kano, A. Textural transition in an aragonite travertine formed under various flow conditions at Pancuran Pitu, Central Java, Indonesia. *Sediment. Geol.* **2012**, *265–266*, 195–209. [CrossRef]
73. Simmons, S.F.; Arehart, G.; Simpson, M.P.; Mauk, J.L. Origin of massive calcite veins in the Golden Cross low-sulfidation, epithermal Au-Ag deposit, New Zealand. *Econ. Geol.* **2000**, *95*, 99–112. [CrossRef]
74. Simmons, S.F.; Browne, P.R.L. Hydrothermal minerals and precious metals in the Broadlands-Ohaaki geothermal system: Implications for understanding low-sulfidation epithermal environments. *Econ. Geol.* **2000**, *95*, 971–999. [CrossRef]
75. Berardi, G.; Vignaroli, G.; Billi, A.; Rossetti, F.; Soligo, M.; Kele, S.; Baykara, M.O.; Bernasconi, S.M.; Castorina, F.; Tecce, F.; et al. Growth of a Pleistocene giant carbonate vein and nearby thermogene travertine deposits at Semproniano, southern Tuscany, Italy: Estimate of CO_2 leakage. *Tectonophysics* **2016**, *690*, 219–239. [CrossRef]

76. Frery, E.; Gratier, J.-P.; Ellouz-Zimmerman, N.; Loiselet, C.; Braun, J.; Deschamps, P.; Blamart, D.; Hamelin, B.; Swennen, R. Evolution of fault permeability during episodic fluid circulation: Evidence for the effects of fluid–rock interactions from travertine studies (Utah–USA). *Tectonophysics* **2015**, *651–652*, 121–137. [CrossRef]
77. Süer, S.; Güleç, N.; Mutlu, H.; Hilton, D.R.; Çifter, C.; Sayin, M. Geochemical Monitoring of Geothermal Waters (2002–2004) along the North Anatolian Fault Zone, Turkey: Spatial and Temporal Variations and Relationship to Seismic Activity. *Pure Appl. Geophys.* **2008**, *165*, 17–43. [CrossRef]
78. Desouky, H.E.; Soete, J.; Claes, H.; Özkul, M.; Vanhaecke, F.; Swennen, R.; Ariztegui, D. Novel applications of fluid inclusions and isotope geochemistry in unravelling the genesis of fossil travertine systems. *Sedimentology* **2015**, *62*, 27–56. [CrossRef]

minerals

Article

Insights into Alpine-Karst-Type Tufa Deposits in Geological Environmental Records: A Case Study of the Calcareous Tufa Profile of the Jiuzhaigou Natural Reserve on the Eastern Margin of the Tibetan Plateau

Congcong Lv [1], Xueqin Zhao [1,*], Yaoxi Jiang [1], Heyan Zhu [1], Hongmin Zhang [2], Fudong Wang [1,*], Qiongfang Li [3] and Keli Hou [1]

1 School of Environment and Resources, Southwest University of Science and Technology, Mianyang 621010, China
2 Sichuan Railway Constration Co., Ltd., Chengdu 610072, China
3 School of Life Science and Engineering, Southwest University of Science and Technology, Mianyang 621010, China
* Correspondence: zhaoxq@swust.edu.cn (X.Z.); wolfdongswust@163.com (F.W.)

Abstract: To study the geological environmental records of alpine-karst-type tufa deposits in the eastern margin of the Tibetan Plateau, the calcareous tufa profile exposed by the "8.8" Jiuzhaigou earthquake was taken as the research object and combined with a field geological investigation. Further, the petrography, sedimentology, chronology, and elemental geochemistry of the calcareous tufa were studied and analyzed. The results show the following. (1) The Sparkling Lake calcareous tufa profile was deposited under the background of a warm and humid climate during the Holocene. The growth pattern follows a bottom-to-top deposition. (2) At 750 ± 30–300 ± 30 aB.P., the calcareous tufa layers were gray-black as a whole, and the changes in mineral composition and elemental geochemistry indicate a fluctuating upward trend for temperature and precipitation during this period. (3) The formation of two sets of black peat layers in the upper part of the tufa calcareous profile is due to the synergistic action of multiple factors caused by strong tectonic activity. In conclusion, the deposition mechanism of the calcareous tufa in Jiuzhaigou was controlled by paleoclimate hydrology and glaciation for a long time, while strong tectonic activity over a short period of time considerably changed the color, structure, element content, and mineral composition of the calcareous tufa.

Keywords: calcareous tufa; carbon-oxygen isotopes; tectonic activity; geological environment; Holocene

Citation: Lv, C.; Zhao, X.; Jiang, Y.; Zhu, H.; Zhang, H.; Wang, F.; Li, Q.; Hou, K. Insights into Alpine-Karst-Type Tufa Deposits in Geological Environmental Records: A Case Study of the Calcareous Tufa Profile of the Jiuzhaigou Natural Reserve on the Eastern Margin of the Tibetan Plateau. *Minerals* **2023**, *13*, 120. https://doi.org/10.3390/min13010120

Academic Editors: Francesca Giustini, Mauro Brilli and Santanu Banerjee

Received: 15 September 2022
Revised: 29 December 2022
Accepted: 11 January 2023
Published: 13 January 2023

1. Introduction

Among terrestrial carbonates (calcareous tufa, stromatolitic tufa, and travertines), calcareous tufa represents a good proxy-indicator of Quaternary climatic and environmental conditions [1–7], widely observed in fluvial areas with calcareous deposits [8]. As tufas are mainly comprised of calcite, they enable direct and precise dating as well as geochemical reconstructions of the past geological environment [7]. Currently, such tufas are being used as a proxy for paleoclimate, paleohydrology, tectonic activity, glaciation, regional geomorphic evolution, human activity, and geoarchaeology studies [2,9–16]. Andrews et al. [2] reviewed paleoclimate records of riverine tufas using stable isotopes, and proposed a synthesis. Tye et al. [17] studied the Long Lake sequence of Marks Tey, Essex, UK, and showed that the $\delta^{18}O$ record of the endogenic carbonates from this sequence records the climatic structure of MIS 11. Matsuoka et al. [18] performed high-resolution stable isotopic analyses of an annually laminated tufa from Shirokawa, SW Japan. Zidi et al. [19] argued that the calcareous tufas of Boulaaba are good indicators of tectonic activity, as they are intimately related to the fault network distribution in the region. Tufa is deposited in open ground-surface environments, which leads to a complex deposition mechanism. Therefore, how to accurately use tufa to reconstruct paleoclimate environment still needs a lot of in-depth research [20].

A large amount of tufa with different genetic types is distributed on the Tibetan Plateau [21]. On the eastern margin of the Tibetan plateau, there is a belt-shaped tufa area from northeast to southwest [22], and its distribution area is mainly located at the turning point before the top of the largest topographic step in the eastern Tibetan Plateau. The regional tufa mainly develops in the periglacial landform area at the lower part of the snow line [23,24]. Because of the special geographical location of the Tibetan Plateau, the tufa in this area is strongly indicative of paleoclimate and paleoenvironment [20,25–30]. Mischke et al. [31] identified the Mid-Holocene tufa section at an altitude of 3815 m in the Qilian Mountains at the northeastern margin of the Tibetan Plateau as a new and possibly important climate archive. Guo et al. [32] reconstructed regional paleoclimatic and paleohydrological changes during the Early and Mid-Holocene by analyzing the rhythmic nature and hydrochemical proxies of tufa in Jiuzhaigou, eastern Tibetan Plateau. Lei et al. [33] discussed the characteristics of major geochemical elements and paleoenvironmental significance of tufa deposited on the ancient lakeshore terrace of the Ngangla Ringsto Lake in the western Tibetan Plateau. Niu et al. [22] and Wang et al. [20] summarized previous studies on the geological information recorded by travertine and tufa on the Tibetan Plateau and proposed the prospect of future studies.

The Jiuzhaigou deposits are typical examples of calcareous tufa with the highest altitude (2200–2900 m) in the world. Jiuzhaigou calcareous tufas are porous deposits containing plants and algal molds or imprints [16,34,35]. The Jiuzhaigou Tufa Scenic Spot, a World Natural Heritage Site (Aba Tibetan and Qiang Autonomous Prefecture, China), is particularly interesting because of its spectacular large-scale plateau tufa lakes, tufa waterfalls, tufa beach flows, and other landforms. On 8 August 2018, an Mw 7.0 earthquake (hereinafter referred to as the "8.8 earthquake" based on the date) occurred in Jiuzhaigou, China. Following the earthquake, many new calcareous tufa profiles were revealed, such as the Nuorilang Waterfall and Sparkling Lake. Before the earthquake, there was a tufa calcareous dam (locally called the Yanacuo Dam) between the Sparkling Lake and the downstream Double-Dragon Lake. The earthquake caused the dam to break in the central part and exposed a rhythmic structure section. Based on mineral composition, thin section observation, major/trace element content, and stable isotope composition ($\delta^{13}C_{V\text{-}PDB}$, $\delta^{18}O_{V\text{-}SMOW}$), combined with previous scholars' research results on calcareous tufa and Jiuzhaigou area, this paper analyzes the mechanism of calcareous tufa deposition in Jiuzhaigou and the influencing factors during the deposition process, and further discusses the geological environment that may be recorded by calcareous tufa deposition.

2. Geographic Setting

The Jiuzhaigou Natural Heritage Scenic Area in Sichuan Province, China (32°54′–33°19′ N, 103°46′–104°14′ N) is located in the conjunction of the Songpan–Ganzi orogenic belt and West Qinling orogenic belt (Figure 1a). The main faults in the surrounding area are the Tazang, Huya, Xueshanliangzi, and Minjiang faults (Figure 1b). The geological structure of the Jiuzhaigou region is complex, with strong neotectonic activity, fold-fault development, and frequent earthquakes (Figure 1b,c). The structural pattern of the scenic spot controls the geomorphological framework in the area and the structural lineaments are mainly distributed in the NW-SE direction (Figure 1c).

Owing to the vulnerability of the aquifer in the Jiuzhaigou area, after it is broken under the action of tectonic stress, rock joints and fractures in the fault are developed, with NW trending dominant directions, which conditions for the formation of a fissure crack. The interaction of groundwater infiltration and rock promotes the further expansion of fissures. In addition, the growth of fissures intensifies the infiltration and circulation of groundwater, alternately, and the fissure–karst process is mutually reinforcing, thereby accelerating the karst process. Identified through a field geological survey and the investigation report of the Sichuan Bureau of Geology and Mineral Resources, the lithofacies of the outcrop strata in the Jiuzhaigou Scenic Spot are marine facies, and the lithology is mainly carbonate rocks, ranging in age from the Paleozoic Devonian to the Mesozoic Triassic. Various limestones (including bioclastic limestone, crystalline limestone, argillaceous limestone, and siliceous limestone) and trace amounts of dolomite, slate, and sandstone constitute the material basis of the Jiuzhaigou karst underground runoff system.

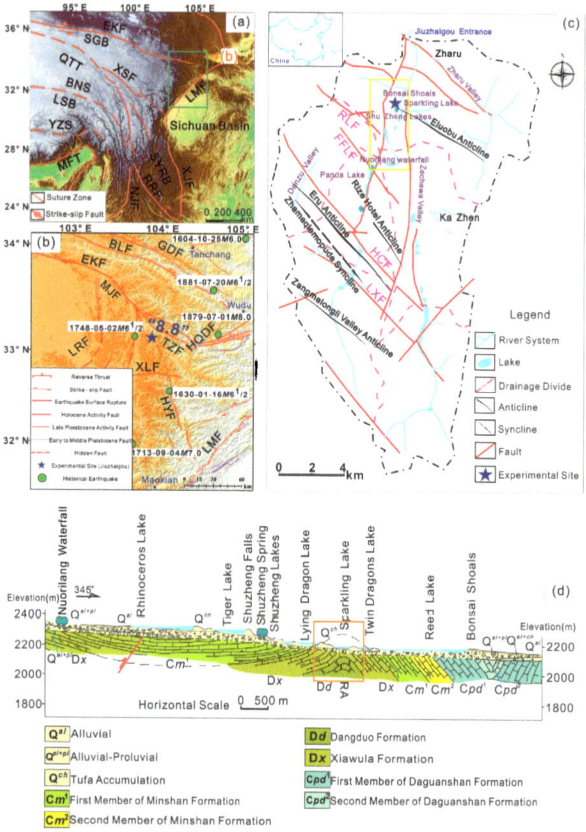

Figure 1. Geological background map of study area. (**a**) Regional tectonic geological map [36]. EKF = East Kunlun fault; XSF = Xianshuihe fault; QTT = QiangTang terrane; BNS = Bangong-Nujiang suture; LSB = Lhasa block; YZS = Yarlung-Zangbo suture; MFT = main frontal thrust; NJF = Nujiang fault; RRF = Red River fault; SYRB = Sichuan-Yunnan rhombic block; XJF = Xiaojiang fault; LMF = Longmenshan fault; SGB = Songpan-Ganzi Block. The green frame is the area in (**b**). (**b**) Regional structure and distribution of magnitude M $\geq$ 6.0 earthquakes in the Jiuzhaigou vicinity between 1630 and 1900 [37]. The blue frame shows the location of the Jiuzhaigou Scenic Spot. GDF = Guanggaishan-Dieshan fault; BLF = Bailongjiang fault; EKF = East Kunlun fault; MJF = MinJiang fault; TZF = Tazang fault; XLF = Xueshanliangzi fault; LMF = Longmenshan fault; HYF = Huya fault; LRF = Longriba fault; HQDF = Hanan-Qingshanwan-Daoqizi fault. The blue five-pointed star is the epicenter of the "8.8" earthquake. (**c**) Outline of the geological structure of the Jiuzhaigou Scenic Spot (from Sichuan Provincial Regional Geological Survey Team 2006). The yellow frame shows the location of Shuzheng Valley. The location of Sparkling Lake is indicated by a blue five-pointed star. LXF = Long Lake-XuanquanValley Fault; HCF = Hawk Claw Fault; FFLF = Five Flower Lake Fault; RLF = Rhinoceros Lake Fault. (**d**) Schematic of the Shuzheng Valley geological profile [38]. RA = Russian Anticline. The orange frame shows the location of the Sparkling Lake in the Shuzheng Valley profile.

The Jiuzhaigou Scenic Spot is distributed in a "Y" shape, with the Shuzheng Valley at the lower branch of the "Y" (Figure 1c). The elevation of the Shuzheng Valley is between 2210 m and 2250 m, and the total length from Entrance to Nuorilang is 13.8 km (Figure 1d). The Jiuzhaigou Scenic Spot has 30 small lakes, including the Sparkling Lake, Lying Dragon Lake, Five Flower Lake, Shuzheng lakes, and so on. The "8.8" earthquake caused serious damage to the two core scenic spots of Jiuzhaigou: the Nuorilang Waterfall and Sparkling Lake. The earthquake caused part of the viewing platform of the Nuorilang Waterfall to collapse. At the downstream of the Shuzheng Valley, the Sparkling Lake burst, forming a trapezoidal gap with a length of approximately 40 m, width of 12 m, and height of 15 m [39] (Figure 2a). The Sparkling Lake dried up almost completely after the "8.8" earthquake, and the calcareous tufa reef and calcareous tufa mound at the bottom of the lake were exposed. Following the collapse of the Sparkling Lake, a layered calcareous tufa profile with a recognizable rhythmic structure was revealed. The exposed profile has clear layers and a continuous deposition (Figure 2b), the color of the layers was mainly off-white, and black peat layers of considerable thickness were sandwiched (Figure 2b,c). The Sparkling Lake calcareous tufa profile provides a good experimental material for this study.

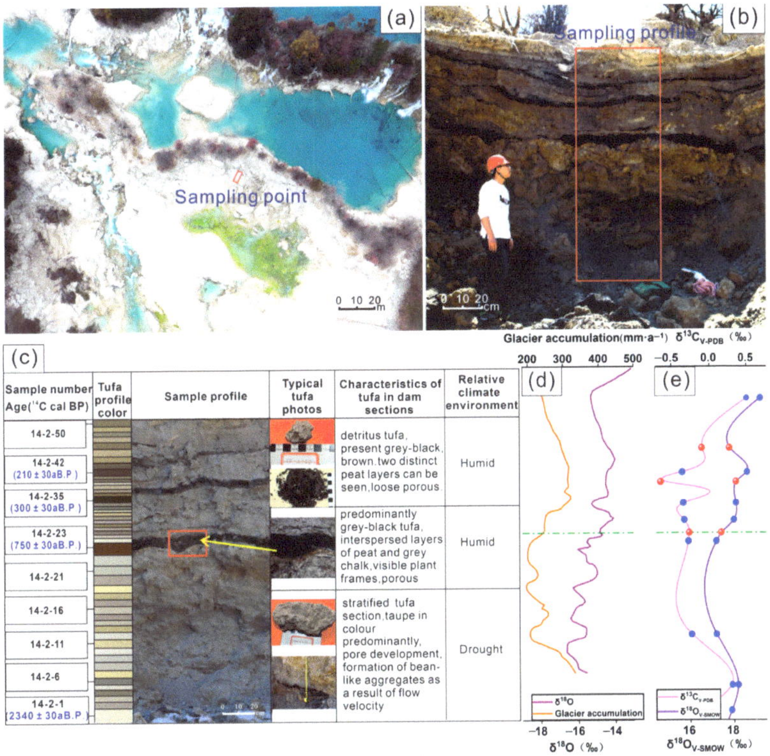

Figure 2. Collected calcareous tufa deposition profile at the Sparkling Lake. (**a**) The lake dried up after the earthquake. The red frame shows the sampling location. (**b**) Sparkling Lake calcareous tufa sampling profile. (**c**) Calcareous tufa characteristics and dating data. (**d**) Changes in ^{18}O and glacier accumulation recorded by Guliya ice cores over the last millennium (data from Yao et al. [40]). (**e**) Carbon-oxygen isotope analysis of the calcareous tufa profile. The blue dots represent experimentally measured carbon-oxygen isotope data. The red dots represent the location of the black peat layer.

3. Materials and Methods

3.1. Sample Collection

After the "8.8 earthquake", the Sparkling Lake essentially dried up (Figure 2a). This study chose the fully exposed and partially collapsed central dam as the research object (Figure 2b). To obtain a fresh sample, the outermost layer of the profile was removed, and a steel pipe with a knife edge was used to knock in the calcareous tufa layer perpendicular to the section. According to the color variation, hardness, and structural characteristics of the calcareous tufa section, the samples are continuously and unevenly stratified from bottom to top. The bit distribution and thickness were marked, and the calcareous tufa sample was put into a clean cotton sample bag for coded storage. A sample was collected from each calcareous tufa layer, with each sample weighing approximately 2 kg. In total, 50 samples were collected, which were recorded as 14-2-1~14-2-50; the codes recorded on the three black peat layers were 14-2-23, 14-2-35, and 14-2-42 (Figure 2c).

Through the field geological survey, we observed and measured the formation occurrence and calcareous tufa profile of the strata in the study area. Further, we determined the attitude of rock formation and structural characteristics of the calcareous tufa. The collected samples were brought to the laboratory for analysis. Based on the establishment of the fine depositional time series of the calcareous tufa profile, the depositional environment changes of calcareous tufa in different periods are analyzed systematically, and the controlling factors affecting the change in depositional environment are also expounded.

3.2. Analytical Methods

The mineral phase analysis of calcareous tufa was completed at the Analytical Testing Center, Southwest University of Science and Technology (Mianyang, China), using an X-ray diffractometer (X'Pert Pro, PANalytical B.V.) under the following conditions: a copper target, 40 kV pipe pressure, 40 mA pipe flow, 0.033° step length, 0.2 °/min integration time, and 3–80 scan range. The stable isotope ($\delta^{13}C_{V\text{-}PDB}$, $\delta^{18}O_{V\text{-}SMOW}$) and major/trace element analysis of calcareous tufa was conducted by the ALS Laboratory Group (Guangzhou, China). The major elements of the rock were analyzed by the X-ray fluorescence spectrometer melting method. The rare earth trace elements were analyzed by an agilent inductively coupled plasma-optical emission spectrometer. Carbon-oxygen isotope analysis was conducted using a Thermo Finnigan Delta^{plus} XP isotope mass spectrometer. The measured results of $\delta^{13}C$ (precisions of 0.1‰, V-PDB standardization) and $\delta^{18}O$ (precisions of 0.5‰, V-SMOW standardization) are expressed in ‰. Radiocarbon dating was conducted on the calcareous tufa samples and performed at the BETA Laboratory (Miami, FL, USA) in accordance with ISO/IEC 17025:2017. The IntCal20 calibration curve was used. To eliminate the influence of "dead carbon", the dating results were calibrated based on the precise dating of carbon in peat layers in the profile. The thin section observation was carried out in the Fundamental Science on Nuclear Wastes and Environmental Safety Laboratory of the Southwest University of Science and Technology (Mianyang, China). The instrument model is the Zeiss upright microscope Axio Scope A1. The technical parameters are as follows: the most advanced IC2S infinity axial, radial double chromatic aberration correction and contrast-enhanced optical system; the analyzer is 360° adjustable, the precision is 0.1°; the magnification is 50~500; and transmission, anti-bireflection light source.

4. Data Analysis

4.1. Calcareous Tufa Mineral Phase Test

Thirteen calcareous tufa samples with layers of different colors were selected for X-ray diffraction (XRD). The diffraction patterns of the obtained samples was compared and analyzed with the powder diffraction file (PDF) card using the MDI Jade 9.0 software. The results show that the diffraction peaks of the calcareous tufa samples are all well fitted with the calcite standard card (PDF#05-0586), indicating that the calcareous tufa was dominated by calcite (Figure 3). The peak position of calcite does not change considerably, with the strongest diffraction peak appearing at 29.4°. The diffraction peak baseline was flat, the peak type was sharp, and the peak was obvious. The corresponding main dominant surfaces are {012}, {104}, {110}, {113}, {202}, {018}, {116}, {122}. After the qualitative information of the mineral is obtained, the Rietveld full-spectrum fitting method is used for semi-quantitative analysis to determine the content of mineral components in calcareous tufa samples of different colors. The analysis results are shown in Table 1. The samples of the Sparkling Lake calcareous tufa profile are shown to contain calcite and quartz. From the 14-2-23 sample upward, the calcite content basically dropped to below 90% and other mineral components appeared (such as Albite, Muscovite, Montmorillonite, and Kaolinite). Table 1 shows that the percentages of calcite in the three sets of black peat calcareous tufa layers are 4.8, 45.9, and 65.8, which are significantly lower than the percentages of about 80 in the gray-brown layers. Silicate minerals are present in all three sets of black peat layers. The content of Muscovite (43.8%) in sample 14-2-23 was even higher than that of calcite.

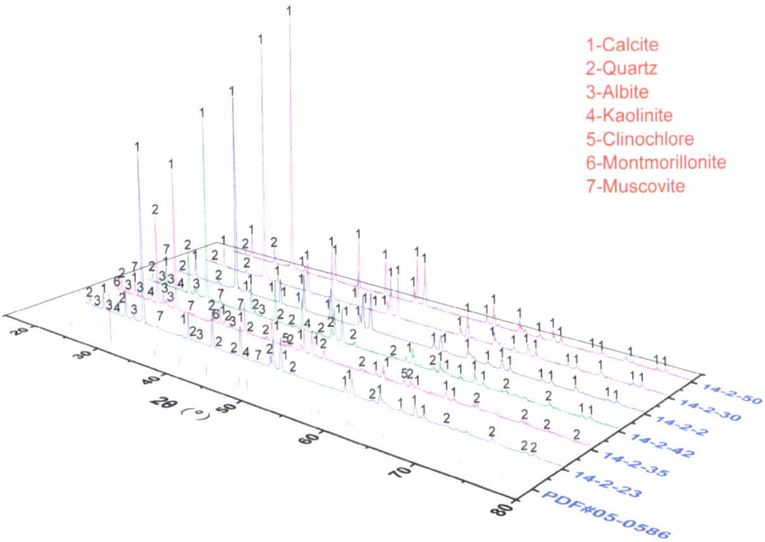

Figure 3. XRD patterns of calcareous tufa profile in Sparkling Lake.

Table 1. Mineral percentage of calcareous tufa profile in Sparkling Lake.

Sample Number	Quartz (%)	Calcite (%)	Clinochlore (%)	Albite (%)	Muscovite (%)	Montmorillonite (%)	Kaolinite (%)
14-2-2	1.8	98.2	-	-	-	-	-
14-2-4	2.0	98	-	-	-	-	-
14-2-5	4.4	95.6	-	-	-	-	-
14-2-11	1.8	98.2	-	-	-	-	-
14-2-22	3.4	96.6	-	-	-	-	-
14-2-23	27.5	4.8	10.8	3.3	43.8	4.6	5.2
14-2-26	8.9	74.3	-	0.4	11.6	-	4.8
14-2-30	2.1	97.9	-	-	-	-	-
14-2-34	5.3	93.2	-	1.5	-	-	-
14-2-35	20.9	45.9	4.7	3.8	19.1	1.8	3.8
14-2-37	3	81.1	-	-	9.4	-	6.5
14-2-42	11.7	65.8	-	1.6	18.3	-	2.6
14-2-50	4.1	95.9	-	-	-	-	-

4.2. Calcareous Tufa Radiocarbon Dating

The chemical index of alteration (CIA), an important index to evaluate the weathering history of sediments, is defined as $Al_2O_3/(Al_2O_3 + CaO^* + Na_2O + K_2O) \times 100$, where the oxide is the molar mass percentage and CaO* is only CaO in silicate [41–43]. Because of the difficulty in accurately separating silicate minerals in the sample, the correction method proposed by McLennan [44] was used to calculate the CIA value. Correlation analysis refers to the measurement of the degree of correlation between variable elements [45,46]. Here, Pearson correlation analysis was performed on the stable isotopes and major/trace elements using the IBM SPSS Statistics 25 software, which intuitively reflected the dependence and correlation degree of each indicator.

The lowest layers and three sets of black peat layers of the calcareous tufa profile were selected for radiocarbon dating. The age of the calcareous tufa profile of the Sparkling Lake was measured to be 2340 ± 30–210 ± 30 aB.P. (Figure 2c), belonging to the Late Holocene. Combined with the warm–humid climate corresponding to the CIA value (60–80) (Table 2), this indicates that the calcareous tufa is the product of deposition in the warm–humid climate of the Late Holocene. Based on the superposition relationship and occurrence measured in the field, the Sparkling Lake calcareous tufa dam shows a bottom-up sequence from old to new, with sedimentary continuity, and the ^{14}C apparent age data are consistent, enabling a more detailed chronological sequence for the calcareous tufa deposition in Jiuzhaigou Scenic Spot.

Table 2. Content and ratios of major/trace elements in calcareous tufa of Sparkling Lake.

Sample Number	K (%)	Na (%)	Ca (%)	Mg (%)	Al (%)	K/Ca	K/Al	Na/Al	Mg/Al	Ca/Al	Al_2O_3 (%)	K_2O (%)	Na_2O (%)	CaO (%)	MgO (%)	CIA
14-2-1	0.05	0.06	38.5	0.24	0.19	0.13	0.26	0.32	1.26	202.63	0.32	0.07	0.08	54.0	0.37	58.18
14-2-2	0.09	0.07	37.5	0.25	0.32	0.24	0.28	0.22	0.78	117.19	0.57	0.11	0.09	52.8	0.42	66.28
14-2-3	0.09	0.07	38.5	0.26	0.31	0.23	0.29	0.23	0.84	124.19	0.53	0.11	0.09	53.7	0.42	64.63
14-2-4	0.17	0.07	37.0	0.28	0.61	0.46	0.28	0.11	0.46	60.66	1.14	0.21	0.09	52.1	0.48	74.51
14-2-5	0.21	0.09	36.9	0.30	0.75	0.57	0.28	0.12	0.40	49.20	1.37	0.26	0.12	50.6	0.50	73.26
14-2-6	0.15	0.07	38.1	0.27	0.53	0.39	0.28	0.13	0.51	71.89	0.97	0.18	0.09	52.4	0.43	72.93
14-2-7	0.09	0.06	37.6	0.25	0.30	0.24	0.30	0.20	0.83	125.33	0.56	0.11	0.08	53.7	0.42	67.47
14-2-8	0.16	0.07	36.9	0.28	0.55	0.43	0.29	0.13	0.51	67.09	1.03	0.19	0.09	52.4	0.47	73.57
14-2-9	0.26	0.09	35.3	0.31	0.98	0.74	0.27	0.09	0.32	36.02	1.81	0.31	0.12	50.2	0.51	76.69
14-2-10	0.18	0.08	36.9	0.29	0.71	0.49	0.25	0.11	0.41	51.97	1.27	0.22	0.11	51.7	0.48	74.27
14-2-11	0.12	0.07	38.8	0.28	0.48	0.31	0.25	0.15	0.58	80.83	0.85	0.15	0.09	53.1	0.45	72.03
14-2-12	0.64	0.16	31.7	0.45	2.55	2.02	0.25	0.06	0.18	12.43	4.76	0.77	0.22	42.7	0.75	79.73
14-2-13	0.22	0.08	36.8	0.29	0.85	0.60	0.26	0.09	0.34	43.29	1.53	0.26	0.11	50.9	0.48	76.12
14-2-14	0.16	0.07	35.7	0.27	0.61	0.45	0.26	0.11	0.44	58.52	1.13	0.20	0.09	52.4	0.46	74.83
14-2-15	0.27	0.09	35.9	0.30	0.92	0.75	0.29	0.10	0.33	39.02	1.73	0.32	0.12	50.3	0.50	75.55
14-2-16	0.26	0.09	35.7	0.31	0.91	0.73	0.29	0.10	0.34	39.23	1.61	0.31	0.12	50.0	0.51	74.54
14-2-17	0.13	0.07	36.2	0.26	0.45	0.36	0.29	0.16	0.58	80.44	0.83	0.16	0.09	52.0	0.43	70.94
14-2-18	0.09	0.06	37.7	0.24	0.31	0.24	0.29	0.19	0.77	121.61	0.54	0.11	0.08	53.3	0.39	66.67
14-2-19	0.17	0.08	36.8	0.26	0.57	0.46	0.30	0.14	0.46	64.56	1.04	0.20	0.11	51.7	0.43	71.23
14-2-20	0.23	0.08	36.0	0.28	0.78	0.64	0.29	0.10	0.36	46.15	1.43	0.27	0.11	50.7	0.47	74.48
14-2-21	0.15	0.07	37.6	0.26	0.51	0.40	0.29	0.14	0.51	73.73	0.92	0.18	0.09	52.5	0.43	71.88
14-2-22	0.35	0.09	33.9	0.30	1.21	1.03	0.29	0.07	0.25	28.02	2.27	0.43	0.12	48.7	0.51	77.21
14-2-23	1.06	0.35	4.64	0.62	3.52	22.84	0.30	0.10	0.18	1.32	7.47	0.70	0.47	6.80	1.14	76.30
14-2-24	0.60	0.23	28.8	0.40	1.98	2.08	0.30	0.12	0.20	14.55	3.74	0.80	0.31	41.5	0.69	73.91
14-2-25	0.67	0.34	28.4	0.45	2.24	2.36	0.30	0.15	0.20	12.68	4.17	0.63	0.46	39.2	0.78	70.80
14-2-26	0.52	0.23	30.1	0.40	1.75	1.73	0.30	0.13	0.23	17.20	3.23	0.34	0.31	42.5	0.69	72.10
14-2-27	0.28	0.11	34.3	0.33	0.96	0.82	0.29	0.11	0.34	35.73	1.81	0.24	0.15	49.5	0.56	73.88
14-2-28	0.20	0.07	36.1	0.30	0.67	0.55	0.30	0.10	0.45	53.88	1.22	0.20	0.09	51.2	0.48	74.39
14-2-29	0.16	0.07	37.5	0.29	0.55	0.43	0.29	0.13	0.53	68.18	0.99	0.26	0.09	52.3	0.48	72.26
14-2-30	0.22	0.08	36.3	0.30	0.76	0.61	0.29	0.11	0.39	47.76	1.39	0.70	0.11	51.2	0.51	74.33
14-2-31	0.29	0.10	34.1	0.31	1.00	0.85	0.29	0.10	0.31	34.10	1.88	0.35	0.13	49.2	0.53	75.50
14-2-32	0.26	0.10	35.8	0.31	0.90	0.73	0.29	0.11	0.34	39.78	1.64	0.31	0.13	50.3	0.53	74.21
14-2-33	0.27	0.09	36.2	0.32	0.95	0.75	0.28	0.09	0.34	38.11	1.74	0.33	0.12	50.0	0.51	75.32
14-2-34	0.39	0.12	33.6	0.35	1.36	1.16	0.29	0.09	0.26	24.71	2.58	0.48	0.16	48.0	0.60	76.33
14-2-35	1.16	0.44	16.15	0.64	3.83	7.18	0.30	0.11	0.17	4.22	7.52	1.44	0.59	22.7	1.16	74.16
14-2-36	0.38	0.14	33.3	0.36	1.24	1.14	0.31	0.11	0.29	26.85	2.31	0.46	0.19	46.9	0.61	73.33
14-2-37	0.28	0.11	35.2	0.32	0.91	0.80	0.31	0.12	0.35	38.68	1.67	0.34	0.15	49.3	0.53	72.29
14-2-38	0.28	0.10	34.5	0.32	0.93	0.81	0.30	0.11	0.34	37.10	1.69	0.34	0.13	48.8	0.52	73.80

Table 2. *Cont.*

Sample Number	K (%)	Na (%)	Ca (%)	Mg (%)	Al (%)	K/Ca	K/Al	Na/Al	Mg/Al	Ca/Al	Al_2O_3 (%)	K_2O (%)	Na_2O (%)	CaO (%)	MgO (%)	CIA
14-2-39	0.30	0.09	34.5	0.32	0.97	0.87	0.31	0.09	0.33	35.57	1.84	0.37	0.12	49.4	0.55	75.10
14-2-40	0.82	0.31	17.55	0.55	2.74	4.67	0.30	0.11	0.20	6.41	5.58	1.05	0.32	25.9	1.02	76.75
14-2-41	0.71	0.27	20.9	0.51	2.39	3.40	0.30	0.11	0.21	8.74	4.67	0.88	0.36	30.3	0.91	74.48
14-2-42	0.69	0.28	23.2	0.50	2.31	2.97	0.30	0.12	0.22	10.04	4.47	0.86	0.38	33.4	0.88	73.40
14-2-43	0.49	0.19	29.6	0.43	1.65	1.66	0.30	0.12	0.26	17.94	3.08	0.60	0.26	42.2	0.75	73.33
14-2-44	0.40	0.16	31.4	0.39	1.34	1.27	0.30	0.12	0.29	23.43	2.53	0.49	0.22	45.2	0.69	73.12
14-2-45	0.20	0.10	36.3	0.33	0.69	0.55	0.29	0.14	0.48	52.61	1.26	0.24	0.13	51.0	0.57	71.59
14-2-46	0.18	0.08	33.7	0.32	0.62	0.53	0.29	0.13	0.52	54.35	1.13	0.22	0.11	50.5	0.56	71.97
14-2-47	0.21	0.10	33.6	0.34	0.75	0.63	0.28	0.13	0.45	44.80	1.38	0.26	0.13	49.6	0.60	72.63
14-2-48	0.08	0.05	36.4	0.30	0.27	0.22	0.30	0.19	1.11	134.81	0.45	0.09	0.07	52.3	0.49	66.18
14-2-49	0.07	0.06	35.8	0.28	0.25	0.20	0.28	0.24	1.12	143.20	0.44	0.09	0.08	52.7	0.46	63.77
14-2-50	0.11	0.08	34.0	0.30	0.37	0.32	0.30	0.22	0.81	91.89	0.70	0.14	0.11	50.8	0.52	66.04

4.3. Thin Section Observation

In the thin-section results, lamellar limestone is evident in Figure 4a. The distribution side around the organic matrix is sparry calcite (Figure 4b–e). Outside, micrite and sparry calcite are distributed around the formed shell. Figure 4f shows sparry calcite and micrite around petioles. In the absence of organisms, there was a mix of micrite and sparry calcite around the void (Figure 4f). The void shape was irregular, which may be related to the recrystallization of micrite calcite. It is speculated that a microbial community exists in the litter accumulation and that the calcareous tufa is preferentially deposited by these organic matrices as a template. Around the organic matrices, bright crystal calcite is formed, and a cryptocrystalline structure dominated by muddy crystal is distributed around it. The calcareous tufa in Jiuzhaigou contains a large amount of algae and microorganisms (Figure 4g). Figure 4h shows the Sparkling Lake red algae population under the light microscope (400 times magnification). The thin-section results show that the tube bundle-shaped calcareous tufa formed by algal filaments and leaves as templates is dominated by splendid calcite on the inside and mainly micrite, microsparkle, and a mixture of the two on the outside. The difference of the crystal structures around the organic matrix indicates that the tubular channels were left during the decomposition of biomass from the inside to the outside. Further, there are gaps between the calcareous tufa inclusions formed by biological templates. In both cases, water flow with slow mobility passes through, which is conducive to recrystallization to form sparry calcite owing to excessive calcite saturation SIc (Figure 4i).

4.4. Results of the Major/Trace Element Analysis of Calcareous Tufa

From the calcareous tufa mineral composition analysis, the major/trace element contents in different color layers were further determined. The peaks of K, Ca, Na, Mg, Al, K/Ca, K/Al, and CIA in the calcareous tufa profile coincide with the black peat layers. In general, their peak values have a good correspondence with the black peat layers (Figure 5). The average values of K (0.97 g/kg), Na (0.36 g/kg), Mg (0.59 g/kg), Al (3.22 g/kg), K/Ca (10.00), K/Al (0.30), and CIA (74.62) in black peat layers were considerably higher than those in other color layers. Furthermore, the average value of Ca (14.66 g/kg) decreased considerably, about 20 g/kg (Table 2). Moreover, the Na/Al, Mg/Al, and Ca/Al peak values are diametrically opposite to the black peat layers (Figure 5). The average values of Na/Al (0.11), Mg/Al (0.19), and Ca/Al (5.19) were considerably lower (Table 2).

The deposition process of calcareous tufa mineral elements is affected by the changes in the geological environment. Therefore, a statistical analysis of major/trace elements can determine the correlation between calcareous tufa mineral elements and the possible origin of the elements. It was observed that the K in the calcareous tufa profile was considerably positively correlated with Na, Mg, Al, K/Ca, K/Al, and CIA ($p < 0.01$) (Table 3 and Figure 6), with a correlation coefficient greater than 0.5, and the variation trend was the closest. This indicates similar sources of the four elements, i.e., silicate minerals, which often occur in soil. Ca showed a considerable negative correlation with K, Na, Mg, Al, CIA, K/Ca, and K/Al, indicating the dilution effect of other elements on Ca. A linear relationship between Al and K, Na, Ca, and Mg was observed, while Al has an exponential relationship with the CIA (Figure 6). These chemical element indexes provide important data for distinguishing different colors of calcareous tufa layers. The results show that the three black peat layers of the calcareous tufa profile have the characteristics of low Ca content and Na/Al, Mg/Al, and Ca/Al ratios, and higher levels of K, Na, Mg, Al, K/Ca, K/Al, and CIA.

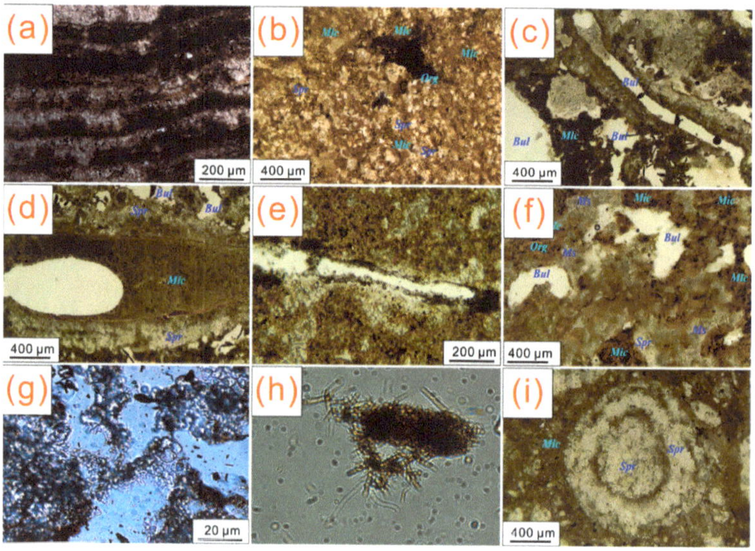

Figure 4. Thin-section results of Sparkling Lake calcareous tufa sample. (**a**) Lamellar limestone. (**b–d**) Surrounding the organic matrix distribution side is sparry calcite. (**e**) Calcite sparry and micrite around petiole. (**f**) Around the void with no living organisms involved. (**g**,**h**) Sparkling Lake red algae population. (**i**) Recrystallization forms sparry calcite. Spr = sparry, Mic = micrite, Ms = microspar, Org = organic.

4.5. Calcareous Tufa Carbon-Oxygen Isotope Test

Three sets of black peat and seven sets of gray-brown calcareous tufa were selected for carbon-oxygen isotope testing (Table 4). The $\delta^{13}C_{V\text{-}PDB}$ values of the calcareous tufa samples range from −0.62‰ to 0.50‰ (average −0.17‰). The $\delta^{18}O_{V\text{-}SMOW}$ values are between 17.78‰ and 19.23‰ (average 18.15‰). The $\delta^{13}C_{V\text{-}PDB}$ values of calcareous tufa black peat layer range from −0.62‰ to −0.10‰ (average −0.36‰). The lowest $\delta^{13}C_{V\text{-}PDB}$ value of the 4-2-35 sample is −0.62‰, corresponding to the second layer of black peat in the calcareous tufa profile. The $\delta^{18}O_{V\text{-}SMOW}$ values of calcareous tufa in black peat layer vary slightly in the range of 17.78‰ to 18.09‰ (average 17.94‰). The correlation diagram shows that $\delta^{18}O_{V\text{-}SMOW}$ has a weak positive correlation or no obvious correlation with $\delta^{13}C_{V\text{-}PDB}$ (Figure 6). The correlation coefficient R^2 = 0.27. Carbon-oxygen isotopes were weakly correlated with K, Na, Ca, Mg, Al, CIA, and their ratios (K/Ca, K/Al, Na/Al, Mg/Al, and Ca/Al) (Table 3 and Figure 6).

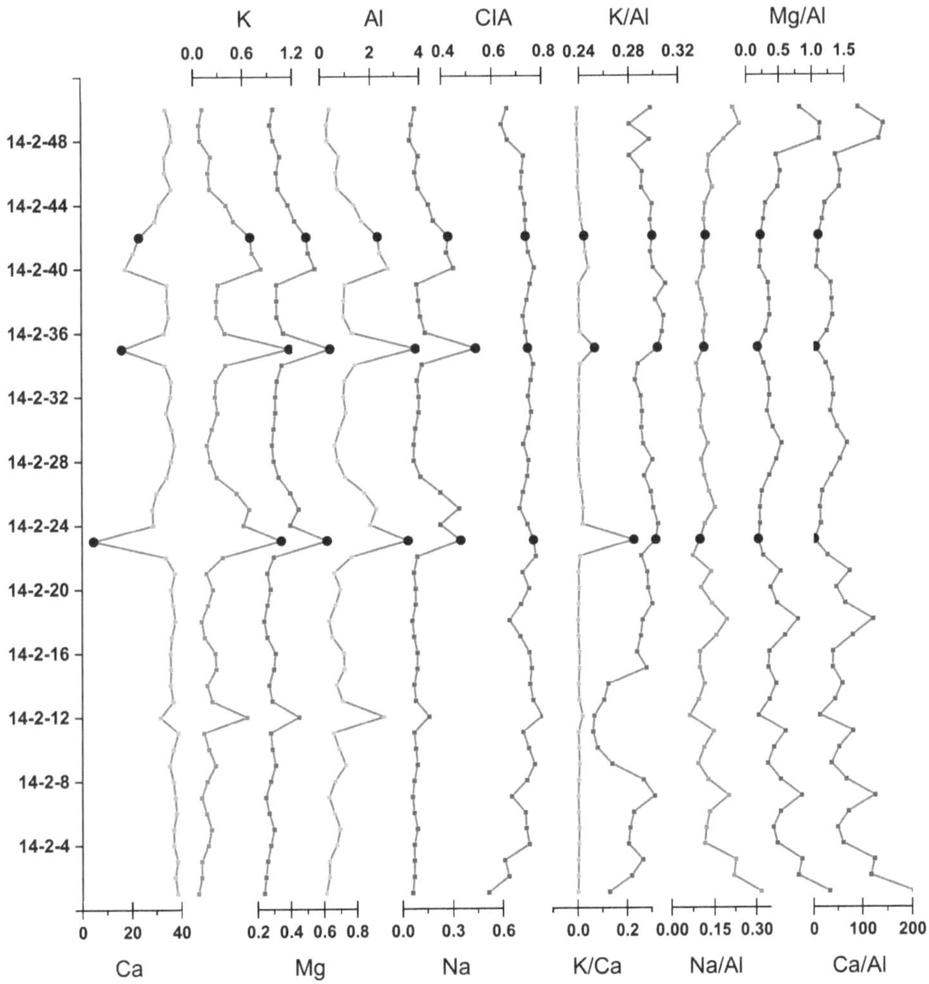

Figure 5. Element (Ca, K, Al, Na, Mg) composition content, element ratios (K/Ca, K/Al, Na/Al, Mg/Al, Ca/Al) and CIA value of calcareous tufa. The black dots indicate the location of the black peat layers.

Table 3. Pearson correlation matrix of main elements, element ratios, and isotopes ($\delta^{13}C_{V\text{-}PDB}$ and $\delta^{18}O_{V\text{-}SMOW}$) in the calcareous tufa profile.

Element	K	Na	Ca	Mg	Al	CIA	K/Ca	K/Al	Na/Al	Ca/Al	Mg/Al	$\delta^{18}O_{V\text{-}SMOW}$	$\delta^{13}C_{V\text{-}PDB}$
K	1												
Na	0.962 **	1											
Ca	−0.927 **	−0.905 **	1										
Mg	0.975 **	0.955 **	−0.954 **	1									
Al	0.997 **	0.949 **	−0.914 **	0.971 **	1								
CIA	0.467 **	0.284 *	−0.320 *	0.391 **	0.492 **	1							
K/Ca	0.727 **	0.670 **	−0.86 6**	0.727 **	0.716 **	0.248	1						
K/Al	0.289 *	0.327 *	−0.333 *	0.279 *	0.231	−0.024	0.211	1					
Na/Al	−0.418 **	−0.237	0.261	−0.342 *	−0.437 **	−0.982 **	−0.195	−0.056	1				
Ca/Al	−0.722 **	−0.621 **	0.573 **	−0.671 **	−0.728 **	−0.869 **	−0.384 **	−0.231	0.865 **	1			
Mg/Al	−0.677 **	0.569 **	0.501 **	−0.594 **	−0.687 **	−0.890	−0.342 *	−0.180	0.880 **	0.980 **	1		
$\delta^{18}O_{V\text{-}SMOW}$	−0.238	−0.561 **	0.125	−0.144	−0.251	−0.561 *	−0.172	0.278	0.515	0.291	0.445	1	
$\delta^{13}C_{V\text{-}PDB}$	−0.529	0.046	0.384	−0.425	−0.529	−0.481	−0.506	−0.138	0.488	0.461	0.566 *	0.529	1

Note: * Correlation is considerable at $p < 0.05$ (two-tailed). ** Correlation is considerable at $p < 0.01$ (two-tailed).

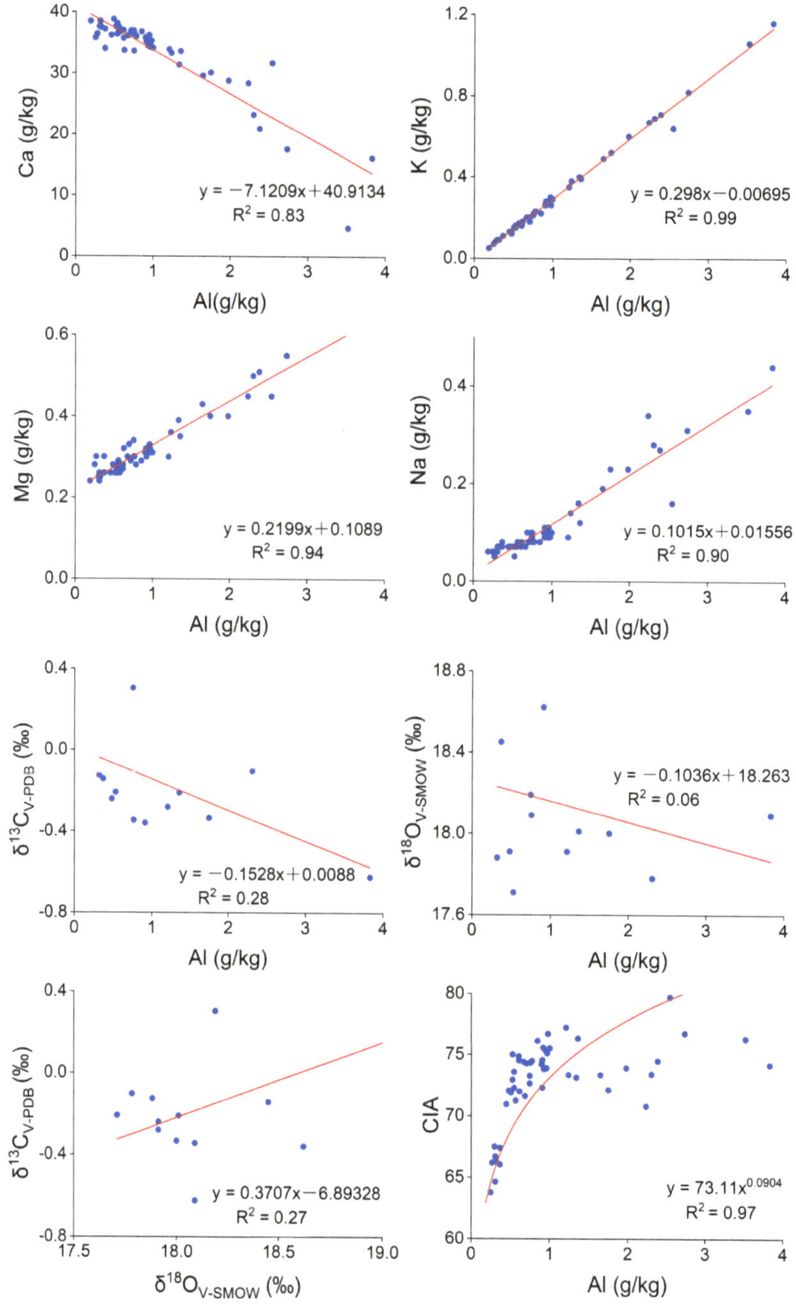

Figure 6. Correlation map of the main elements and carbon-oxygen isotopes of the calcareous tufa profile in the Sparkling Lake. It shows the correlation between elements (K, Na, Al, Mg, Ca, and CIA) and the plots of $\delta^{13}C_{V\text{-}PDB}$ and $\delta^{18}O_{V\text{-}SMOW}$.

Table 4. Carbon-oxygen isotope data of Sparkling Lake calcareous tufa profile.

Sample Number	$\delta^{13}C_{V\text{-}PDB}$ (‰)	$\delta^{18}O_{V\text{-}SMOW}$ (‰)
14-2-2	−0.13	17.88
14-2-5	0.30	18.19
14-2-11	−0.24	17.91
14-2-22	−0.28	17.91
14-2-23	the content of carbonate is low enough to measure C isotope	the content of carbonate is low enough to measure O isotope
14-2-26	−0.33	18.00
14-2-30	−0.35	18.09
14-2-34	−0.21	18.01
14-2-35	−0.62	18.09
14-2-37	−0.36	18.62
14-2-42	−0.10	17.78
14-2-50	0.50	19.23

5. Discussion

5.1. Response of Tufa Deposition to Paleohydrological Changes and Glaciation

Several authors have investigated the water source of the Jiuzhaigou Scenic Spot [38,47,48]. The Jiuzhaigou Scenic Spot is considered an independent karst hydrogeological structure unit where atmospheric precipitation (including atmospheric snowfall) is the only source of water supply. Therefore, isotope composition of the calcareous tufa in the study area reflects the changes in rainfall and temperature (Figure 2d,e). During 750 ± 30–300 ± 30 aB.P., the carbon isotope value of calcareous tufa showed a continuous decreasing trend. However, there is a weak positive correlation or no obvious correlation between carbon-oxygen isotopes in calcareous tufa (Table 3 and Figure 6), which is the same as the record of cave carbonate deposits in Southwest China [49,50]. Carbon-oxygen isotopes are consistent with precipitation and temperature changes recorded in the Guliya Ice Core [40] (before 750 ± 30 aB.P., it was a long cold and dry period, after which the precipitation and temperature showed a fluctuating upward trend) (Figure 2e). Carbon-oxygen isotopes records are consistent with the hydrological changes recorded in a compilation of 2000-year flood historical data in Sichuan [51]. Therefore, the analysis suggests that the information recorded by the calcareous tufa profile in the Sparkling Lake clearly reflects the heavy rainfall and the ablation of glaciers since the Holocene.

On the one hand, the isotope"dilution effect"caused by high rainfall conditions is extremely obvious in the Jiuzhaigou Valley [52–55]. That is, the light carbon isotope in the rain has played a "dilution role" in the heavy carbon isotope in calcareous tufa, making the carbon isotope in calcareous tufa lower. Before 750 ± 30 aB.P., the calcareous tufa carbon isotope values of Sparkling Lake were higher (average −0.09‰) (Table 4). It shows that the rainfall is relatively stable at this stage, and the water and soil are kept in good condition. At 750 ± 30 aB.P., the content of calcareous tufa carbonate dropped sharply, which was insufficient for the determination of carbon isotope. The content of calcite and Ca elements reached the lowest value (Tables 1 and 2). This period may result from flooding caused by heavy rains over a long period of time. After 750 ± 30 aB.P., the calcite and Ca contents decreased as a whole. However, the contents of other minerals and elements such as K, Na, Mg, and Al have increased to different degrees (Table 2 and Figure 5). The color of the calcareous tufa layers deepens, mainly to gray-black and brown (Figure 2b,c). The $\delta^{13}C_{V\text{-}PDB}$ value showed a decreasing trend. Around 300 ± 30 aB.P., the carbon isotope value reaches the lowest (−0.624‰) (Table 4). The occurrence of this phenomenon may be related to soil leaching. Heavy rainfall in the Jiuzhaigou area is speculated to have led to increased erosion; as a result, the water body that washed the soil carried large amounts of impurities (plants, organisms, minerals, and organic debris) (Figure 4). These impurity-carrying water bodies reduce the deposition rate of calcite and the increase in colored elements in its impurities, which makes the calcareous tufa layers at this stage darker than the calcareous tufa formed in the cold and dry period, showing a gray-black or brown color. After 300 ± 30 aB.P., The $\delta^{13}C_{V\text{-}PDB}$ value continues to rise (Table 4). This indicates that rainfall decreases at this stage. At 210 ± 30 aB.P., a thinner layer of black peat appears. It is presumed to be related to the influence of tectonic activity or intensified human activity. During this period, vegetation was destroyed, soil erosion intensified, and the environment deteriorated.

On the other hand, the calcareous tufa in Jiuzhaigou was formed by multi-period accumulation. Previous studies suggested that the early tufas were accumulation landforms formed by alpine karstification in multiple periglacial environments [56,57]. Early trufa in Jiuzhaigou formed in a periglacial environment 520,000~30,000 years ago. After 30,000 years, the temperature in the region showed a fluctuating increase. Jiuzhaigou gradually withdrew from the periglacial environment. However, it was still covered by seasonal ice and snow. Currently, the runoff replenishment source of Jiuzhaigou is mainly rainfall, supplemented by seasonal snow melt water, which has obvious periodicity. However, the change in snow and ice is more sensitive to temperature than rainfall change, so ice and snow change has a better response to temperature change. Melting ice and snow increases free CO_2 in runoff, so the dissolution capacity is enhanced [58]. Part of the meltwater from ice and snow forms surface runoff and flows to the low-altitude Shuzheng Valley (Figure 1d). As water seeps further underground to become pore water, crevice water migrates along the fault direction. The strong dissolution of snowmelt runoff causes soil erosion, and the introduction of impurities such as soil particles and plant organic debris darkens the calcareous tufa layers (Figure 4). At the same time, the impurities deposited on the snow surface are washed out of the snow layer during the snowmelt period and migrate with the surface runoff and groundwater, which can replenish and dilute the elements in groundwater and surface water. The meltwater impurities contain more K, Na, Mg, and Al, which dilute the Ca content, affecting the elemental composition in the calcareous tufa layers (Table 2 and Figure 5). Due to the apparent isotopic fractionation during snow and ice melting, the light isotopic composition of the snow melt water is abundant [59], and the magnitude of contribution to the Jiuzhaigou Valley affects the stable isotope in the calcareous tufa. In summary, it can be concluded that the change in the meltwater flow of ice and snow caused by temperature changes directly or indirectly affects the deposition of calcareous tufa layers with different colors.

5.2. *Response of Calcareous Tufa Deposition to Tectonic Activity*

Calcareous tufa is highly sensitive to climate change. Therefore, calcareous tufas are often used as reliable climate proxies in many geological settings. Conversely, the research on the response of calcareous tufa to tectonic activity and earthquakes is rather limited. Pazzaglia et al. [60] concluded that the presence of calcareous tufas in the Ellera basin successfully allowed the disclosure of a wide array of paleoenvironmental and tectonic information and could be potentially extended to similar depositional contexts. The findings of Martini et al. [61] indicate that calcareous tufa deposits could serve as a sensitive proxy for tectonics when such deposits are associated with coarse-grained clastic deposits.

Jiuzhaigou is located in the transition zone of the deep and large fault zone, where tectonic activity is intense and earthquakes occur frequently (Figure 1a,b). Tufa development is closely related to this strong tectonic activity. Hence, calcareous tufa can be used to determine the occurrence time of faults or seismic activities and reveal more tectonic activity characteristics and evolution [20,22,23,62,63]. Three sets of black peat layers with a thickness of 5–20 cm developed in the calcareous tufa profile, aged 750, 300, and 210 ± 30 aB.P. Based on historical records [64,65], during the formation of the upper two sets of black peat layers, a series of large earthquakes of magnitude above 6.0 occurred around the area where Jiuzhaigou is located (Figure 1b). The analysis suggests that the formation of the black peat layers in the calcareous tufa profile may be related to strong seismic activities in the study area. The strong fault activity led to the fragmentation of underground rocks and the development of underground fractures and broken zones in addition to inducing surface landslides and rock collapse and enhancing chemical weathering. The dissolution amount of the related minerals increased, affording an increase in K, Na, Mg, and Al contents in the black peat layers (Table 2 and Figure 5). Moreover, the strong tectonic activity triggered mountain collapse and soil erosion in the valleys, resulting in a large number of sources (broken rocks, vegetation, and particles of soil) mixed into the surface runoff. As a result, the hydrochemical environment was affected, and a large number of sources were accumulated during the deposition of calcareous tufa (Figure 4). Zheng et al. [66] compared the water chemistry and calcareous tufa deposition before and after the "8.8" earthquake in Jiuzhaigou, revealing the algal enrichment caused by enhanced soil erosion, which also caused the calcareous tufa deposits to become loose and porous. Moreover, the strong tectonic activity led to an abrupt change in the calcareous tufa deposition environment, the collapse of calcareous tufa dams, and the diversion of water bodies. A large amount of surface water leakage, aquatic algae death, the growth of airborne cyanobacteria, and weathering continues to increase, surface calcareous tufa will then disintegrate and desertification occurs, resulting in the loss of water and black calcareous tufa. The effects of strong earthquakes can persist for 10~20 years or more, after which calcareous tufa deposits enter the state of "normal deposits" (no longer affected by strong earthquakes to deposit) with a rhythmic structure. Dang et al. [67] and Qiao et al. [68] clarified that the "8.8" earthquake increased the degree of soil erosion and changed the water regulation effect of vegetation in the watershed. Soil erosion in the watershed has increased, a substantial number of sources such as sediment and wood have entered the body of water, and aquatic plants such as fir-leaf algae have grown in large numbers, resulting in turbid lakes and silted water bodies. Finally, the hydrochemical environment of calcareous tufa deposition is affected. Level changes as well as disappearances of groundwater in tectonically active regions due to earthquakes have been reported [69,70], suggesting that hydrological changes and depositional breaks in calcareous tufa in this area may also be caused by earthquakes. Qiu et al. [71] also proved the influence of the tectonic activity and earthquake on the Dawan travertine deposition in Huanglong (100 km away from the Jiuzhaigou Scenic Spot).

In summary, the formation of black peat layers in the calcareous tufa profile is the result of many factors caused by strong tectonic activity. Due to the lack of relevant studies on similar calcareous tufa profiles, a systematic comparative analysis cannot be conducted. Therefore, further studies are required to verify the above conclusions.

6. Conclusions

From the analysis of sedimentology, petrography and elemental geochemical characteristics of the differently colored layers of calcareous tufa in the Jiuzhaigou Scenic spot and ^{14}C dating results, combined with regional tectonic characteristics and geological environment changes, the following conclusions are drawn.

(1) The radiocarbon age of the calcareous tufa profile in the Sparkling Lake is 2340 ± 30–210 ± 30 aB.P. Therefore, the calcareous tufa of the Sparkling Lake belongs to the Late Holocene with a warm and humid climate. The sedimentary sequence of the calcareous tufa profile is the normal bottom-up sequence from old to new with good sedimentary continuity.

(2) Based on the dating results and geochemical analysis, it is believed that the deposition process of the calcareous tufa profile was controlled by paleoclimate and glaciation. The layers' color in the upper part of the tufa profile is significantly darker, suggesting a weak positive correlation of carbon-oxygen isotopes and the trend of change at or above centennial scale. The paleoclimate and glaciation changes reflected by these features are consistent with climate changes recorded in the ice cores on the Tibetan Plateau, carbonate cave deposits in Southwest China, and literature records. Therefore, this indicates that the information recorded in the calcareous tufa profile of the Sparkling Lake clearly reflects the heavy rainfall and the ablation of glaciers since the Holocene.

(3) The anomalous black peat layers in the calcareous tufa profile are related to strong regional tectonic activity (strong seismic activity). The ages of the three sets of black

peat layers in the upper part of the Sparkling Lake calcareous tufa profile are 750, 300, and 210 ± 30 aB.P., respectively, measured by ^{14}C dating. Among them, the ages of 300 and 210 ± 30 aB.P. are consistent with the time when a series of earthquakes with a magnitude above 6.0 occurred around Jiuzhaigou (1630~1900). The analysis shows that under the synergism effect of various factors caused by strong tectonic activities, the dissolution amount of related minerals is increased, and large amounts of vegetation, sediment, and other sources are mixed into the calcareous tufa for accumulation, affecting the hydrochemical environment of calcareous tufa deposition and forming a large amount of plant debris accumulation. In addition, strong tectonic activity led to the collapse of the calcareous tufa dam; moreover, water bodies were diverted, the calcareous tufa lost water and oxidized, the vegetation grew, and finally, black peat layers were formed.

Author Contributions: Conceptualization, C.L., X.Z. and F.W.; software, C.L., Y.J. and K.H.; validation, C.L., X.Z. and F.W.; formal analysis, C.L. and H.Z. (Hongmin Zhang); investigation, C.L., X.Z., Y.J., H.Z. (Hongmin Zhang) and F.W.; resources, X.Z., F.W. and Q.L.; data curation, C.L. and H.Z. (Hongmin Zhang); writing—original draft preparation, C.L. and H.Z. (Hongmin Zhang); writing—review and editing, C.L., X.Z., H.Z. (Heyan Zhu) and K.H.; visualization, C.L., Y.J. and H.Z. (Heyan Zhu); supervision, X.Z., F.W. and Q.L.; project administration, X.Z. and F.W.; funding acquisition, X.Z. and F.W. All authors have read and agreed to the published version of the manuscript.

Funding: This work was supported by the National Natural Science Foundation of China (grant nos. 41672206, 41973053, 42042030, U21A2016); the State Key Laboratory of Earthquake Dynamics, Institute of Geology, China Earthquake Administration (grant no. LED2016B07); and the Sichuan Province Overseas High-end Talent Introduction Project (grant no. 22RCYJ0060).

Data Availability Statement: Not applicable.

Acknowledgments: We greatly appreciate the help of the Jiuzhaigou Administration Bureau PetroChina for the fieldwork. We would also like to express our sincere gratitude to Qunwei Dai and Lanjie Hou from the Southwest University of Science and Technology for their helpful discussion and guidance on the early conception of this manuscript. Thanks are also owed to Lujia Yu for his help with the fieldwork and his comments in the process. Many thanks to the academic editors and the three peer reviewers for their constructive suggestions to improve the quality of the article.

Conflicts of Interest: The authors declare no conflict of interest.

References

1. Andrew, J.E.; Pedley, M.; Dennis, P.F. Palaeoenvironmental records in Holocene Spanish tufas; a stable isotope approach in search of reliable climatic archives. *Sedimentology* **2000**, *47*, 961–978. [CrossRef]
2. Andrews, J.E. Palaeoclimatic records from stable isotopes in riverine tufas: Synthesis and review. *Earth-Sci. Rev.* **2006**, *75*, 85–104. [CrossRef]
3. Viles, H.A.; Taylor, M.P.; Nicoll, K.; Neumann, S.; Nash, D.J.; Bullard, J.; North, C.P. Facies evidence of hydroclimatic regime shifts in tufa depositional sequences from the arid Naukluft Mountains, Namibia. *Sediment. Geol.* **2007**, *195*, 39–53. [CrossRef]
4. Bertini, A.; Minissale, A.; Ricci, M. Use of Quaternary travertines of central-southern Italy as archives of paleoclimate, paleohydrology and neotectonics. *Alp. Mediterr. Quat.* **2008**, *21*, 99–112.
5. Capezzuoli, E.; Gandin, A.; Sandrelli, F.; Pedley, H.M.; Rogerson, M. Calcareous tufa as indicators of climatic variability; a case study from southern Tuscany (Italy). *Geol. Soc. Spec. Publ.* **2010**, *336*, 263–281. [CrossRef]
6. Cremaschi, M.; Zerboni, A.; Spoetl, C.; Felletti, F. The calcareous tufa in the Tadrart Acacus Mt. (SW Fezzan, Libya); an early Holocene palaeoclimate archive in the central Sahara. *Palaeogeogr. Palaeoclimatol. Palaeoecol.* **2010**, *287*, 81–94. [CrossRef]
7. Dabkowski, J. High potential of calcareous tufas for integrative multidisciplinary studies and prospects for archaeology in Europe. *J. Archaeol. Sci.* **2014**, *52*, 72–83. [CrossRef]
8. Pentecost, A. The Quaternary travertine deposits of Europe and Asia Minor. *Quat. Sci. Rev.* **1995**, *14*, 1005–1028. [CrossRef]
9. Liu, Z.; Li, H.; You, C.; Wan, N.; Sun, H. Thickness and stable isotopic characteristics of modern seasonal climate-controlled sub-annual travertine laminas in a travertine-depositing stream at Baishuitai, SW China; implications for paleoclimate reconstruction. *Environ. Geol.* **2006**, *51*, 257–265. [CrossRef]
10. Luigi, D.F.; Claudio, F.; Andrea, B.; Erlisiana, A.; Mauro, B.; Michele, S.; Paola, T. Plateau versus fissure ridge travertines from Quaternary geothermal springs of Italy and Turkey: Interactions and feedbacks between fluid discharge, paleoclimate, and tectonics. *Earth Sci. Rev.* **2013**, *123*, 35–52. [CrossRef]
11. Gradzinski, M.; Wroblewski, W.; Dulinski, M.; Hercman, H.; Capezzuoli, E.; Capezzuoli, E. Earthquake-affected development of a travertine ridge. *Sedimentology* **2014**, *61*, 238–263. [CrossRef]

12. Kokh, S.N.; Sokol, E.V.; Deev, E.V.; Ryapolova, Y.M.; Rusanov, G.G.; Tomilenko, A.A.; Bul'Bak, T.A. Post-late glacial calcareous tufas from the Kurai fault zone (southeastern Gorny Altai, Russia). *Sediment. Geol.* **2017**, *355*, 1–19. [CrossRef]
13. Priestley, S.C.; Karlstrom, K.E.; Love, A.J.; Crossey, L.J.; Polyak, V.J.; Asmerom, Y.; Meredith, K.T.; Crow, R.; Keppel, M.N.; Habermehl, M.A. Uranium series dating of Great Artesian Basin travertine deposits: Implications for palaeohydrogeology and palaeoclimate. *Palaeogeogr. Palaeoclimatol. Palaeoecol.* **2017**, *490*, 163–177. [CrossRef]
14. Meyer, M.C.; Aldenderfer, M.S.; Wang, Z.; Hoffmann, D.L.; Dahl, J.A.; Degering, D.; Haas, W.R.; Schluetz, F. Permanent human occupation of the central Tibetan Plateau in the early Holocene. *Science* **2017**, *357*, 64–67. [CrossRef] [PubMed]
15. Andrea, S.; Paola, M.; Michele, S.; Paola, T.; Erlisiana, A.; Andrea, B.; Stefania, F.; Massimo, R.; Luca, T. The uplift of the Adriatic flank of the Apennines since the Middle Pleistocene: New insights from the Tronto River basin and the Acquasanta Terme Travertine (central Italy). *Geomorphology* **2020**, *352*, 106990. [CrossRef]
16. Wang, F.; Zhao, X.; Dong, F.; Enrico, C.; Alexander, I.M.; Du, J. Classification of Alpine-type travertine in Jiuzhaigou valley on the eastern margin of the Qinghai-Tibet Plateau. *Carsol. Sin.* **2021**, *40*, 112–124. [CrossRef]
17. Tye, G.J.; Sherriff, J.; Candy, I.; Coxon, P.; Palmer, A.; McClymont, E.L.; Schreve, D.C. The $\delta^{18}O$ stratigraphy of the Hoxnian lacustrine sequence at Marks Tey, Essex, UK: Implications for the climatic structure of MIS 11 in Britain. *J. Quat. Sci.* **2016**, *31*, 75–92. [CrossRef]
18. Matsuoka, J.; Kano, A.; Oba, T.; Watanabe, T.; Sakai, S.; Seto, K. Seasonal variation of stable isotopic compositions recorded in a laminated tufa, SW Japan. *Earth Planet. Sci. Lett.* **2001**, *192*, 31–44. [CrossRef]
19. Zidi, M.K.; Ben Ahmed, W.; Henchiri, M. Late Pleistocene Boulaaba travertine and calcareous tufa, Kasserine, Central Tunisia: Implications for North African spring-fed Ca-carbonates. *Quatern. Int.* **2022**, *637*, 57. [CrossRef]
20. Wang, Z.; Yin, J.; Yuan, D. Possibilities and problems associated with travertines and tufas in Quaternary studies: A case of the Tibetan Plateau. *Chin. Sci. Bull.* **2018**, *63*, 1012–1023. [CrossRef]
21. Geological Survey of China University of Geosciences. *The Instruction of Quaternary Geological and Geomorphologic Map of Qinghai-Tibetan Plateau and Its Adjacent Area*; Geological Survey of China University of Geosciences: Wuhan, China, 2010.
22. Niu, X.; Zheng, M.; Liu, X.; Qi, L. Sedimentary property and the geological significance of travertines in Qinghai-Tibetan Plateau. *Sci. Technol. Rev.* **2017**, *35*, 59–64.
23. Zhou, L. Characteristics of the Typical Hot Springs in the Central Tibet. Master's Thesis, China University of Geosciences, Beijing, China, 2012.
24. Zhou, X. On Periglacial Karst in the Minshan Mountains. *Acta Geol. Sichuan* **2013**, *33*, 479–487.
25. Zentmyer, R.; Myrow, P.M.; Newell, D.L. Travertine deposits from along the South Tibetan Fault System near Nyalam, Tibet. *Geol. Mag.* **2008**, *145*, 753–765. [CrossRef]
26. Liu, Z.; Sun, H.; Baoying, L.; Xiangling, L.; Wenbing, Y.; Cheng, Z. Wet-dry seasonal variations of hydrochemistry and carbonate precipitation rates in a travertine-depositing canal at Baishuitai, Yunnan, SW China; implications for the formation of biannual laminae in travertine and for climatic reconstruction. *Chem. Geol.* **2010**, *273*, 258–266. [CrossRef]
27. Gao, J.; Zhou, X.; Fang, B.; Li, T.; Tang, L.; Catto, N. U-series dating of the travertine depositing near the Rongma Hot Springs in northern Tibet, China, and its paleoclimatic implication. *Quatern. Int.* **2013**, *298*, 98–106. [CrossRef]
28. Qin, J.; Han, P.; Che, X.; Yuan, G.; Fang, B.; Wang, G. Resuming the Holocene paleoclimate using $\delta^{18}O$ and trace elements of travertine in Rongma area, Tibet. *Earth Sci. Front.* **2014**, *21*, 312–322. [CrossRef]
29. Wang, Z.; Meyer, M.C.; Hoffmann, D.L. Sedimentology, petrography and early diagenesis of a travertine-colluvium succession from Chusang (southern Tibet). *Sediment. Geol.* **2016**, *342*, 218–236. [CrossRef]
30. Wang, Z.; Meyer, M.C.; Gliganic, L.A.; Hoffmann, D.L.; May, J. Timing of fluvial terrace formation and concomitant travertine deposition in the upper Sutlej River (Tirthapuri, southwestern Tibet) and paleoclimatic implications. *Quat. Sci. Rev.* **2017**, *169*, 357–377. [CrossRef]
31. Mischke, S.; Chengjun, Z. A laminated tufa carbonate from the mid Holocene of the Qilian Mountains and its potential for palaeoclimate inferences. *Episodes* **2008**, *31*, 401–407. [CrossRef]
32. Guo, Y.; Ge, Y.; Cui, P.; Chen, X.; Mao, P.; Liu, T.; Zhou, L. Early and mid-Holocene hydroclimate change recorded in tufa deposits in the Jiuzhaigou gully, eastern Tibetan Plateau. *Catena* **2021**, *196*, 104834. [CrossRef]
33. Lei, G.L.; Zhang, H.C.; Li, Z.Z.; Adam, M.H.; Zhu, Y.; Jiang, X.Y.; Chen, X.L.; Chang, F.Q.; Li, H.Y. Geochemical features and significance of shoreline tufa from a closed-basin lake ngangla ring tso in the western tibetan plateau. *Quat. Sci.* **2013**, *33*, 839–847. [CrossRef]
34. Florsheim, J.L.; Ustin, S.L.; Tang, Y.; Di, B.; Huang, C.; Qiao, X.; Peng, H.; Zhang, M.; Cai, Y. Basin-scale and travertine dam-scale controls on fluvial travertine, Jiuzhaigou, southwestern China. *Geomorphology* **2013**, *180–181*, 267–280. [CrossRef]
35. Lugli, S.; Tang, Y.; Reghizzi, M.; Qiao, X.; Schreiber, B.C.; Deng, G. Seasonal Pattern In the High-Elevation Fluvial Travertine From the Jiuzhaigou National Nature Reserve, Sichuan, Southwestern China. *J. Sediment. Res.* **2017**, *87*, 253–271. [CrossRef]
36. Bai, D.; Unsworth, M.J.; Meju, M.A.; Ma, X.; Teng, J.; Kong, X.; Sun, Y.; Sun, J.; Wang, L.; Jiang, C.; et al. Crustal deformation of the eastern Tibetan Plateau revealed by magnetotelluric imaging. *Nat. Geosci.* **2010**, *3*, 358–362. [CrossRef]
37. Yi, G.; Long, F.; Liang, M.; Zhang, H.; Zhao, M.; Ye, Y.; Zhang, Z.; Qi, Y.; Wang, S.; Gong, Y.; et al. Focal mechanism solutions and seismogenic structure of the 8 August 2017 M7.0 Jiuzhaigou earthquake and its aftershocks, northern Sichuan. *Chin. J. Geophys.* **2017**, *60*, 4083–4097.

38. Gan, J.; Liu, M.; Huang, R.; Fan, C.; Li, Q.; Wang, C. A study of hydrological cycle in the Jiuzhaigou core scenic area. *Hydrogeol. Eng. Geol.* **2010**, *37*, e39. [CrossRef]
39. Wang, D.; Zhou, Y.; Pei, X.; Ouyang, C.; Du, J.; Gianvito, S. Dam-break dynamics at Huohua Lake following the 2017 M_w 6.5 Jiuzhaigou earthquake in Sichuan, China. *Eng. Geol.* **2021**, *289*, 106145. [CrossRef]
40. Yao, T.; Qing, D.; Tian, L.; Jiao, K.; Yang, Z.; Xie, C.; Thompson, L.G. Changes of temperature and precipitation over the Qinghai-Tibet Plateau during the past 2ka—Guliya Ice Core record. *Sci. Sin. Terrae* **1996**, *26*, 348–353.
41. Nesbitt, H.W.; Young, G.M. Prediction of some weathering trends of plutonic and volcanic rocks based on thermodynamic and kinetic considerations. *Geochim. Cosmochim. Acta* **1984**, *48*, 1523–1534. [CrossRef]
42. Nesbitt, H.W.; Young, G.M. Formation and diagenesis of weathering profiles. *J. Geol.* **1989**, *97*, 129–147. [CrossRef]
43. Kautz, C.Q.; Martin, C.E. Chemical and physical weathering in New Zealand's Southern Alps monitored by bedload sediment major element composition. *Appl. Geochem.* **2007**, *22*, 1715–1735. [CrossRef]
44. McLennan, S.M. Weathering and global denudation. *J. Geol.* **1993**, *101*, 295–303. [CrossRef]
45. Zhang, C.; Yang, B. *Fundamentals of Quantitative Geography*, 2nd ed.; Higher Education Press: Beijing, China, 2007; p. 202.
46. Xu, J. *Mathematical Methods in Contemporary Geography*, 3rd ed.; Higher Education Press: Beijing, China, 2017; p. 432.
47. Yin, G.; Fan, X.; Guo, J.; Yang, J. Isotope Tracer on Water Cycly System in Jiuzaigou, Sichuan. *Acta Geogr. Sin.* **2000**, *55*, 487–494. [CrossRef]
48. Yin, G.; Ni, S.; Zhang, Q. Deuterium excess parameter and geohydrology significance-Taking the geohydrology researches in Jiuzaigou and Yele, Sichuan for example. *J. Chengdu Univ. Technol.* **2001**, *28*, 251–254.
49. Dykoski, C.A.; Edwards, R.L.; Cheng, H.; Yuan, D.; Cai, Y.; Zhang, M.; Lin, Y.; Qing, J.; An, Z.; Revenaugh, J. A high-resolution, absolute-dated Holocene and deglacial Asian monsoon record from Dongge Cave, China. *Earth Planet. Sci. Lett.* **2005**, *233*, 71–86. [CrossRef]
50. Hu, C.; Henderson, G.M.; Huang, J.; Xie, S.; Sun, Y.; Johnson, K.R. Quantification of Holocene Asian monsoon rainfall from spatially separated cave records. *Earth Planet. Sci. Lett.* **2008**, *266*, 221–232. [CrossRef]
51. Changjiang Water Resources Commission of the Ministry of Water Resources. *Compilation of 2000-Year Flood Historical Data in Sichuan*; Cultural Relics Press: Beijing, China, 1993; p. 594.
52. Liu, Z.; Li, Q.; Sun, H.; Wang, J. Seasonal, diurnal and storm-scale hydrochemical variations of typical epikarst springs in subtropical karst areas of SW China: Soil CO_2 and dilution effects. *J. Hydrol.* **2007**, *337*, 207–223. [CrossRef]
53. Sun, H.; Liu, Z. Wet–dry seasonal and spatial variations in the $\delta^{13}C$ and $\delta^{18}O$ values of the modern endogenic travertine at Baishuitai, Yunnan, SW China and their paleoclimatic and paleoenvironmental implications. *Geochim. Cosmochim. Acta* **2010**, *74*, 1016–1029. [CrossRef]
54. Wang, H.; Yan, H.; Liu, Z. Contrasts in variations of the carbon and oxygen isotopic composition of travertines formed in pools and a ramp stream at Huanglong Ravine, China: Implications for paleoclimatic interpretations. *Geochim. Cosmochim. Acta* **2014**, *125*, 34–48. [CrossRef]
55. Yan, H.; Liu, Z.; Sun, H. Effect of in-stream physicochemical processes on the seasonal variations in $\delta^{13}C$ and $\delta^{18}O$ values in laminated travertine deposits in a mountain stream channel. *Geochim. Cosmochim. Acta* **2017**, *202*, 179–189. [CrossRef]
56. Li, H.; Zhou, X.; Li, G. Thermoluminescence age of sedimentary calcium carbonate in late geological history. *Chin. Sci. Bull.* **1988**, *33*, 775–778.
57. Dong, F. China's travertine-the most natural asset in karst landscape worthy of world research and protection. *Carsol. Sin.* **2021**, *40*, 1–3.
58. Ocampo, J.; Klinger, J. Adsorption of N_2 and CO_2 on ice. *J. Colloid Interface Sci.* **1982**, *86*, 377–383. [CrossRef]
59. Qian, Z.; Yuan, Z.; Xu, J. Climatic characteristics of the last glacial period of Quaternary and Holocene climate recorded by oxygen and carbon isotopes in YaO2 hole, Yanhai, Yimeng, Inner Mongolia. *Geol. Chem. Miner.* **2002**, *24*, 96–100. [CrossRef]
60. Pazzaglia, F.; Barchi, M.R.; Buratti, N.; Cherin, M.; Pandolfi, L.; Ricci, M. Pleistocene calcareous tufa from the Ellera basin (Umbria, central Italy) as a key for an integrated paleoenvironmental and tectonic reconstruction. *Quatern. Int.* **2013**, *292*, 59–70. [CrossRef]
61. Martini, I.; Capezzuoli, E. Interdigitated fluvial clastic deposits and calcareous tufa testifying an uplift of the catchment area: An example from the Pianizzoli area (southern Tuscany, Italy). *Sediment. Geol.* **2014**, *299*, 60–73. [CrossRef]
62. Peng, G.; Jiao, W. Sedimentary age of travertine sinter in Jianchuan and new activity of Jianchuan fault. *J. Seismol. Res.* **1990**, *13*, 166–172.
63. Yuan, D.; Liu, Z.; Lin, Y.; Shen, J.; He, S.; Xu, S. *Karst Dynamical System in China*; Geology Press: Beijing, China, 2002; p. 275.
64. China Earthquake Administration. *Catalogue of Historical Strong Earthquakes in China*; Seismological Press: Beijing, China, 1995.
65. Sun, C. *Sichuan Earthquake Full Record*; Sichuan People's Publishing House: Chengdu, China, 2010.
66. Zheng, X.; Tang, Y.; Du, J.; Lugli, S.; Xiao, Y.; Yang, Q.; Song, H.; Qiao, X. Enhanced soil erosion threatens fluvial tufa landscapes after an Ms 7.0 earthquake in the Jiuzhaigou World Heritage Site, southwestern China. *Sci. Total Environ.* **2022**, *848*, 157632. [CrossRef]
67. Dang, Z.; Ren, J.; An, C.; Dai, Q.; Dong, F.; Deng, Y.; Yang, Q.; Zhuo, M. Effect of Ms 7.0 earthquake on travertine landscapes and hydrochmistry of Jiuzhaigou core scenic spots. *Carsol. Sin.* **2019**, *38*, 186–192.
68. Qiao, X.; Xiao, Y.; Du, J.; Tang, Y.; Xiao, W.; Zheng, X.; Zhang, M. Tufa Landscapes in the Key Scenic Areas of the Jiuzhaigou Natural World Heritage Site: A ritical Review and Future Research Needs. *Earth Environ.* **2022**, *50*, 202–218.

69. Hosono, T.; Yamada, C.; Shibata, T.; Tawara, Y.; Wang, C.Y.; Manga, M.; Rahman, A.T.M.S.; Shimada, J. Coseismic Groundwater Drawdown Along Crustal Ruptures During the 2016 Mw 7.0 Kumamoto Earthquake. *Water Resour. Res.* **2019**, *55*, 5891–5903. [CrossRef]
70. Ichiyanagi, K.; Imatsu, M.; Ide, K.; Shimada, J. Effects on fluvial discharges of the 2016 Kumamoto earthquakes, Japan. *J. Hydrol.* **2020**, *583*, 124600. [CrossRef]
71. Qiu, S.; Wang, F.; Dong, F.; Tian, F.; Zhao, X.; Dai, Q.; Li, Q.; Zhu, Y.; Wang, Y. Sedimentary evolution of the Dawan travertines and their geological environmental significance, Huanglong, China. *Depos. Rec.* **2022**, *8*, 251–265. [CrossRef]

minerals

MDPI

Article

Geochemical Characterization of Laminated Crystalline Crust Travertines Formed by Ca^{2+}-Deficient Hot Springs at Sobcha (China)

Zhipeng Lu [1,2], Huaguo Wen [1,2,3,4,*], Lianchao Luo [5], Liang Li [6] and Ying Nie [7]

1. State Key Laboratory of Oil and Gas Reservoir Geology and Exploitation, Chengdu University of Technology, Chengdu 610059, China
2. Institute of Sedimentary Geology, Chengdu University of Technology, Chengdu 610059, China
3. CNPC Key Laboratory of Carbonate Reservoir, Chengdu University of Technology, Chengdu 610059, China
4. Key Laboratory of Deep-Time Geographical Environment Reconstruction and Application, Ministry of Natural Resources, Chengdu 610059, China
5. Department of Earth Sciences, University of Florence, 50121 Firenze, Italy
6. College of Earth Sciences, Chengdu University of Technology, Chengdu 610059, China
7. School of Geoscience and Technology, Southwest Petroleum University, Chengdu 610500, China

* Correspondence: wenhuaguo08@cdut.edu.cn

Abstract: Travertines formed of crystalline crust have been widely reported, but there has not been focus on their geochemical characteristics. We therefore carefully conducted a series of geochemical investigations and U-Th dating on a travertine mound mainly composed of crystalline crust from Sobcha (southwest China) to determine their geochemical features and geological implications. The Sobcha travertines dominantly consist of granular crystals and fan crystals and show $\delta^{13}C$ from 3.4‰ to 4.9‰ V-PDB, $\delta^{18}O$ from −26.7‰ to −23.7‰ V-PDB, and $^{87}Sr/^{86}Sr$ from 0.712458 to 0.712951. When normalized to PASS, the Sobcha travertines exhibit MREE enrichment relative to HREE and LREE, HREE enrichment relative to LREE, and positive Eu anomalies. The $\delta^{13}C$ signatures and mother CO_2 evaluation of the Sobcha travertines show that the Sobcha travertines were thermogene travertines largely receiving mother CO_2 from (upper) mantle (i.e., magmatic CO_2) or a mixture of soil-derived CO_2 and CO_2 related to carbonate decarbonation. The $^{87}Sr/^{86}Sr$ of the Sobcha travertines is out of the $^{87}Sr/^{86}Sr$ ranges of local deposits exposed at Sobcha and surrounding areas but is well matched with the mean $^{87}Sr/^{86}Sr$ of Nadi Kangri volcanic rocks which cropped out to the northeast of the studied travertines (over 20 km away). This might indicate the important role of the Nadi Kangri volcanic rocks in suppling Sr to the studied travertines, but more studies are required. The LREE depletion compared to MREE and HREE in the Sobcha travertines was interpreted to be caused by the difference in geochemical mobility between LREEs and HREEs during water–rock interaction at depth, while the MREE enrichment compared to HREE was considered to be most likely inherited from reservoir/aquifer rocks. The positive Eu anomalies of the Sobcha travertines may result from very high reservoir temperatures and/or preferential dissolution of Eu-rich minerals/rocks (especially plagioclase). The Sobcha travertine mounds displays no or very slight vertical variations in $\delta^{13}C$, $^{87}Sr/^{86}Sr$, and REE patterns, indicating the compositional stability of mother CO_2 and paleo-fluids. However, a significant vertical increase in $\delta^{18}O$ was observed and was explained as the result of gradual water temperature decrease related to climate cooling, self-closure of the vents, or mound vertical growth. The findings in this study might help us better understand the deposition of crystalline crust in Ca^{2+}-deficient hot spring systems.

Keywords: travertine mound; crystalline crust; C-O-Sr isotopes; rare-earth elements; Tibet

Citation: Lu, Z.; Wen, H.; Luo, L.; Li, L.; Nie, Y. Geochemical Characterization of Laminated Crystalline Crust Travertines Formed by Ca^{2+}-Deficient Hot Springs at Sobcha (China). *Minerals* **2023**, *13*, 220. https://doi.org/10.3390/min13020220

Academic Editors: Francesca Giustini, Mauro Brilli and Giovanni Ruggieri

Received: 18 November 2022
Revised: 29 January 2023
Accepted: 30 January 2023
Published: 2 February 2023

1. Introduction

Travertines (i.e., thermogene/hypogean travertines) are typical terrestrial carbonate rocks/deposits associated with warm-hot springs on the earth's surface [1]. Travertine systems are often characterized by diverse morphologies, rapid changes of lithofacies and environments, rapid precipitation rates, and the deficiency of macrophytes and faunas [1–4]. More attention has recently been paid to travertines largely owing to their great potentials in enhancing our understanding of tectonic activity [5–15], paleo-climate [16–19], paleo-environment [16,17,20,21], thermal fluid circulation in geothermal systems [8,22], $CaCO_3$ polymorphism and crystallization [23–25], microbe-mediated carbonate precipitation [26–40], simulation of engineered geological storage and leakage of CO_2 [41,42], and pre-salt carbonate hydrocarbon reservoirs [43–45].

The lithofacies composition of travertines is quite complex [46,47], but one attractive lithofacies in many travertine systems is abiotic crystalline crust, which is formed of abundant well-packed bright calcite and/or aragonite crystals and often has a low porosity. Similar lithofacies in carbonate deposits from other environments (e.g., marine, lake) is commonly treated as diagenetic products. In contrast, abiotic crystalline crust in travertine systems is commonly considered to be primary deposits mainly formed by rapid passive CO_2 degassing [47–53]. However, calcite/aragonite crystals constituting abiotic crystalline crust shows various morphologies and sizes, such as dendritic crystals over 1 cm high, and raft-like crystals more than 5 cm long [49–52,54]. Therefore, abiotic crystalline crust is often further subdivided according to the morphology and size of the main fabrics. Gandin and Capezzuoli [47], for example, split abiotic crystalline crust in tufa and travertine systems up into five subtypes: feather-like/dendritic crystals, fan/ray crystals, banded palisade crystals, and foam rock and calcite rafts, which can be formed under numerous deposition conditions.

Many works on the sedimentological, petrological, and mineralogical characteristics of abiotic crystalline crust have been carried out over the past few decades [47–52,55–57]. Geochemical characteristics of travertines have also been investigated in some studies [58–61]. However, most of them only focused on the stable carbon and oxygen isotope compositions of travertines [52,61–67]. The $^{87}Sr/^{86}Sr$ and rare-earth elements signatures of travertines were poorly examined and interpreted in past studies. This study thus carefully characterized the geochemical compositions ($\delta^{13}C$, $\delta^{18}O$, $^{87}Sr/^{86}Sr$, and rare-earth elements) of laminated crystalline crust travertines at Sobcha, Tibet, southwestern China. Based on the geochemical results, this study attempted to determine the genesis, (paleo-)fluid source(s), and fluid evolution of the Sobcha travertine system. Some geochronological data were also obtained in this study. Considering that the Sobcha travertine system was likely deposited by hot springs deficient in Ca^{2+} (Ca^{2+} concentrations of active hot springs at Sobcha are 0.17 mM) [68], the findings obtained in this study might aid in the geochemical interpretation of crystalline crust travertines, especially those formed in Ca^{2+}-deficient hot spring systems.

2. Geological Setting

Sobcha is located in central Tibet, southwestern China (32°31′36.7″ N, 89°56′39″ E; altitude: ca. 4735 m) (Figure 1) and lies to the north of Qixiang Co Lake (ca. 2 km). The sedimentary sequences cropping out at Sobcha and surrounding areas mainly include (1) Upper Triassic–Lower Jurassic Sobcha Formation (also known as Xiaochaka Formation and Suobucha Formation) marine deposits mainly consisting of limestone, (2) Lower Jurassic Quse Formation and Middle Jurassic Sewa Formation marine deposits mainly composed of clastic deposits (e.g., shale, sandstone, and claystone), and (3) Neogene and Quaternary terrestrial clastic deposits [69–71]. A series of E–W faults developed at Sobcha and surrounding areas. The studied travertines were formed on the Upper Triassic–Lower Jurassic Sobcha Formation marine deposits and are situated very close to two faults (Figure 1).

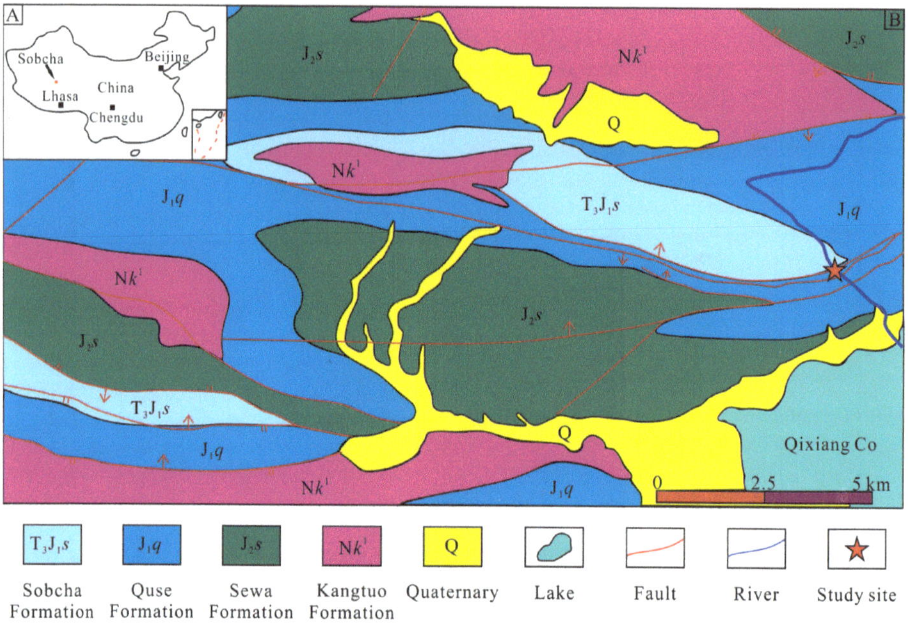

Figure 1. (**A**) The location of Sobcha in the Tibetan Plateau, southwestern China. (**B**) Simplified geological map of the study area (modified from Fu et al. [70]).

The studied Sobcha travertines are fossil travertine deposits, but modern hot springs and associated deposits, though very limited, are still present near the studied travertines. Liao [68] called this place Quse or Sobcha Hot Fountain. Specifically, this place is a hot spring area composed of 12 vents [68]. The location information of these hot spring vents given by Liao [68] is very close to that of the Sobcha travertines (89°56′25″ E, 32°31′35″ N; altitude: 4770 m). A pioneering study by Liao [68] showed that most of the hot springs at Sobcha have vent temperatures from 45 °C to 50 °C and only two hot springs at Sobcha show vent temperatures near 60 °C. The water composition of one hot spring was also reported by Liao [68]: pH = 8.62, Ca^{2+} concentration = 0.17 mM, Mg^{2+} concentration = 0.69 mM, K^+ concentration = 0.87 mM, Na^+ concentration = 22.17 mM, HCO_3^- concentration = 14.95 mM, SO_4^{2-} concentration = 1.58 mM, Cl^- concentration = 3.55 mM, and CO_3^{2-} concentration = 1.68 mM. Na^+ is absolutely dominant in the cation composition. Thus, the studied fossil travertines were anticipated to be formed by Ca^{2+}-deficient hot springs, similar to those in Tengchong (China) [72], Lake Bogoria (Kenya) [73], and Waikite (New Zealand) [74].

3. Methods

Fifty-eight fossil travertines samples were collected in the field. Specific sampling sites for geochemical analysis are shown in Figure 2D. To evaluate their mineralogical composition, thirty powder samples were made using an agate motor and a pestle and were then analyzed with a DX-2700 X-ray diffractometer (XRD) (Cu-Kα radiation; 2θ from 5° to 60°). Fifty-eight thin sections were prepared and were observed with a Nikon LV100POL polarizing microscope to check the petrographic and mineralogical features of the Sobcha travertines. Both the XRD and thin section analyses were conducted at Chengdu University of Technology, China.

Figure 2. (**A**) General view of the studied travertine mound at Sobcha. (**B**) Nearly horizontal laminated abiotic crystalline crust travertines; (**C**) Paleo-vent situated at the top of the studied travertine mound (its location can be found in panel (**A**)). (**D**,**E**) Schematic diagram of sampling sites of the Sobcha travertines for geochemical analysis. The red dots represent the sample site for geochemical analysis, the number represents the sample number, and the same number represents the same sample.

$\delta^{13}C$, $\delta^{18}O$, $^{87}Sr/^{86}Sr$, and concentrations of some trace elements (including rare-earth elements, manganese, strontium, zirconium, copper) of the Sobcha travertines were examined to determine the genesis of the travertines and the characteristics of paleo-fluids. Travertine samples used for geochemical analyses were ground into powder using an agate motor and a pestle and the powder samples were then sifted using a 200 mesh sieve. $\delta^{13}C$ and $\delta^{18}O$ signatures were determined for thirty samples using a Thermo Fisher Scientific DELTA V Advantage isotope ratio mass spectrometer and their values were reported relative to the Vienna Pee Dee Belemnite standard (V-PDB). The analytical precision is commonly better than 0.2‰ for $\delta^{13}C$ and 0.3‰ for $\delta^{18}O$. $^{87}Sr/^{86}Sr$ values were determined for thirty travertine samples using a Thermo Fisher Scientific Triton Plus mass spectrometer. The final $^{87}Sr/^{86}Sr$ results were corrected by assuming a non-radiogenic $^{86}Sr/^{88}Sr$ isotopic ratio of 0.1194. Rare-earth elements (i.e., REE), Mn, Sr, Zr, and Cu concentrations were determined for twenty travertine samples using a Jena Plasma Quant MS ICP-MS. U-Th dating of three travertine samples collected from the bottom (Sample 1), middle (Sample 21), and top (Sample 30) of the sampling profile at Sobcha, respectively, was also conducted using a Neptune Plus MC-ICP-MS. The $\delta^{13}C$-$\delta^{18}O$ analyses, $^{87}Sr/^{86}Sr$ analyses, trace element analyses, and U-Th dating were completed at Yangtze University (China), Chengdu University of Technology (China), Beijing Createch Testing Technology Co., Ltd. (Beijing, China), and the Institute of Geology and Geophysics (Chinese Academy of Sciences), respectively.

In this study, rare-earth elements were divided into three groups: light REE (i.e., LREE: La, Ce, Pr, and Nd), middle REE (i.e., MREE: Sm, Eu, Gd, Tb, Dy, and Ho), and heavy REE (i.e., HREE: Er, Tm, Yb, and Lu) [75]. To better interpret REE characteristics of the Sobcha travertines, REE of Post-Archean Australian Shale (PAAS) [76] was utilized to normalize the REE values of the studied travertines. Additionally, anomalies of Eu were presented as $(Eu/Eu^*)_N$, and were calculated using the following equations: $(Eu/Eu^*)_N = Eu_N/(Sm_N{}^2 \times Tb_N)^{1/3}$ [77]. The relative enrichments of LREE, MREE, and HREE were presented as $(Pr/Yb)_N$ (LREE compared to HREE), $(Pr/Tb)_N$ (LREE compared to MREE), and $(Tb/Yb)_N$ (MREE compared to HREE) [75].

4. Results

4.1. Description of the Sobcha Travertines

The studied travertine system is a domical mound system developed on a slightly inclined slope (Figure 2), morphologically similar to the travertine mounds found in Chusang (southwestern China) [78]. It is characterized by an asymmetric outline with a nearly flat surface at the upslope side and a relatively steep surface at the downslope side. An orifice which had been nearly fully sealed was found on the top surface of the mound (Figure 2C). Determining the exact scale of the mound system was handicapped by the fact that some of the travertine mound was covered by clastic deposits and/or soils and that fluvial incision eroded parts of the travertine mound (Figure 2A). However, the residual travertines were observed to be at least over twenty meters wide. The fluvial incision was largely caused by a north-flowing river (i.e., Sobcha River). Due to the fluvial incision, a four-meter-high travertine profile was exposed, providing a good place for our field observation and sampling.

4.2. Petrology and Mineralogy

The outcropping travertines at Sobcha show great laminated structures, which are characterized by the alternation of wavy dark laminae and wavy light laminae (Figure 2B). Both dark laminae and wavy light laminae have highly variable thicknesses, but in general, light laminae are thicker and more dominant. Furthermore, the Sobcha travertines are consolidated and show visible large calcite crystals ranging from 1 to 10 mm in hand specimens. Large cavities (up to 2 cm in diameter) are also visible in the outcrop, but their development is very limited in fresh samples. Therefore, these cavities are largely secondary (probably generated by the dissolution of meteoric water or temporary flood water).

According to the lithofacies classification of Gandin and Capezzuoli [47], most of the Sobcha travertines are identified as (abiotic) crystalline crust. In this study, the Sobcha travertines were termed 'laminated crystalline crust' to clearly indicate their textural and petrological characteristics. Apart from the laminated crystalline crust, a small amount of clotted peloidal boundstone was also observed in the studied travertine system. XRD and thin-section analyses shows the predominance of calcite in both the laminated crystalline crust and the clotted peloidal boundstone.

The clotted peloidal boundstone is formed of dark micrite peloids and sparite (Figure 3A), while components constituting the laminated crystalline crust include fan crystals and granular crystals (Figure 3B,C). The fan crystals range from 0.2 to 2 mm long and usually develop as thin layers ca. 30 mm thick (Figure 3B,C). The granular crystals are the most widely developed crystals in the Sobcha travertines and have highly variable sizes between 0.1 and 0.5 mm in diameter (Figure 3B). The granular crystals have a more irregular morphology and are subhedral to anhedral in shape (Figure 3B).

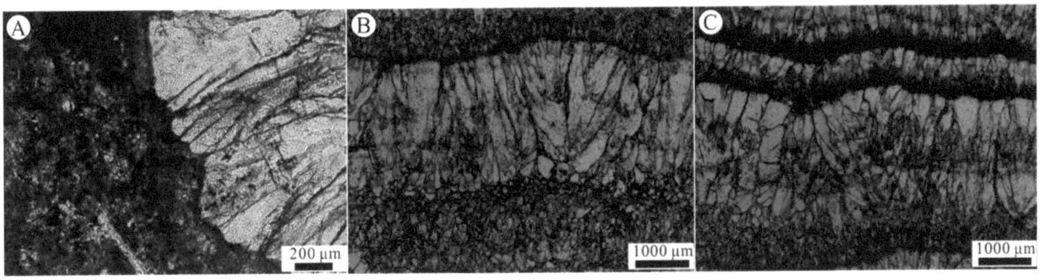

Figure 3. Thin-section microphotographs of the Sobcha travertines (**A–C**); plane-polarized light. (**A**) Clotted peloidal boundstone and fan crystals. (**B**) Fan crystals and granular crystals. (**C**) Fan crystals.

4.3. $\delta^{13}C$, $\delta^{18}O$, and $^{87}Sr/^{86}Sr$

The δ^{13}C, δ^{18}O, and ^{87}Sr/^{86}Sr signatures of the Sobcha travertines are listed in Table S1. The Sobcha travertines show δ^{13}C from 3.4‰ to 4.9‰ V-PDB (average = 4.0‰ V-PDB). δ^{18}O of the Sobcha travertines has a range larger than δ^{13}C (from −26.7‰ to −23.7‰ V-PDB, average = −25.6‰ V-PDB). A positive δ^{13}C-δ^{18}O correlation was found in the Sobcha travertines, but the correlation is not very strong (R^2 = 0.50, n = 30) (Figure 4). Additionally, it is notable that there is a gradual positive δ^{18}O shift from the bottom to the top of the studied profile, although a few anomalous values are also present (Figure 5A). Such increase was also found in δ^{13}C of the travertines, but the δ^{13}C excursion is very slight (Figure 5B). Unlike visible δ^{13}C and δ^{18}O variations, ^{87}Sr/^{86}Sr of the Sobcha travertines are nearly unchanged and show a very narrow range from 0.712458 to 0.712951 (average = 0.712737, n = 30) (Figure 5C).

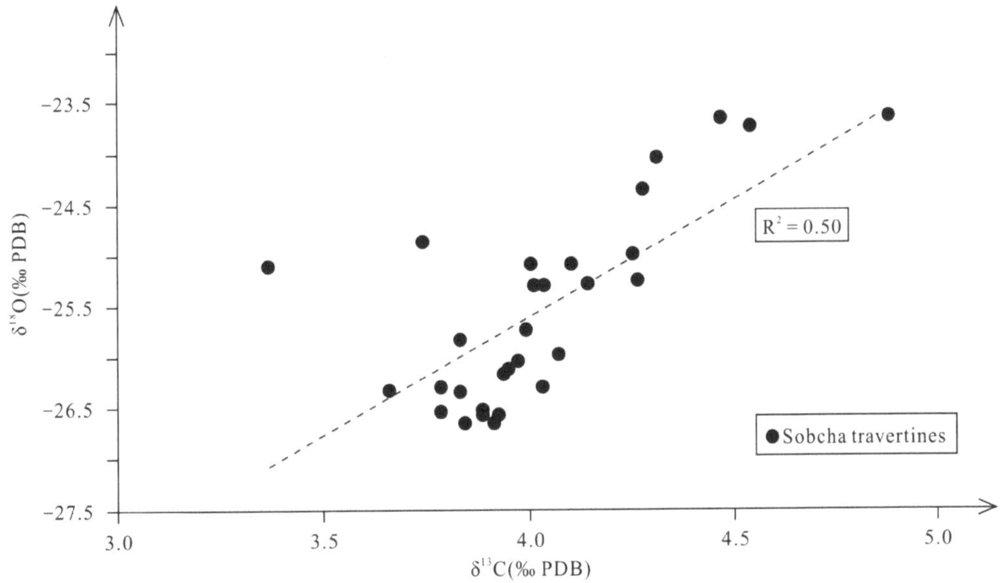

Figure 4. Bivariant graph of δ^{18}O and δ^{13}C of travertines from Sobcha.

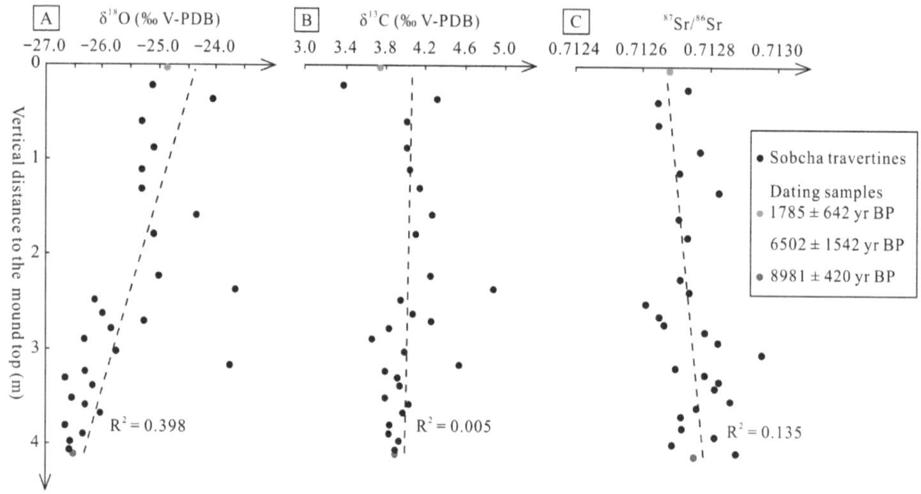

Figure 5. (**A–C**) Vertical variations of $\delta^{18}O$, $\delta^{13}C$, and $^{87}Sr/^{86}Sr$ of the Sobcha travertines in the studied profile.

4.4. U-Th Dating

The U-Th dating results of three travertine samples (i.e., Samples 1, 21, and 30) are listed in Table S2. Samples 1, 21, and 30 were collected from the bottom, middle, and top of the sampling profile at Sobcha, respectively. Sample 1 has a measured depositional age of 8981 ± 420 yr BP, while Sample 21 exhibits a depositional age of 6502 ± 1542 yr BP. Sample 30 is the youngest of the three samples and its corrected age is 1785 ± 642 yr BP. The dating results show that travertine deposition of the studied travertine mound system began, at least, in the early Holocene and ceased in the late Holocene.

4.5. Rare-Earth Elements, Mn, Sr, Zr, and Cu

Rare-earth elements, Mn, Sr, Zr, and Cu concentrations and some main calculated geochemical parameters of the Sobcha travertines are listed in Tables S3 and S4. ΣREE of the Sobcha travertines is highly variable (ranging from 0.61 to 23.05 $\mu g\ g^{-1}$, average = 8.04 $\mu g\ g^{-1}$), but most of the samples show ΣREE below 13.00 $\mu g\ g^{-1}$. The Sobcha travertines show significant MREE and HREE enrichment relative to LREE, as shown by their very low $(Pr/Tb)_N$ (from 0.10 to 0.55, mostly between 0.10 and 0.25, average = 0.16) and $(Pr/Yb)_N$ (from 0.12 to 0.60, commonly between 0.12 and 0.24, average = 0.19). $(Tb/Yb)_N$ ratios of the Sobcha travertines vary from 1.09 to 1.21 (average = 1.14), indicating the Sobcha travertines are weakly MREE-enriched compared to HREE.

Elemental anomaly calculations show that the Sobcha travertines have $(Eu/Eu^*)_N$ ratios range from 1.05 to 1.68 (average = 1.36) and most of the $(Eu/Eu^*)_N$ data are between 1.28 and 1.43. This reflects unneglectable positive Eu anomalies in the Sobcha travertines.

The Sobcha travertines have very low Zr and Mn concentrations and Mn/Sr ratios. Their Zr concentrations are from 0.38 to 3.24 $\mu g\ g^{-1}$ (average = 1.73 $\mu g\ g^{-1}$), but most of the values are lower than 2 $\mu g\ g^{-1}$. Mn concentrations of the Sobcha travertines are between 279 and 1280 $\mu g\ g^{-1}$ and have an average value of 825.75 $\mu g\ g^{-1}$. The calculated Mn/Sr ratios are within a small range from 0.53 to 2.48 (average = 1.63). Cu concentrations of the Sobcha travertines are between 0.19 and 1.17 $\mu g\ g^{-1}$ and have an average value of 0.52 $\mu g\ g^{-1}$.

5. Interpretation and Discussion

5.1. Interpretation of $\delta^{13}C$: Source of Mother CO_2

$\delta^{13}C$ analyses of travertines are good tools in the determination of mother CO_2 sources [52,62,79]. $\delta^{13}C$ of mother CO_2 (i.e., $\delta^{13}C_{mother\text{-}CO2}$) of spring-related carbonates have been used to divide them into thermogene travertine and meteogene travertine [79–83]. For example, according to Pentecost and Viles [83], $\delta^{13}C$ of thermogene travertines ranges from −4‰ to 8‰ V-PDB, whereas $\delta^{13}C$ of meteogene travertines ranges from −11‰ to 0 V-PDB. $\delta^{13}C$ values of the Sobcha travertines are between 3.4‰ and 4.9‰ V-PDB, which is obviously in the $\delta^{13}C$ range of thermogene travertines. In addition, Figure 6 also shows that the Sobcha travertines belongs to hypogean CATT (calcitic or aragonitic travertine and tufa) [60]. However, spring-related carbonates with $\delta^{13}C$ in the travertine range may show highly different CO_2 sources. Thus, $\delta^{13}C_{mother\text{-}CO2}$ of the Sobcha travertines were calculated and compared with $\delta^{13}C$ of potential CO_2 sources to determine their genesis and mother CO_2 source(s).

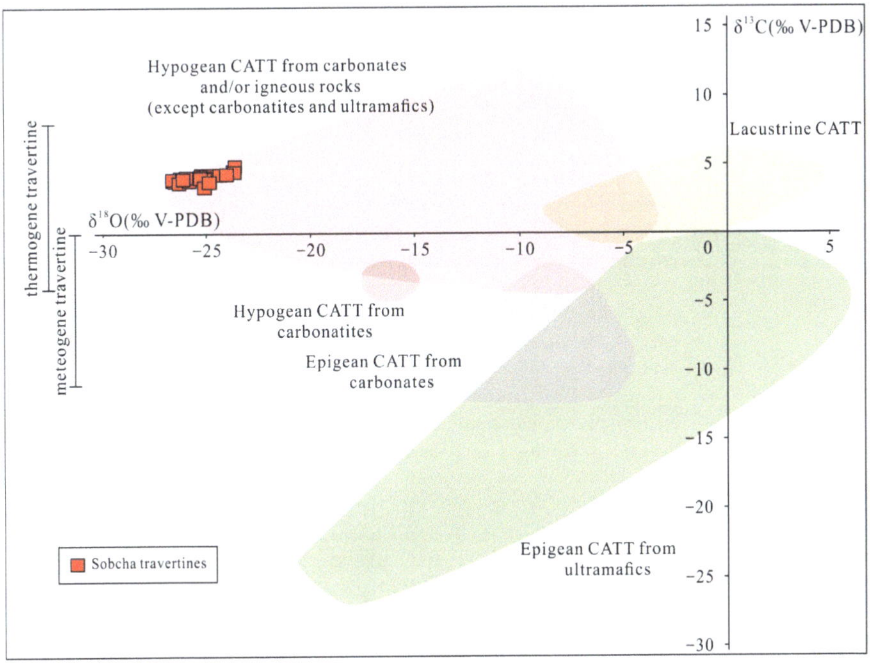

Figure 6. $\delta^{13}C$ and $\delta^{18}O$ composition for the Sobcha travertines. $\delta^{13}C$ ranges for thermogene and meteogene travertines were from Pentecost and Viles [64]. $\delta^{13}C$ and $\delta^{18}O$ ranges of different CATT (i.e., calcitic or aragonitic travertine and tufa) were from Teboul et al. [65].

The empirical equation developed by Panichi and Tongiorgi [84] was adapted to calculate $\delta^{13}C_{mother\text{-}CO2}$ of the Sobcha travertines. The calculated $\delta^{13}C_{mother\text{-}CO2}$ varies from −6.5‰ and −4.6‰ V-PDB (Table S1). A simple comparison in Figure 7 shows that the calculated $\delta^{13}C_{mother\text{-}CO2}$ overlaps the $\delta^{13}C$ range of mantle-derived CO_2, and is significantly different from $\delta^{13}C$ of soil-related CO_2 and $\delta^{13}C$ of marine carbonate rocks in the study area and surrounding areas [69,85–88]. This might be indicative of the close relationship between the Sobcha travertines and mantle-derived CO_2. Such magmatic CO_2-containing volatiles has been found in hot springs from Naqu (close to the study area) and was interpreted as the product of magma bodies in the shallow crust [89,90]. Unfortunately, detailed analyses of hot spring gases from the study area have not been performed. Thus, such a mantle-derived CO_2 origin is possible but is not the only interpretation.

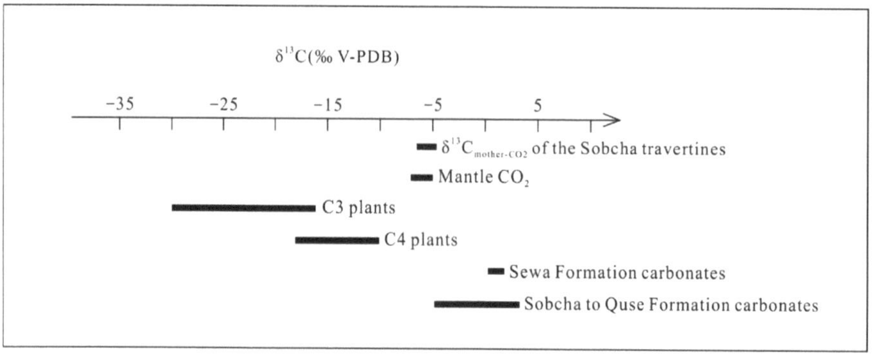

Figure 7. Calculated $\delta^{13}C_{mother\text{-}CO2}$ of the Sobcha travertines in comparison with $\delta^{13}C$ of potential carbon sources. $\delta^{13}C$ ranges of mantle-derived CO_2 and CO_2 related to C3 plants and C4 plants were from Vogel [69] and Prokoph et al. [68]. $\delta^{13}C$ ranges of carbonate rocks of the Sewa Formation and Sobcha to Quse Formation were from Yi [70] and Yi [49]; Fu et al. [67]; Yi [70].

An alternative explanation of the similarity between the calculated $\delta^{13}C_{mother\text{-}CO2}$ range and $\delta^{13}C$ of mantle-derived CO_2 is underground CO_2 mixing. Apart from magmatic CO_2, there are at least two main CO_2 sources for travertine systems: soil CO_2 (C3 or C4 plants) and carbonate-related CO_2 (either by the dissolution or decarbonation of carbonate rocks) [82]. In our study area, the distribution of C4 plants is very limited, probably due to its high altitude [91]. Therefore, the plants in the study area are mainly C3 plants. For the carbonate-related CO_2, given the low Ca^{2+} concentration (0.17 mM) and high HCO_3^- concentrations (14.95 mM) of the modern spring in the study area [68], the dissolution of carbonate rocks is considered to only supply minor CO_2 to the spring waters and travertines at Sobcha. However, this cannot exclude the CO_2 contribution of carbonate decarbonization because this process can provide CO_2 to the spring water (in other words, elevating HCO_3^- concentration) without increasing the concentrations of Ca^{2+} and Mg^{2+}. If there was an underground mixing between soil CO_2 and carbonate-related CO_2, their mixture may yield similar $\delta^{13}C$ to mantle-derived CO_2. Thus, mixing, if present, might happen between soil CO_2 (C3 plants) and CO_2 related to carbonate decarbonation.

5.2. Interpretation of $^{87}Sr/^{86}Sr$: Sr Source Rocks

The $^{87}Sr/^{86}Sr$ analysis of travertines can be used to determine their Sr sources [72]. With respect to the studied travertines, all the rocks exposing in the study area are their potential Sr source rocks. A simple comparison between $^{87}Sr/^{86}Sr$ of the studied travertines and $^{87}Sr/^{86}Sr$ of potential Sr source rocks [92–94] was made and is given in Figure 8. The comparison shows that the $^{87}Sr/^{86}Sr$ range of the Sobcha travertines is significantly higher than $^{87}Sr/^{86}Sr$ ranges of carbonate rocks and clastic rocks exposed in/near the study area (mainly the Quse Formation) and Cambrian to Cenozoic marine carbonates (Figure 8). This suggests that the studied travertines must acquire more radiogenic Sr from other Sr source rocks, instead of from the local rocks cropping out in the study area. Interestingly, the studied travertines display $^{87}Sr/^{86}Sr$ very close to the average $^{87}Sr/^{86}Sr$ value of Nadi Kangri volcanic rocks. However, the nearest exposed Nadi Kangri volcanic rocks are located in the Bilong Co area and are ca. 20 km away from the study area [95]. Thus, simply treating the Nadi Kangri volcanic rocks as the main Sr source rocks of the studied travertines is not very appropriate. Conclusively, further studies are required to uncover the main Sr source rocks of the Sobcha travertines.

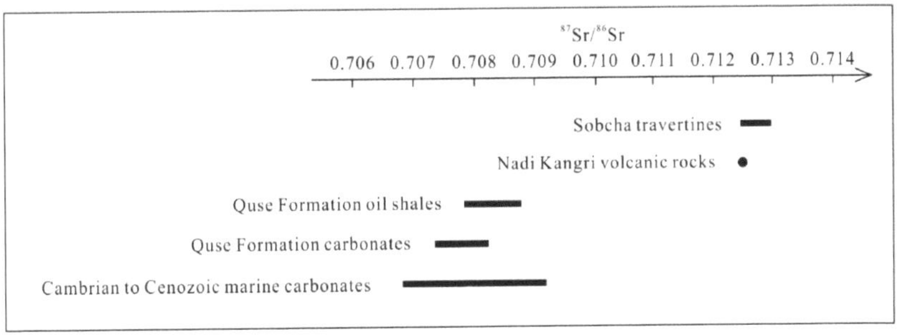

Figure 8. $^{87}Sr/^{86}Sr$ in the studied travertines in comparison with $^{87}Sr/^{86}Sr$ ranges of potential Sr source rocks: oil shales and micritic limestones of the Quse Formation (data from Fu et al. [76]), average values of Nadi Kangri volcanic rocks (data from Fu et al. [75]), and Cambrian to Cenozoic marine carbonates (data from McArthur et al. [77]).

5.3. Interpretation of REE Patterns and Eu Anomalies

REE of carbonate rocks might not be pristine, because they can be easily contaminated by other materials, such as clastic detritus [96–103] and Fe-Mn (oxyhydr) oxides [104] during or after deposition. Thus, prior to the analysis of REE of carbonates, a contaminant evaluation is often required. The Sobcha travertines shows Zr concentrations lower than 4 $\mu g\ g^{-1}$ (Table S3). This indicates that there is no significant contamination from clastic detritus [98]. In addition, the Sobcha travertines display no Cu–ΣREE correlations and have very low Mn concentrations (279 to 1660 $\mu g\ g^{-1}$) (Table S3, Figure 9), reflecting that the contamination of Fe-Mn (oxyhydr)oxides is not important [105]. The REE modification of post-depositional processes can be also neglected because the Mn/Sr ratios of the Sobcha travertines are all less than 3 (Table S4) [106–108]. These show that our REE data are not significantly contaminated and can provide original geological information.

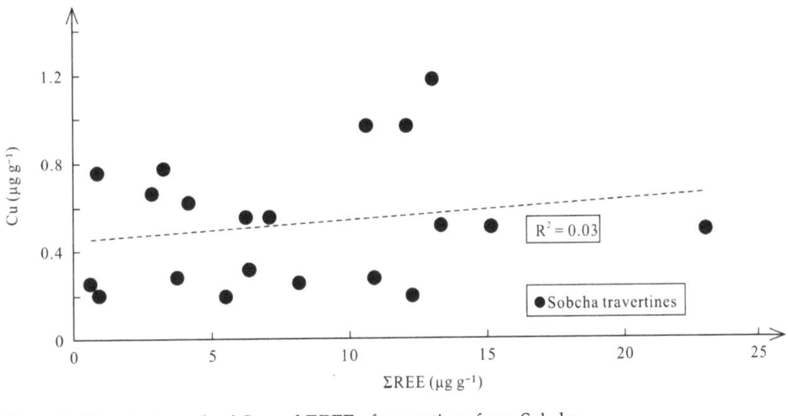

Figure 9. Bivariant graph of Cu and ΣREE of travertines from Sobcha.

There are two striking REE characteristics of the Sobcha travertines: (1) MREE-enrichment relative to both HREE and LREE and HREE-enrichment relative to LREE, and (2) positive Eu anomalies (1.28 and 1.43; Figure 10 and Table S4). There are two main factors which might influence REE patterns of the studied travertine: solution complexation and REE of aquifer rocks. Paleo-fluids depositing the Sobcha travertines might be compositionally similar to active hot springs at Sobcha (i.e., bicarbonate- and carbonate-rich) [68]. Such fluids could form strong HREE-complexes (i.e., preferential HREE incorporation into fluids) and the resulting fluids and associated deposits would be HREE-enriched compared to REE of their aquifer rocks [109]. However, it is strange that although the Sobcha travertines are HREE-enriched compared to LREE, their HREE are depleted compared to MREE. This indicates that their MREE-enrichment was largely inherited from aquifer rocks, instead of caused by solution complexation. However, this does not mean that the influence of solution complexation on the REE pattern of the Sobcha travertines is very weak. It is here believed that solution complexation might indeed cause the HREE-enrichment in the Sobcha travertines, but such HREE-enrichment did not fully mask the aquifer rock REE information recorded in the travertines (i.e., MREE-enrichment). In Figure 11 the REE patterns of the Sobcha travertines are compared with those of rocks cropping out in the study area, such as shale, calci-mudstone, and marl of the Quse Formation and Nadi Kangri volcanic rocks (including basalt, rhyolite, dacite, and tuff) [70,110–115]. It is clear that none of the surface rocks display MREE-enrichment, excluding the possibility that these rocks are aquifer rocks. Such inference is consistent with the results of the $^{87}Sr/^{86}Sr$ analyses in Section 5.2 because aquifer rocks were often considered as predominant Sr source rocks [116–118].

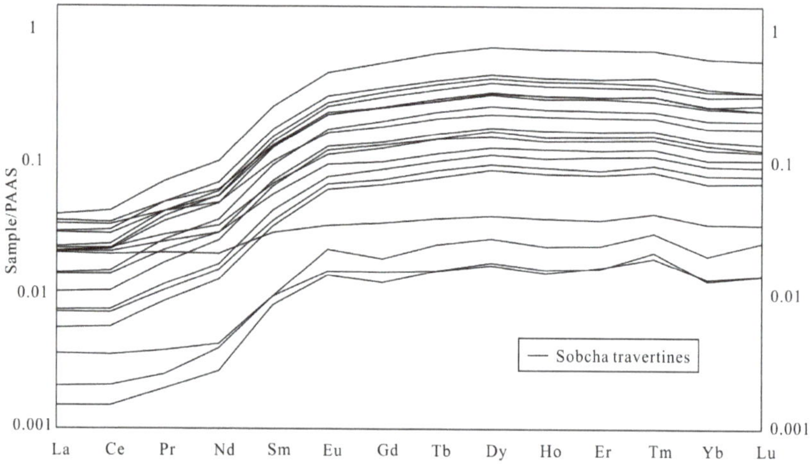

Figure 10. PAAS-normalized REE patterns for the travertine samples of Sobcha.

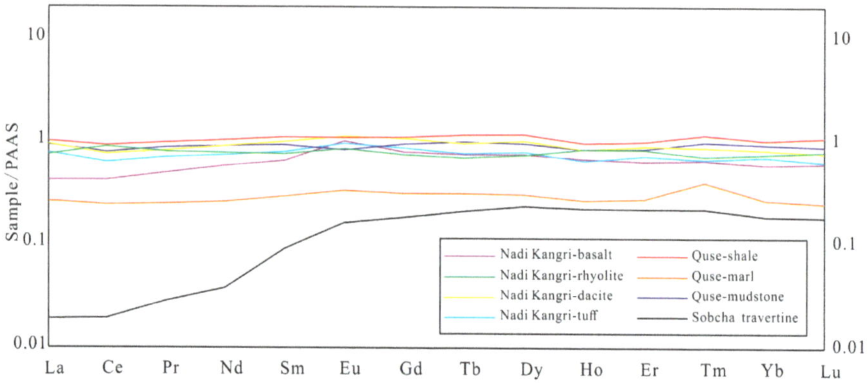

Figure 11. Comparison of PAAS-normalization patterns between Sobcha travertines and potential source rocks: basalt, rhyolite, and dacite of the Nadi Kangri volcanic rocks (data from Fu et al. [93]); tuff of the Nadi Kangri volcanic rocks (data from Wang et al. [96]; Fu et al. [97]; Wang et al. [98]); shale of the Quse Formation (data from Fu et al. [50]); mudstone of the Quse Formation (data from Nie et al. [95]); and marl of the Quse Formation (data from Fu et al. [94]).

Positive Eu anomalies in marine carbonate rocks are often indicative of a hydrothermal origin of their parent fluids [75,119]. At high temperatures (often >250 °C), Eu might be fractionated from other REE because of the reduction of Eu into Eu^{2+} [120], which would finally cause Eu enrichment in high-temperature fluids. As a result, carbonate rocks deposited from these fluids, or their cooled products might show positive Eu anomalies. At Sobcha, the temperature of its reservoir(s) has not been investigated. However, geothermal systems with reservoir temperatures over 250 °C have been found in Central Tibet (e.g., Yangbajing geothermal field) [121]. Therefore, it is possible that positive Eu anomalies of the Sobcha travertines might be related to undiscovered high-temperature reservoirs beneath Sobcha. An alternative explanation of the positive Eu anomalies in the Sobcha travertines might be the preferential dissolution of Eu-rich minerals/rocks (especially plagioclase). Indeed, such a process is not uncommon in groundwater systems [122]. Therefore, positive Eu anomalies of the studied travertines might be caused by high-temperature geothermal reservoir(s), preferential dissolution of Eu-rich minerals/rocks, or both, and more studies are still necessary in this region.

5.4. Evolution of Paleo-Fluids

Travertine deposition may be affected by various factors, such as tectonic activity, climate, hydrodynamics, and hydrochemistry [10,12,17,123] and is thus not always steady. For example, some of these complex controlling factors (either internal or external), such as climate change, may lead to the changes in the flow path and/or discharge. However, changes in (paleo-)fluids (e.g., the source and physicochemical properties of the mother water) might be reflected by the geochemical features of travertine deposits, such as $\delta^{13}C$, $^{87}Sr/^{86}Sr$, and REE. The $\delta^{13}C$ of the Sobcha travertines shows a great stability (Table S1, Figure 5B). The small range of $\delta^{13}C$ indicates that the CO_2 source of the Sobcha travertines did not change. The $^{87}Sr/^{86}Sr$ and REE signatures of the Sobcha travertines also display little changes in the sampling profile (Table S1 and Figure 5C), indicating the source and physicochemical stability of the mother water of the Sobcha travertines

The $\delta^{18}O$ of the Sobcha travertines has a larger variation than $\delta^{13}C$ and shows a striking gradual increase from bottom to top of the sampling profile (Table S1 and Figure 5A). The $\delta^{18}O$ signature of travertines is important to the recovery of paleo-temperature of the mother water in travertine deposition systems [52,62,79]. We were unable to recover the exact temperature changes of the mother water depositing the Sobcha travertines. However, according to common traditional oxygen isotope thermometers, there is an inverse relationship between $\delta^{18}O$ of carbonate deposits and water temperature [124]. Thus, the positive $\delta^{18}O$ excursion in the studied travertine profile of Sobcha, at least, indicates that paleo-fluid temperature forming the Sobcha travertines decreased gradually. However, the factors that cause such a $\delta^{18}O$ increase or temperature decrease remain unknown.

Climate (mainly atmospheric temperature and rainfall) may influence the temperature of mother water of travertines [63,64,125]. The Sobcha travertine deposition began, at least, in the early Holocene and ceased in the late Holocene. From the early Holocene to the middle Holocene and to the late Holocene, the southern Qinghai–Tibet Plateau was generally in a climate background from drought to humid and to drought, as recorded in the $\delta^{18}O$ composition of cave deposits [126–128]. In the Holocene, the temperature of the Qinghai–Tibet Plateau also experienced a gradual increase from the early to the middle period and a gradual decrease from the middle to the late period [126]. In general, the trend of $\delta^{18}O$ during the deposition period of the Sobcha travertines are similar to that of $\delta^{18}O$ recorded in cave deposits of the southern Qinghai–Tibet Plateau at the same time. This seems to indicate that the $\delta^{18}O$ of the Sobcha travertines, or rather the temperature of the mother water, is affected by climate.

In addition to climate, a gradual decrease in the discharge of the mother water may also lead to a faster decrease in the temperature of the mother water (assuming the air temperature is stable). This would in turn lead to a gradual increase in the $\delta^{18}O$ of the deposits. The gradual decrease in the discharge of the mother water of the Sobcha travertines may also be due to the gradual drought of the climate during the deposition period. However, the self-closure of the vents and mound vertical growth might also lead to a gradual decrease in water discharge of mound springs. For the Sobcha travertines, we found that the paleo-vent was completely closed due to the growth of deposit (Figure 2C). Therefore, it is possible that water discharge of paleo-fluids depositing the Sobcha travertines gradually decreased due to the gradual closure of the vent(s) and/or mound vertical growth.

6. Conclusions

Through the geochemical and geochronological study on the Sobcha fossil travertine mound, we mainly draw the following conclusions:

(1) The Sobcha travertines are dominantly composed of laminated crystalline crust mainly composed of granular crystals and fan crystals. In addition, a small amount of clotted peloidal boundstone was also observed. Calcite has an absolute predominance in the Sobcha travertines.

(2) $\delta^{13}C$ and $\delta^{18}O$ analyses show that the Sobcha travertines belong to thermogene travertines, and its parent CO_2 may be derived from mantle-derived CO_2 or the mixture of soil-derived CO_2 and CO_2 related to carbonate decarbonation.

(3) The Sobcha travertines display $^{87}Sr/^{86}Sr$ very close to the average $^{87}Sr/^{86}Sr$ value of Nadi Kangri volcanic rocks. However, further studies are required to uncover the main Sr source rocks of the Sobcha travertines.

(4) For the PAAS-normalized REE patterns of the Sobcha travertines, the LREE depletion relative to HREE of the Sobcha travertines may be due to the differences in geochemical mobility between LREEs and HREEs during water–rock interaction at depth, while the MREE enrichment of the Sobcha travertines may be inherited from aquifer rocks. Positive Eu anomalies were also observed in the Sobcha travertines and may result from the hydrothermal property of the mother water and/or preferential dissolution of Eu-rich minerals/rocks.

(5) Travertine deposition in the studied travertine mound began, at least, in the early Holocene and ceased in the late Holocene. During the whole deposition period, the source of CO_2 and the source of the mother water of Sobcha travertines remained stable. However, $\delta^{18}O$ of the studied travertines gradually decreased, probably because of climate drying, self-closure of the vents, or mound vertical growth.

Supplementary Materials: The following supporting information can be downloaded at https://www.mdpi.com/article/10.3390/min13020220/s1. Table S1: $\delta^{13}C$, $\delta^{18}O$, and $^{87}Sr/^{86}Sr$ values of the Sobcha travertines and their calculated $\delta^{13}C_{mother\text{-}CO2}$ (i.e., $\delta^{13}C$ of mother CO_2). $\delta^{13}C_{mother\text{-}CO2}$ was evaluated using the equation from Panichi and Tongiorgi [84]. Table S2: U-Th dating results of the Sobcha travertines. Table S3: Concentrations ($\mu g\ g^{-}1$) of trace elements in the travertine samples from Sobcha. Table S4: Mn/Sr, enrichment indexes, and elemental anomalies of the studied travertines from Sobcha.

Author Contributions: Conceptualization, Z.L., H.W. and L.L. (Lianchao Luo); methodology, Z.L., H.W. and L.L. (Lianchao Luo); software, Z.L. and L.L. (Lianchao Luo); validation, Z.L., H.W. and L.L. (Lianchao Luo); formal analysis, Z.L. and L.L. (Lianchao Luo); investigation, Z.L., L.L. (Liang Li), and Y.N.; resources, H.W.; data curation, H.W.; writing—original draft preparation, Z.L.; writing—review and editing, Z.L., H.W., L.L. (Lianchao Luo), L.L. (Liang Li), and Y.N.; visualization, Z.L. and L.L. (Lianchao Luo); supervision, H.W.; project administration, H.W.; funding acquisition, H.W. All authors have read and agreed to the published version of the manuscript.

Funding: This research was funded by the National Natural Science Foundation of China (grant number 41972116 and 41572097 to H. Wen).

Data Availability Statement: Not applicable.

Acknowledgments: We are grateful to Mingshi Fen (Chengdu University of Technology) for his help in SEM analyses, Di Yang (Chengdu University of Technology) for her help in Sr isotope analyses, and Miao Deng (Chengdu University of Technology) for his help in XRD analyses, Xiugen Fu (Southwest Petroleum University) for his help in field survey, and Xuefeng Wang (Uranium chronology laboratory, Institute of Geology and Geophysics, Chinese Academy of Sciences) for his help in U-Th dating analyses.

Conflicts of Interest: The authors declare no conflict of interest.

References

1. Capezzuoli, E.; Gandin, A.; Pedley, M. Decoding tufa and travertine (fresh water carbonates) in the sedimentary record: The state of the art. *Sedimentology* **2014**, *61*, 1–21. [CrossRef]
2. Ford, T.D.; Pedley, H.M. A review of tufa and travertine deposits of the world. *Earth-Sci. Rev.* **1996**, *41*, 117–175. [CrossRef]
3. Pedley, M. Tufas and travertines of the Mediterranean region: A testing ground for freshwater carbonate concepts and developments. *Sedimentology* **2009**, *56*, 221–246. [CrossRef]
4. Wen, H.; Luo, L.; Luo, X.; You, Y.; Du, L. Advances and Prospects of Terrestrial Thermal Spring Travertine Research. *Acta Sedimentol. Sin.* **2019**, *37*, 1162–1180, (In Chinese with English Abstract).
5. Altunel, E.; Hancock, P.L. Morphology and structural setting of Quaternary travertines at Pamukkale, Turkey. *Geol. J.* **1993**, *28*, 335–346. [CrossRef]
6. Altunel, E.; Hancock, P.L. Structural Attributes of Travertine-Filled Extensional Fissures in the Pamukkale Plateau, Western Turkey. *Int. Geol. Rev.* **1996**, *38*, 768–777. [CrossRef]
7. Altunel, E.; Karabacak, V. Determination of horizontal extension from fissure-ridge travertines: A case study from the Denizli Basin, southwestern Turkey. *Geodin. Acta* **2005**, *18*, 333–342. [CrossRef]
8. Brogi, A.; Alcicek, M.C.; Yalciner, C.C.; Capezzuoli, E.; Liotta, D.; Meccheri, M.; Rimondi, V.; Ruggieri, G.; Gandin, A.; Boschi, C.; et al. Hydrothermal fluids circulation and travertine deposition in an active tectonic setting: Insights from the Kamara geothermal area (western Anatolia, Turkey). *Tectonophysics* **2016**, *680*, 211–232. [CrossRef]
9. Brogi, A.; Capezzuoli, E. Travertine deposition and faulting: The fault-related travertine fissure-ridge at Terme S. Giovanni, Rapolano Terme (Italy). *Int. J. Earth Sci.* **2009**, *98*, 931–947. [CrossRef]
10. Brogi, A.; Capezzuoli, E. Earthquake impact on fissure-ridge type travertine deposition. *Geol. Mag.* **2014**, *151*, 1135–1143. [CrossRef]
11. Brogi, A.; Liotta, D.; Capezzuoli, E.; Matera, P.F.; Kele, S.; Soligo, M.; Tuccimei, P.; Ruggieri, G.; Yu, T.L.; Shen, C.C.; et al. Travertine deposits constraining transfer zone neotectonics in geothermal areas: An example from the inner Northern Apennines (Bagno Vignoni-Val d'Orcia area, Italy). *Geothermics* **2020**, *85*, 22. [CrossRef]
12. Gradzinski, M.; Wroblewski, W.; Dulinski, M.; Hercman, H. Earthquake-affected development of a travertine ridge. *Sedimentology* **2014**, *61*, 238–263. [CrossRef]
13. Hancock, P.L.; Chalmers, R.M.L.; Altunel, E.; Cakir, Z. Travitonics: Using travertines in active fault studies. *J. Struct. Geol.* **1999**, *21*, 903–916. [CrossRef]
14. Selim, H.H.; Yanik, G. Development of the Cambazli (Turgutlu/MANISA) fissure-ridge-type travertine and relationship with active tectonics, Gediz Graben, Turkey. *Quat. Int.* **2009**, *199*, 157–163. [CrossRef]
15. Shiraishi, F.; Morikawa, A.; Kuroshima, K.; Amekawa, S.; Yu, T.-L.; Shen, C.-C.; Kakizaki, Y.; Kano, A.; Asada, J.; Bahniuk, A.M. Genesis and diagenesis of travertine, Futamata hot spring, Japan. *Sediment. Geol.* **2020**, *405*, 105706. [CrossRef]
16. Prado-Perez, A.J.; Huertas, A.D.; Crespo, M.T.; Sanchez, A.M.; Del Villar, L.P. Late Pleistocene and Holocene mid-latitude palaeoclimatic and palaeoenvironmental reconstruction: An approach based on the isotopic record from a travertine formation in the Guadix-Baza basin, Spain. *Geol. Mag.* **2013**, *150*, 602–625. [CrossRef]
17. Rodriguez-Berriguete, A.; Alonso-Zarza, A.M. Controlling factors and implications for travertine and tufa deposition in a volcanic setting. *Sediment. Geol.* **2019**, *381*, 13–28. [CrossRef]
18. Toker, E.; Kayseri-Ozer, M.S.; Ozkul, M.; Kele, S. Depositional system and palaeoclimatic interpretations of Middle to Late Pleistocene travertines: Kocaba, Denizli, south-west Turkey. *Sedimentology* **2015**, *62*, 1360–1383. [CrossRef]
19. Uysal, I.T.; Unal-Imer, E.; Shulmeister, J.; Zhao, J.X.; Karabacak, V.; Feng, Y.X.; Bolhar, R. Linking CO_2 degassing in active fault zones to long-term changes in water balance and surface water circulation, an example from SW Turkey. *Quat. Sci. Rev.* **2019**, *214*, 164–177. [CrossRef]
20. Camuera, J.; Alonso-Zarza, A.M.; Rodriguez-Berriguete, A.; Rodriguez-Gonzalez, A. Origin and palaeo-environmental significance of the Berrazales carbonate spring deposit, North of Gran Canaria Island, Spain. *Sediment. Geol.* **2014**, *308*, 32–43. [CrossRef]
21. Kele, S.; Korpás, L.; Demény, A.; Kovács-Pálffy, P.; Bajnóczi, B.; Medzihradszky, Z. Paleoenvironmental evaluation of the Tata Travertine Complex (Hungary), based on stable isotopic and petrographic studies. *Acta Geol. Hung.* **2006**, *49*, 1–31. [CrossRef]

22. Capezzuoli, E.; Ruggieri, G.; Rimondi, V.; Brogi, A.; Liotta, D.; Alcicek, M.C.; Alcicek, H.; Bulbul, A.; Gandin, A.; Meccheri, M.; et al. Calcite veining and feeding conduits in a hydrothermal system: Insights from a natural section across the Pleistocene Golemezli travertine depositional system (western Anatolia, Turkey). *Sediment. Geol.* **2018**, *364*, 180–203. [CrossRef]
23. Jones, B. Review of aragonite and calcite crystal morphogenesis in thermal spring systems. *Sediment. Geol.* **2017**, *354*, 9–23. [CrossRef]
24. Jones, B. Review of calcium carbonate polymorph precipitation in spring systems. *Sediment. Geol.* **2017**, *353*, 64–75. [CrossRef]
25. Luo, L.C.; Capezzuoli, E.; Rogerson, M.; Vaselli, O.; Wen, H.G.; Lu, Z.P. Precipitation of carbonate minerals in travertine-depositing hot springs: Driving forces, microenvironments, and mechanisms. *Sediment. Geol.* **2022**, *438*, 27. [CrossRef]
26. Fouke, B.W. Hot-spring Systems Geobiology: Abiotic and biotic influences on travertine formation at Mammoth Hot Springs, Yellowstone National Park, USA. *Sedimentology* **2011**, *58*, 170–219. [CrossRef]
27. Guo, L.I.; Riding, R. Origin and diagenesis of Quaternary travertine shrub fabrics, Rapolano Terme, central Italy. *Sedimentology* **1994**, *41*, 499–520. [CrossRef]
28. Jones, B.; Peng, X.T. Signatures of biologically influenced CaCo3 and Mg-Fe silicate precipitation in hot springs: Case study from the Ruidian geothermal area, western Yunnan Province, China. *Sedimentology* **2014**, *61*, 56–89. [CrossRef]
29. Rogerson, M.; Pedley, H.M.; Kelham, A.; Wadhawan, J.D. Linking mineralisation process and sedimentary product in terrestrial carbonates using a solution thermodynamic approach. *Earth Surf. Dyn.* **2014**, *2*, 197–216. [CrossRef]
30. Shiraishi, F.; Eno, Y.; Nakamura, Y.; Hanzawa, Y.; Asada, J.; Bahniuk, A.M. Relative influence of biotic and abiotic processes on travertine fabrics, Satono-yu hot spring, Japan. *Sedimentology* **2019**, *66*, 459–479. [CrossRef]
31. Sugihara, C.; Yanagawa, K.; Okumura, T.; Takashima, C.; Harijoko, A.; Kano, A. Transition of microbiological and sedimentological features associated with the geochemical gradient in a travertine mound in northern Sumatra, Indonesia. *Sediment. Geol.* **2016**, *343*, 85–98. [CrossRef]
32. Takashima, C.; Kano, A. Microbial processes forming daily lamination in a stromatolitic travertine. *Sediment. Geol.* **2008**, *208*, 114–119. [CrossRef]
33. Della Porta, G.; Hoppert, M.; Hallmann, C.; Schneider, D.; Reitner, J. The influence of microbial mats on travertine precipitation in active hydrothermal systems (Central Italy). *Depos. Rec.* **2022**, *8*, 165–209. [CrossRef]
34. Norris, T.B.; Castenholz, R.W. Endolithic photosynthetic communities within ancient and recent travertine deposits in Yellowstone National Park. *FEMS Microbiol. Ecol.* **2006**, *57*, 470–483. [CrossRef] [PubMed]
35. Okumura, T.; Takashima, C.; Shiraishi, F.; Nishida, S.; Yukimura, K.; Naganuma, T.; Koike, H.; Arp, G.; Kano, A. Microbial Processes Forming Daily Lamination in an Aragonite Travertine, Nagano-yu Hot Spring, Southwest Japan. *Geomicrobiol. J.* **2011**, *28*, 135–148. [CrossRef]
36. Okumura, T.; Takashima, C.; Shiraishi, F.; Nishida, S.; Kano, A. Processes Forming Daily Lamination in a Microbe-Rich Travertine Under Low Flow Condition at the Nagano-yu Hot Spring, Southwestern Japan. *Geomicrobiol. J.* **2013**, *30*, 910–927. [CrossRef]
37. Okumura, T.; Takashima, C.; Kano, A. Textures and processes of laminated travertines formed by unicellular cyanobacteria in Myoken hot spring, southwestern Japan. *Island Arc.* **2013**, *22*, 410–426. [CrossRef]
38. Okumura, T.; Takashima, C.; Shiraishi, F.; Akmaluddin; Kano, A. Textural transition in an aragonite travertine formed under various flow conditions at Pancuran Pitu, Central Java, Indonesia. *Sediment. Geol.* **2012**, *265–266*, 195–209. [CrossRef]
39. Takashima, C.; Okumura, T.; Nishida, S.; Shimamoto, T.; Koike, H.; Kano, A. Microbial Control on Lamina Formation in a Travertine of Crystal Geyser, Utah. In *Advances in Stromatolite Geobiology*; Reitner, J., Quéric, N.-V., Arp, G., Eds.; Springer: Berlin/Heidelberg, Germany, 2011; pp. 123–133.
40. Valeriani, F.; Crognale, S.; Protano, C.; Gianfranceschi, G.; Orsini, M.; Vitali, M.; Spica, V.R. Metagenomic analysis of bacterial community in a travertine depositing hot spring. *New Microbiol.* **2018**, *41*, 126–135.
41. Burnside, N.M.; Shipton, Z.K.; Dockrill, B.; Ellam, R.M. Man-made versus natural CO_2 leakage: A 400 k.y. history of an analogue for engineered geological storage of CO_2. *Geology* **2013**, *41*, 471–474. [CrossRef]
42. Moore, J.; Adams, M.; Allis, R.; Lutz, S.; Rauzi, S. Mineralogical and geochemical consequences of the long-term presence of CO_2 in natural reservoirs: An example from the Springerville-St. Johns Field, Arizona, and New Mexico, USA. *Chem. Geol.* **2005**, *217*, 365–385. [CrossRef]
43. Chafetz, H.; Barth, J.; Cook, M.; Guo, X.; Zhou, J. Origins of carbonate spherulites: Implications for Brazilian Aptian pre-salt reservoir. *Sediment. Geol.* **2018**, *365*, 21–33. [CrossRef]
44. Mancini, A.; Capezzuoli, E.; Erthal, M.; Swennen, R. Hierarchical approach to define travertine depositional systems: 3D conceptual morphological model and possible applications. *Mar. Pet. Geol.* **2019**, *103*, 549–563. [CrossRef]
45. Ronchi, P.; Cruciani, F. Continental carbonates as a hydrocarbon reservoir, an analog case study from the travertine of Saturnia, Italy. *AAPG Bull.* **2015**, *99*, 711–734. [CrossRef]
46. Della Porta, G. Carbonate build-ups in lacustrine, hydrothermal and fluvial settings: Comparing depositional geometry, fabric types and geochemical signature. *Geol. Soc. Lond. Spec. Publ.* **2015**, *418*, 17–68. [CrossRef]
47. Gandin, A.; Capezzuoli, E. Travertine: Distinctive depositional fabrics of carbonates from thermal spring systems. *Sedimentology* **2014**, *61*, 264–290. [CrossRef]

48. Alcicek, M.C.; Alcicek, H.; Altunel, E.; Arenas, C.; Bons, P.; Brogi, A.; Capezzuoli, E.; de Riese, T.; Della Porta, G.; Gandin, A.; et al. Comment on "First records of syn-diagenetic non-tectonic folding in Quaternary thermogene travertines caused by hydrothermal incremental veining" by Billi et al. Tectonophysics 700–701 (2017) 60–79. *Tectonophysics* **2017**, *721*, 491–500. [CrossRef]
49. Jones, B.; Peng, X.T. Intrinsic versus extrinsic controls on the development of calcite dendrite bushes, Shuzhishi Spring, Rehai geothermal area, Tengchong, Yunnan Province, China. *Sediment. Geol.* **2012**, *249*, 45–62. [CrossRef]
50. Jones, B.; Renaut, R.W. Cyclic development of large, complex, calcite dendrite crystals in the Clinton travertine, Interior British Columbia, Canada. *Sediment. Geol.* **2008**, *203*, 17–35. [CrossRef]
51. Jones, B.; Renaut, R.W.; Owen, R.B.; Torfason, H. Growth patterns and implications of complex dendrites in calcite travertines from Lysuholl, Sn ae fellsnes, Iceland. *Sedimentology* **2005**, *52*, 1277–1301. [CrossRef]
52. Luo, L.C.; Wen, H.G.; Li, Y.; You, Y.X.; Luo, X.T. Mineralogical, crystal morphological, and isotopic characteristics of smooth slope travertine deposits at Reshuitang, Tengchong, China. *Sediment. Geol.* **2019**, *381*, 29–45. [CrossRef]
53. Kano, A.; Okumura, T.; Takashima, C.; Shiraishi, F. *Geomicrobiological Properties and Processes of Travertine: With a Focus on Japanese Sites*; Springer: Singapore, 2019.
54. Jones, B.; Peng, X.T. Hot spring deposits on a cliff face: A case study from Jifei, Yunnan Province, China. *Sediment. Geol.* **2014**, *302*, 1–28. [CrossRef]
55. Chafetz, H.S.; Folk, R.L. Travertines; depositional morphology and the bacterially constructed constituents. *J. Sediment. Res.* **1984**, *54*, 289–316. [CrossRef]
56. Chafetz, H.S.; Guidry, S.A. Deposition and diagenesis of Mammoth Hot Springs travertine, Yellowstone National Park, Wyoming, U.S.A. *Can. J. Earth Sci.* **2003**, *40*, 1515–1529. [CrossRef]
57. Jones, B.; Renaut, R.W. Noncrystallographic Calcite Dendrites from Hot-Spring Deposits at Lake Bogoria, Kenya. *J. Sediment. Res. Sect. A-Sediment. Petrol. Process.* **1995**, *65*, 154–169.
58. Olsson, J.; Stipp, S.L.S.; Gislason, S.R. Element scavenging by recently formed travertine deposits in the alkaline springs from the Oman Semail Ophiolite. *Mineral. Mag.* **2014**, *78*, 1479–1490. [CrossRef]
59. Qin, J.; Han, P.; Che, X.; Yuan, G.; Fang, B.; Wang, G. Resuming the Holocene paleoclimate using delta~(18)O and trace elements of travertine in Rongma area, Tibet. *Earth Sci. Front.* **2014**, *21*, 312–322.
60. Teboul, P.A.; Durlet, C.; Gaucher, E.C.; Virgone, A.; Girard, J.P.; Curie, J.; Lopez, B.; Camoin, G.F. Origins of elements building travertine and tufa: New perspectives provided by isotopic and geochemical tracers. *Sediment. Geol.* **2016**, *334*, 97–114. [CrossRef]
61. Porras-Toribio, I.; Pi-Puig, T.; Villanueva-Estrada, R.E.; Rubio-Ramos, M.A.; Sole, J. Mineralogy, Geochemistry, and Stable Isotopes (C, O, S) of Hot Spring Waters and Associated Travertines near Tamiahua Lagoon, Veracruz, Gulf of Mexico (Mexico). *Minerals* **2022**, *12*, 822. [CrossRef]
62. Jones, B.; Peng, X.T. Mineralogical, crystallographic, and isotopic constraints on the precipitation of aragonite and calcite at Shigiang and other hot springs in Yunnan Province, China. *Sediment. Geol.* **2016**, *345*, 103–125. [CrossRef]
63. Liu, Z.H.; Li, H.C.; You, C.F.; Wan, N.J.; Sun, H.L. Thickness and stable isotopic characteristics of modern seasonal climate-controlled sub-annual travertine laminas in a travertine-depositing stream at Baishuitai, SW China: Implications for paleoclimate reconstruction. *Environ. Geol.* **2006**, *51*, 257–265. [CrossRef]
64. Sun, H.; Liu, Z. Wet–dry seasonal and spatial variations in the $\delta^{13}C$ and $\delta^{18}O$ values of the modern endogenic travertine at Baishuitai, Yunnan, SW China and their paleoclimatic and paleoenvironmental implications. *Geochim. Cosmochim. Acta* **2010**, *74*, 1016–1029. [CrossRef]
65. Blasco, M.; Auque, L.F.; Gimeno, M.J.; Asta, M.P.; Mandado, J. Stable isotope characterisation of recent aragonite travertine deposits associated with the Fitero thermal waters (Spain). *Int. J. Earth Sci.* **2020**, *109*, 877–892. [CrossRef]
66. Kele, S.; Ozkul, M.; Forizs, I.; Gokgoz, A.; Baykara, M.O.; Alcicek, M.C.; Nemeth, T. Stable isotope geochemical study of Pamukkale travertines: New evidences of low-temperature non-equilibrium calcite-water fractionation. *Sediment. Geol.* **2011**, *238*, 191–212. [CrossRef]
67. Mohammadzadeh, H.; Daneshvar, M.R.M. A comparison of hydro-geochemistry and stable isotope composition of travertine-depositing springs, Garab in NE Iran and Pamukkale in SW Turkey. *Carbonates Evaporites* **2020**, *35*, 23. [CrossRef]
68. Liao, Z.J. *Thermal Springs and Geothermal Energy in the Qinghai-Tibetan Plateau and the Surroundings*; Springer: Singapore, 2018.
69. Yi, F. Carbon Isotopic Records and Its Paleoenvironmental Significance during the Triassic-jurassic Transition. in Qiangtang Area, Tibet. Ph.D. Thesis, Chengdu University of Technology, Chengdu, China, 2018. (In Chinese with English Abstract).
70. Fu, X.G.; Wang, J.; Chen, W.B.; Feng, X.L.; Wang, D.; Song, C.Y.; Zeng, S.Q. Elemental geochemistry of the early Jurassic black shales in the Qiangtang Basin, eastern Tethys: Constraints for palaeoenvironment conditions. *Geol. J.* **2016**, *51*, 443–454. [CrossRef]
71. Wang, Y.S.; Zheng, C.Z. Lithostratigraphy, Sequence Stratigraphy, and Biostratigraphy of the Su-Obucha and Quse Formations and the Triassic-Jurassic Boundary in The Sewa Area on The South Margin of the Qiangtang Basin, Northern Tibet. *J. Stratigr.* **2007**, *31*, 377–384, (In Chinese with English Abstract).
72. Luo, L.C.; Wen, H.G.; Capezzuoli, E. Travertine deposition and diagenesis in Ca-deficiency perched hot spring systems: A case from Shihuadong, Tengchong, China. *Sediment. Geol.* **2021**, *414*, 17. [CrossRef]
73. Jones, B.; Renaut, R.W.; Rosen, M.R. High-temperature (>90 degrees C) calcite precipitation at Waikite Hot Springs, North Island, New Zealand. *J. Geol. Soc.* **1996**, *153*, 481–496. [CrossRef]

74. Renaut, R.W.; Owen, R.B.; Jones, B.; Tiercelin, J.J.; Tarits, C.; Ego, J.K.; Konhauser, K.O. Impact of lake-level changes on the formation of thermogene travertine in continental rifts: Evidence from Lake Bogoria, Kenya Rift Valley. *Sedimentology* **2013**, *60*, 428–468. [CrossRef]
75. Zhao, Y.Y.; Wei, W.; Santosh, M.; Hu, J.; Wei, H.T.; Yang, J.; Liu, S.; Zhang, G.L.; Yang, D.D.; Li, S.Z. A review of retrieving pristine rare earth element signatures from carbonates. *Paleogeogr. Paleoclimatol. Paleoecol.* **2022**, *586*, 18. [CrossRef]
76. McLennan, S.M. *Rare Earth Elements in Sedimentary Rocks: Influence of Provenance and Sedimentary Processes*; De Gruyter: Berlin, Germany, 1989. [CrossRef]
77. Lawrence, M.G.; Greig, A.; Collerson, K.D.; Kamber, B.S. Rare earth element and yttrium variability in South East Queensland waterways. *Aquat. Geochem.* **2006**, *12*, 39–72. [CrossRef]
78. Wang, Z.J.; Meyer, M.C.; Hoffmann, D.L. Sedimentology, petrography and early diagenesis of a travertine-colluvium succession from Chusang (southern Tibet). *Sediment. Geol.* **2016**, *342*, 218–236. [CrossRef]
79. Jones, B.; Peng, X.T. Growth and development of spring towers at Shiqiang, Yunnan Province, China. *Sediment. Geol.* **2017**, *347*, 183–209. [CrossRef]
80. Kele, S.; Vaselli, O.; Szabó, C.; Minissale, A. Stable isotope geochemistry of Pleistocene travertine from Budakalász (Buda Mts, Hungary). *Acta Geol. Hung.* **2003**, *46*, 161–175. [CrossRef]
81. Pentecost, A. The quaternary travertine deposits of Europe and Asia Minor. *Quat. Sci. Rev.* **1995**, *14*, 1005–1028. [CrossRef]
82. Pentecost, A. *Travertine*; Springer: Berlin/Heidelberg, Germany, 2005.
83. Pentecost, A.; Viles, H. A Review and Reassessment of Travertine Classification. *Géographie Phys. et Quat.* **1994**, *48*, 305–314. [CrossRef]
84. Panichi, C.; Tongiorgi, E.; Panichi, C.; Tongiorgi, E. Carbon isotopic composition of CO_2 from springs, fumaroles, mofettes and travertines of central and southern Italy: A preliminary prospection method of geothermal areas. In Proceedings of the Second United Nations Symposium on the Development and Use of Geothermal, San Francisco, CA, USA, 20 May 1975; Volume 2, pp. 815–825.
85. Fu, X.G.; Wang, J.; Wen, H.G.; Song, C.Y.; Wang, Z.W.; Zeng, S.Q.; Feng, X.L.; Wei, H.Y. A Toarcian Ocean Anoxic Event record from an open-ocean setting in the eastern Tethys: Implications for global climatic change and regional environmental pertur-bation. *Sci. China Earth Sci.* **2021**, *64*, 1860–1872, (In Chinese with English Abstract). [CrossRef]
86. Prokoph, A.; Shields, G.A.; Veizer, J. Compilation and time-series analysis of a marine carbonate $\delta^{18}O$, $\delta^{13}C$, $^{87}Sr/^{86}Sr$ and $\delta^{34}S$ database through Earth history. *Earth-Sci. Rev.* **2008**, *87*, 113–133. [CrossRef]
87. Vogel, J.C. 4—Variability of Carbon Isotope Fractionation during Photosynthesis. In *Stable Isotopes and Plant Carbon-Water Relations*; Ehleringer, J.R., Hall, A.E., Farquhar, G.D., Eds.; Academic Press: San Diego, CA, USA, 1993; pp. 29–46.
88. Yi, H.S. The carbon isotope fluctuation and its origin interpretion during the Early to Middle Jurassic transition period in the Shuanghu area of Qiangtang Basin. *Sediment. Geol. Tethyan Geol.* **2021**, *41*, 505–511, (In Chinese with English Abstract).
89. Luo, M.; Huang, H.G.; Zhang, P.; Wu, Q.H.; Chen, D.F. Origins of gas discharging from the Qiangtang Basin in the northern Qinghai-Tibet Plateau, China: Evidence from gas compositions, helium, and carbon isotopes. *J. Geochem. Explor.* **2014**, *146*, 119–126. [CrossRef]
90. Yokoyama, T.; Nakai, S.i.; Wakita, H. Helium and carbon isotopic compositions of hot spring gases in the Tibetan Plateau. *J. Volcanol. Geotherm. Res.* **1999**, *88*, 99–107. [CrossRef]
91. Wang, R.Z.; Ma, L.N. Climate-driven C4 plant distributions in China: Divergence in C4 taxa. *Sci. Rep.* **2016**, *6*, 27977. [CrossRef] [PubMed]
92. Fu, X.G.; Wang, J.; Tan, F.W.; Feng, X.L.; Chen, W.B.; Song, C.Y.; Zeng, S.Q. Sr and Nd isotopic systematics of mid-cretaceous organic-rich rocks (oil shales) from the qiangtang basin: Implications for source regions and sedimentary paleoenvironment. *Oil Shale* **2015**, *32*, 109–123. [CrossRef]
93. Fu, X.G.; Wang, J.; Zeng, Y.H.; Tan, F.W.; Feng, X.L. Source regions and the sedimentary paleoenvironment of marine oil shale from the bilong co area, Northern Tibet, China: An SR-ND isotopic study. *Oil Shale* **2012**, *29*, 306–321. [CrossRef]
94. McArthur, J.M.; Howarth, R.J.; Shields, G.A.; Zhou, Y. Chapter 7—Strontium Isotope Stratigraphy. In *The Geologic Time Scale*; Gradstein, F.M., Ogg, J.G., Schmitz, M.D., Ogg, G.M., Eds.; Elsevier: Boston, MA, USA, 2012; pp. 127–144.
95. Hu, F.Z.; Fu, X.G.; Lin, L.; Song, C.Y.; Wang, Z.W.; Tian, K.Z. Marine Late Triassic-Jurassic carbon-isotope excursion and biological extinction records: New evidence from the Qiangtang Basin, eastern Tethys. *Glob. Planet. Change* **2020**, *185*, 103093. [CrossRef]
96. Bolhar, R.; Van Kranendonk, M.J. A non-marine depositional setting for the northern Fortescue Group, Pilbara Craton, inferred from trace element geochemistry of stromatolitic carbonates. *Precambrian Res.* **2007**, *155*, 229–250. [CrossRef]
97. Elderfield, H.; Upstill-Goddard, R.; Sholkovitz, E.R. The rare earth elements in rivers, estuaries, and coastal seas and their significance to the composition of ocean waters. *Geochim. et Cosmochim. Acta* **1990**, *54*, 971–991. [CrossRef]
98. Frimmel, H.E. Trace element distribution in Neoproterozoic carbonates as palaeoenvironmental indicator. *Chem. Geol.* **2009**, *258*, 338–353. [CrossRef]
99. Goldstein, S.J.; Jacobsen, S.B. Rare earth elements in river waters. *Earth Planet. Sci. Lett.* **1988**, *89*, 35–47. [CrossRef]
100. Nothdurft, L.D.; Webb, G.E.; Kamber, B.S. Rare earth element geochemistry of Late Devonian reefal carbonates, canning basin, Western Australia: Confirmation of a seawater REE proxy in ancient limestones. *Geochim. Et Cosmochim. Acta* **2004**, *68*, 263–283. [CrossRef]

101. Olivier, N.; Boyet, M. Rare earth and trace elements of microbialites in Upper Jurassic coral- and sponge-microbialite reefs. *Chem. Geol.* **2006**, *230*, 105–123. [CrossRef]
102. Van Kranendonk, M.J.; Webb, G.E.; Kamber, B.S. Geological and trace element evidence for a marine sedimentary environment of deposition and biogenicity of 3.45 Ga stromatolitic carbonates in the Pilbara Craton, and support for a reducing Archaean ocean. *Geobiology* **2003**, *1*, 91–108. [CrossRef]
103. Zhao, Y.Y.; Zheng, Y.F.; Chen, F.K. Trace element and strontium isotope constraints on sedimentary environment of Ediacaran carbonates in southern Anhui, South China. *Chem. Geol.* **2009**, *265*, 345–362. [CrossRef]
104. Bayon, G.; German, C.R.; Burton, K.W.; Nesbitt, R.W.; Rogers, N. Sedimentary Fe-Mn oxyhydroxides as paleoceanographic archives and the role of aeolian flux in regulating oceanic dissolved REE. *Earth Planet. Sci. Lett.* **2004**, *224*, 477–492. [CrossRef]
105. Zhao, Y.Y.; Wei, W.; Li, S.Z.; Yang, T.; Zhang, R.X.; Somerville, I.; Santosh, M.; Wei, H.T.; Wu, J.Q.; Yang, J.; et al. Rare earth element geochemistry of carbonates as a proxy for deep-time environmental reconstruction. *Paleogeogr. Paleoclimatol. Paleoecol.* **2021**, *574*, 21. [CrossRef]
106. Dehler, C.M.; Elrick, M.; Bloch, J.D.; Crossey, L.J.; Karlstrom, K.E.; Marais, D.J.D. High-resolution $\delta^{13}C$ stratigraphy of the Chuar Group (ca. 770–742 Ma), Grand Canyon: Implications for mid-Neoproterozoic climate change. *GSA Bull.* **2005**, *117*, 32–45. [CrossRef]
107. Kaufman, A.J.; Knoll, A.H. Neoproterozoic variations in the C-isotopic composition of seawater: Stratigraphic and biogeochemical implications. *Precambrian Res.* **1995**, *73*, 27–49. [CrossRef]
108. Le Guerroué, E.; Allen, P.A.; Cozzi, A. Chemostratigraphic and sedimentological framework of the largest negative carbon isotopic excursion in Earth history: The Neoproterozoic Shuram Formation (Nafun Group, Oman). *Precambrian Res.* **2006**, *146*, 68–92. [CrossRef]
109. Debruyne, D.; Hulsbosch, N.; Muchez, P. Unraveling rare earth element signatures in hydrothermal carbonate minerals using a source-sink system. *Ore Geol. Rev.* **2016**, *72*, 232–252. [CrossRef]
110. Fu, X.G.; Wang, J.; Tan, F.W.; Chen, M.; Chen, W.B. The Late Triassic rift-related volcanic rocks from eastern Qiangtang, northern Tibet (China): Age and tectonic implications. *Gondwana Res.* **2010**, *17*, 135–144. [CrossRef]
111. Fu, X.G.; Wang, M.; Zeng, S.Q.; Feng, X.L.; Wang, D.; Song, C.Y. Continental weathering and palaeoclimatic changes through the onset of the Early Toarcian oceanic anoxic event in the Qiangtang Basin, eastern Tethys. *Paleogeogr. Paleoclimatol. Paleoecol.* **2017**, *487*, 241–250. [CrossRef]
112. Nie, Y.; Fu, X.G.; Xu, W.L.; Wen, H.G.; Wang, Z.W.; Song, C.Y. Redox conditions and climate control on organic matter accumulation and depletion during the Toarcian in the Qiangtang Basin, eastern Tethys. *Int. J. Earth Sci.* **2020**, *109*, 1977–1990. [CrossRef]
113. Wang, J.A.; Fu, X.G.; Chen, W.X.; Wang, Z.J.; Tan, F.W.; Chen, M.; Zhuo, J.W. Chronology and geochemistry of the volcanic rocks in Woruo Mountain region, northern Qiangtang depression: Implications to the Late Triassic volcanic-sedimentary events. *Sci. China Ser. D-Earth Sci.* **2008**, *51*, 194–205. [CrossRef]
114. Fu, X.G.; Wang, J.; Wang, Z.J.; Chen, W.X. U-Pb Zircon Age and Geochemical Characteristics of Volcanic Rocks from the Juhua Mountain Area in the northern Qiangtang Basin, northern Xizang(Tibet). *Geol. Rev.* **2008**, *54*, 232–242, (In Chinese with English Abstract).
115. Wang, Z.J.; Wang, J.; Tan, F.W.; Fu, X.G.; Chen, W.X.; Chen, M. Geochemical characteristics of volcanic rocks of the Upper Triassic Nadi Kangri Formation in the north Qiangtang basin, Qinghai-Tibet Plateau. *Geol. Bull. China* **2008**, *27*, 83–91, (In Chinese with English Abstract).
116. Barbieri, M.; Boschetti, T.; Petitta, M.; Tallini, M. Stable isotope (^{2}H, ^{18}O and $^{87}Sr/^{86}Sr$) and hydrochemistry monitoring for groundwater hydrodynamics analysis in a karst aquifer (Gran Sasso, Central Italy). *Appl. Geochem.* **2005**, *20*, 2063–2081. [CrossRef]
117. Hao, Y.L.; Pang, Z.H.; Kong, Y.L.; Tian, J.; Wang, Y.C.; Liao, D.W.; Fan, Y.F. Chemical and isotopic constraints on the origin of saline waters from a hot spring in the eastern coastal area of China. *Hydrogeol. J.* **2020**, *28*, 2457–2475. [CrossRef]
118. Vuataz, F.-D.; Goff, F.; Fouillac, C.; Calvez, J.-Y. A strontium isotope study of the VC-1 core hole and associated hydrothermal fluids and rocks from Valles Caldera, Jemez Mountains, New Mexico. *J. Geophys. Res. Solid Earth* **1988**, *93*, 6059–6067. [CrossRef]
119. Bau, M.; Dulski, P. Distribution of yttrium and rare-earth elements in the Penge and Kuruman iron-formations, Transvaal Supergroup, South Africa. *Precambrian Res.* **1996**, *79*, 37–55. [CrossRef]
120. Bau, M. Rare-earth element mobility during hydrothermal and metamorphic fluid-rock interaction and the significance of the oxidation state of europium. *Chem. Geol.* **1991**, *93*, 219–230. [CrossRef]
121. Wang, Y.C.; Li, L.; Wen, H.G.; Hao, Y.L. Geochemical evidence for the nonexistence of supercritical geothermal fluids at the Yangbajing geothermal field, southern Tibet. *J. Hydrol.* **2022**, *604*, 22. [CrossRef]
122. Möller, P.; Dulski, P.; De Lucia, M. REY Patterns and Their Natural Anomalies in Waters and Brines: The Correlation of Gd and Y Anomalies. *Hydrology* **2021**, *8*, 116. [CrossRef]
123. Priestley, S.C.; Karlstrom, K.E.; Love, A.J.; Crossey, L.J.; Polyak, V.J.; Asmerom, Y.; Meredith, K.T.; Crow, R.; Keppel, M.N.; Habermehl, M.A. Uranium series dating of Great Artesian Basin travertine deposits: Implications for palaeohydrogeology and palaeoclimate. *Paleogeogr. Paleoclimatol. Paleoecol.* **2018**, *490*, 163–177. [CrossRef]
124. Urey, H.C. The thermodynamic properties of isotopic substances. *J. Chem. Soc.* **1947**, 562–581. [CrossRef] [PubMed]
125. Kano, A.; Kawai, T.; Matsuoka, J.; Ihara, T. High-resolution records of rainfall events from clay bands in tufa. *Geology* **2004**, *32*, 793–796. [CrossRef]

126. Chen, F.H.; Zhang, J.F.; Liu, J.B.; Cao, X.Y.; Hou, J.Z.; Zhu, L.P.; Xu, X.K.; Liu, X.J.; Wang, M.D.; Wu, D.; et al. Climate change, vegetation history, and landscape responses on the Tibetan Plateau during the Holocene: A comprehensive review. *Quat. Sci. Rev.* **2020**, *243*, 21. [CrossRef]
127. Dykoski, C.A.; Edwards, R.L.; Cheng, H.; Yuan, D.; Cai, Y.; Zhang, M.; Lin, Y.; Qing, J.; An, Z.; Revenaugh, J. A high-resolution, absolute-dated Holocene and deglacial Asian monsoon record from Dongge Cave, China. *Earth Planet. Sci. Lett.* **2005**, *233*, 71–86. [CrossRef]
128. Wang, Y.; Cheng, H.; Edwards, R.L.; He, Y.; Kong, X.; An, Z.; Wu, J.; Kelly, M.J.; Dykoski, C.A.; Li, X. The Holocene Asian Monsoon: Links to Solar Changes and North Atlantic Climate. *Science* **2005**, *308*, 854–857. [CrossRef]

minerals

MDPI

Article

Travertines of the South-Eastern Gorny Altai (Russia): Implications for Paleoseismology and Paleoenvironmental Conditions

Evgeny V. Deev [1,2,*], Svetlana N. Kokh [1,3], Yuri Dublyansky [1,4], Ella V. Sokol [3], Denis Scholz [5], Gennady G. Rusanov [6] and Vadim N. Reutsky [1,3]

1 Department of Geology and Geophysics, Novosibirsk State University, 2 Pirogov Street, Novosibirsk 630090, Russia
2 A.A. Trofimuk Institute of Petroleum Geology and Geophysics, Siberian Branch of the Russian Academy of Sciences, 3 Acad. Koptyug Avenue, Novosibirsk 630090, Russia
3 V.S. Sobolev Institute of Geology and Mineralogy, Siberian Branch of the Russian Academy of Sciences, 3 Acad. Koptyug Avenue, Novosibirsk 630090, Russia
4 Institut für Geologie, Leopold-Franzens-Universität, Innrain 52, A-6020 Innsbruck, Austria
5 Institute of Geosciences, Johannes Gutenberg-Universität Mainz, J.-J.-Becher-Weg 21, 55128 Mainz, Germany
6 Gorno-Altaisk Expedition JSC, 15 Sovetskaya Street, Maloyeniseiskoe Village, Altai Territory 659370, Russia
* Correspondence: deevev@ngs.ru

Abstract: The south-eastern Gorny Altai is one of the most hazardous seismogenic area in the north of Central Asia. We present a synthesis of field, ^{230}Th-U geochronological, mineralogical and geochemical data collected on seven Quaternary travertines. All travertines occur within the zones of active faults that border the Chuya and Kurai intermontane basins. Travertine cement mainly comprises calcite (with minor amounts of aragonite), which cements alluvial, alluvial fan, and colluvial deposits. The results of ^{230}Th-U dating suggest that deposition of the travertines was triggered by large paleoearthquakes in the last eight thousand years. Several stages of travertine formation with ages 9–11 ka BP correspond to the known period of strong paleoseismicity in the region (8–16 ka BP). The 123 ka BP travertine resulted from a slip triggered by the Middle Pleistocene deglaciation, while that of 400 ka BP represents seismic motions likely associated with the main Cenozoic orogenic phase. All travertine forming events fall within warm and wet climatic phases (interglacials). Large earthquakes activated faults and caused a rapid rise along them of ambient-temperature bicarbonate groundwater, which was previously sealed in deep-seated Upper Neoproterozoic–Paleozoic limestone-dolostone aquifers. Rapid CO_2 degassing of the spring water was the most important control of calcite or aragonite precipitation. Such travertines represent an important tool for paleoseismological research in seismically active regions.

Keywords: large paleoearthquake; active fault; travertine; paleoenvironmental conditions; ^{230}Th-U dating; Gorny Altai

Citation: Deev, E.V.; Kokh, S.N.; Dublyansky, Y.; Sokol, E.V.; Scholz, D.; Rusanov, G.G.; Reutsky, V.N. Travertines of the South-Eastern Gorny Altai (Russia): Implications for Paleoseismology and Paleoenvironmental Conditions. *Minerals* **2023**, *13*, 259. https://doi.org/10.3390/min13020259

Academic Editors: Francesca Giustini, Mauro Brilli and János Haas

Received: 19 December 2022
Revised: 4 February 2023
Accepted: 7 February 2023
Published: 12 February 2023

1. Introduction

The term 'travertine' refers to a large group of freshwater carbonate rocks that precipitate to form bicarbonate water upon or near the surface. As authors of many overviews note [1–3], there is no unified classification system for the calcareous fresh-water spring deposits, and differentiation between travertine and tufa is problematic. Pedley [4] and Ford and Pedley [5] defined travertines as carbonate precipitates from warm-to-hot hydrothermal water and tufas as precipitates from ambient-temperature water. Pentecost [6] and Pentecost and Viles [7] introduced the term 'cemented rudite' to describe surface-cemented rudites, consisting of cemented screes, alluvium, breccia, gravel, etc. We do not use the term 'cemented rudite' in our study.

'Travertine-cemented gravel' [8,9], 'conglomerate carbonate cements' [10], or 'carbonate cement in gravels' [11,12] are considered as an independent lithofacies genetically related to a travertine-deposition system. This carbonate cement precipitated from ascending bicarbonate waters discharged into the permeable colluvium, till, alluvium, and mixed debris sediments. This process leads to the deposition of variably inclined large bodies (up to several square kilometres). Obviously, erosion is the reason why we often are dealing with the remnants of the originally large bodies and have not found travertine lithofacies associated with vent or slope environments, feeding channels along active faults, calcite veins associated with fractures, fluid escape features, etc. The incorrect definition of the 'travertine-cemented gravel' lithofacies leads to the loss of a huge archive of paleoseismological and paleoclimatic information on mountainous areas. According to Pentecost [1], the term 'travertine', in its broadest sense, refers to all chemically precipitated continental carbonate precipitates formed in or near seepages, springs, along streams, rivers, and occasionally in lakes. It is in this way that the term 'travertine' is applied for both travertine and tufa fossil deposits by Claes et al. [13,14]. In our study, the term 'travertine' is always used in its broadest sense.

Travertines can be classified according to several criteria, the most important being the stable isotope composition, geochemistry, fabric and morphology, and the depth at which the travertine-forming fluids circulated [1,7,14–16]. Pentecost [1] divided travertines into 'meteogene' and 'thermogene' according to the carrier of CO_2 incorporated in travertine formation. The CO_2 involved in 'meteogene' travertine precipitation has meteoric origin (vegetation, soil, and atmosphere). Waters of 'meteogene' travertines are often characterized by ambient temperatures. The CO_2 for 'thermogene' travertines originates from deeply seated thermal processes (magmatic CO_2, decarbonation of sedimentary calcareous parent rocks during regional and contact metamorphism). Travertines of thermogenic origin are enriched in ^{13}C. 'Thermogene' travertines often precipitated from hot spring water, but thermogene source waters are not necessarily hot [1]. According to Crossey et al. [15] waters containing deeply derived CO_2 (regardless of temperature) are classified as 'endogenic'.

The origin of travertine-forming fluids is frequently associated with different source rocks dissolved in non-hydrothermal superficial ('epigean') or hydrothermal deep ('hypogean') hydrogeological systems. The classification of travertines as 'epigean' or 'hypogean' was based on the stable isotope composition of $CaCO_3$ [16]. The Ba/Sr ratio was used in the reconstruction of source rock lithology prevailing in the hydrogeological or paleo-hydrogeological reservoir. In this study, the travertine classification is based on Teboult et al. [16].

Travertines may be relevant for paleoclimate studies, providing information on, e.g., the climate-controlled temperature of groundwater; provenance of atmospheric precipitation, recharging groundwater; characteristics of soils and intensity of biological activity in them; etc. [17–22]. Numerous studies on Quaternary travertines worldwide have demonstrated the close link between the circulation of travertine-forming fluids and active faulting [23–41]. Attempts have been made to link the mobilisation of CO_2-rich fluids to seismic triggers, to use travertines, which source fluids have deep origin, to determine the age and recurrence periods of paleoearthquakes, and to estimate slip rates and coseismic slip along seismogenic faults [42–49]. These travertines are also used for the estimation of the CO_2 flux from tectonically active regions in the Earth's atmosphere, necessary to understand the global geological carbon cycle [50].

Gorny Altai (49–52° N, 82–90° E) represents the northern end of the fan of the Mongolian Altai structures, and together with the latter is part of the system of intracontinental Cenozoic orogens in northern Central Asia, formed as a far-field response to the India–Eurasia convergence and collision [51]. The neotectonic crustal deformation and mountain growth in the Gorny Altai, especially intensified during the Quaternary period, produced 6 km of structural relief (between high mountain ranges and the bottoms of the large intermontane basins), based on the displacement of the Mesozoic pre-orogenic erosion surface [52–55]. They were accompanied by strong seismicity and occurred at times of significant climate change [56–62].

To date, several dozen travertine deposits have been documented in various areas of the Gorny Altai [63,64]. Importantly, in the south-eastern part of the Gorny Altai (Figure 1) all known travertine deposits are spatially associated with zones of active faults [41,48,65,66]. The goals of this study are: (1) the generalization of the available data on the structures of travertine deposits of the south-eastern part of the Gorny Altai, and their petrographic, mineralogical and geochemical features and ^{230}Th-U geochronology; (2) the correlation of episodes of travertine formation with movements along active faults, including the ages of paleoearthquakes; (3) the evaluation of the possible connection between travertine formation and the climate.

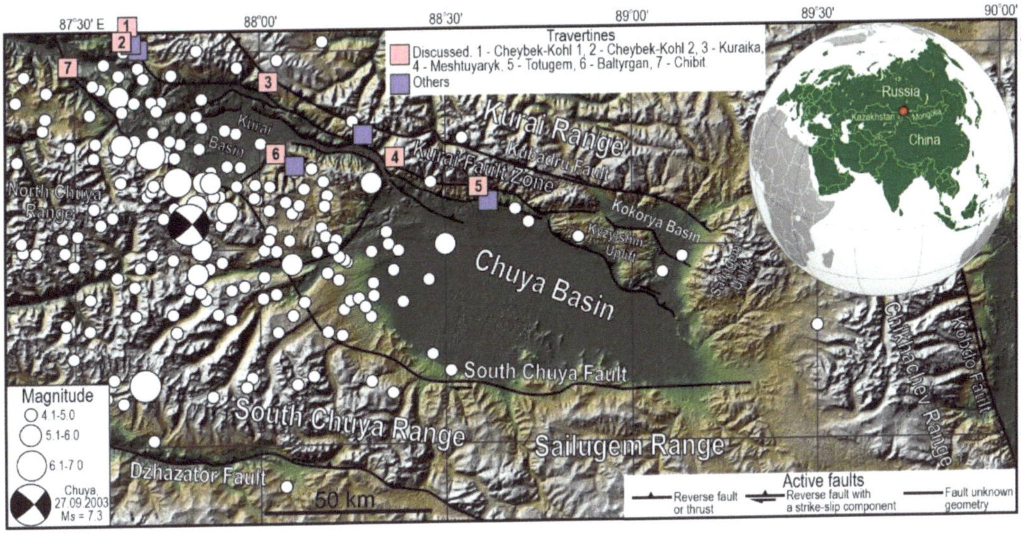

Figure 1. Main active faults, earthquakes, and travertines in the south-eastern Gorny Altai. Base map was made from GeoMapApp (http://www.geomapapp.org, accessed on 5 December 2022) [67].

2. Geological Setting

2.1. Morphostructures of the South-Western Gorny Altai

The main morphostructures in the south-western Gorny Altai are mountain ranges with altitudes of 3000–4000 m a.s.l.: North Chuya, South Chuya, Sayluygem, Chikhachev, and Kurai ranges. The ranges frame the Chuya (70 × 35 km), Kurai (35 km × 20 km), and Kokorya (23 km × 9 km) intermontane basins (Figure 1). The Chuya and Kurai basins are separated from each other by the Chagan Uplift (2900 m a.s.l.), and the Chuya and Kokorya basins by the Kyzylshin Uplift (2560 m a.s.l.).

Cenozoic siliciclastic infill of the basins overlies the deformed Late Neoproterozoic–Carboniferous and Jurassic volcanogenic, siliciclastic, and carbonate rocks (Figure 2) as well as relics of the Late Cretaceous–Paleogene weathering crust [68]. Controlled-source resistivity results show that the thickness of the infill reaches 1.0–1.6 km [53,54,56,69]. The basins were formed in the Paleogene–Neogene period as pull-apart structures, in which 600–800 m of fine-grained lacustrine (in the central parts of the basins) and coarser near-shore lacustrine and alluvial (closer to basin margins) sediments have accumulated [53,54,70,71]. In the Early to the earliest Middle Pleistocene period the main phase of the Cenozoic orogenesis and the growth of mountain ranges occurred. At that time the pull-apart structures transformed into basins bounded by thrusts and reverse faults. Up to 300 m of coarse brown intermountain molasse have accumulated in the basins [52–54,71]. The upper part of the sedimentary fill of the basins comprises Middle Pleistocene–Holocene glacial, fluvioglacial, alluvial, limnic, mud-debris, colluvial, and aeolian sediments [53,56,63,72].

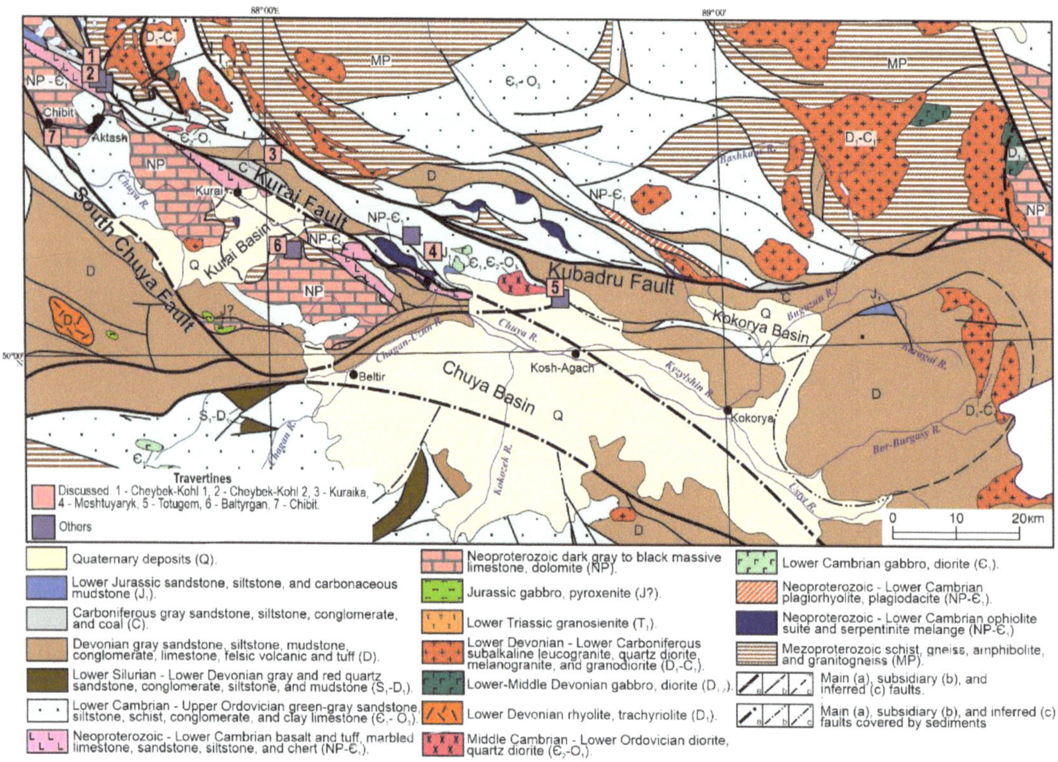

Figure 2. Geological map of the south-western Gorny Altai (simplified from [73]), showing locations of travertine occurrences.

2.2. Active Faults and Seismicity

The Kurai and Chuya basins are bordered to the south by the active South Chuya Fault (Figure 1), separating them from the North- and South Chuya ranges [74,75]. The Chuya (the Altai) earthquake (27 September 2003; Ms = 7.3, I = IX) with its epicenter at the back of the Chagan Uplift, the largest of the historic and instrumental Gorny Altai earthquakes, was associated with the South Chuya Fault (Figure 1). Its largest aftershocks occurred on 27 September 2003 (Ms = 6.4) and 1 October 2003 (Ms = 6.6). The main shock and the largest aftershocks had a dextral strike-slip focal mechanism with a NW trending plane [76]. The earthquake caused the formation of surface ruptures exceeding 70 km in length. In the vast areas of the Chuya and Kurai basins, multiple secondary seismic deformations, triggered by this large seismic event, have been detected [77,78].

The structure at the northern margin of the Chuya, Kurai, and Kokorya basins (Figure 1) is controlled by the active Kurai Fault Zone (KFZ). The positive flower structure of the KFZ with the Kubadru master fault is the result of transpression [52,71,74]. The Kubadru Fault dips towards the N-NE reverse fault with a dextral strike-slip component. To the south, there is a series of sub-parallel reverse faults and thrusts along which the Kurai Range thrusts over the basins. The backthrusts with a southern dip are pinnate to them. The displacements along the two fault systems lead to the growth of forebergs at the ridge front. Forebergs separate narrow (from a few meters to a few kilometres) negative morphostructures from the Chuya, Kurai, and Kokorya basins [57,71,74,79,80]. The regional seismological catalogues contain records only $M \leq 5$ instrumental and historic events along the KFZ. Morphostructural and seismological data also indicate that the faults associated with the Chagan and Kyzylshin uplifts are still active [48,71,80].

Numerous fault scarps and large landslides known in the area indicate that abundant large Late Pleistocene–Holocene paleoearthquakes are associated with the South Chuya Fault and the KFZ [57,74,77]. Trenching studies, supported by radiocarbon, optically stimulated luminescence, and infrared-stimulated luminescence dating revealed that at least 12 large paleoearthquakes with Mw of 6.6 to 7.6 occurred along the Kurai faults over the past 16 thousand years [49,57,71,74,79,81].

2.3. Climate, Permafrost, and Hydrogeology

The south-eastern Gorny Altai is characterized by a sharp continental climate and desert-steppe landscapes. The mean temperatures in January are from −26 to −30 °C, and in July 10 to 14 °C; the mean annual temperature is −6 °C. The areal distribution of precipitation is heterogeneous: the annual amounts range from 67 to 300 mm in the intermontane depressions and 600 to 800 mm on the slopes of the mountain ranges [82].

The Chuya and Kurai depressions and the surrounding mountain ranges are affected by permafrost, whose thickness ranges from 6 to more than 100 m [82,83]. On the surface, permafrost manifests as patterned ground and pingos. Numerous thermokarst lakes are associated with permafrost degradation.

The Chuya and Kurai intermontane basins host the largest artesian basins of the Gorny Altai, whose aquifers are associated with Neogene-Quaternary sediments. Fissure waters are ubiquitous in the bedrock of the basin rim, often associated with fault zones. Such waters discharge as descending low-yield springs, most of which are known in the Kurai Range. Fissure waters are recharged primarily by the infiltration of meteoric precipitation and meltwater. Fresh (salinity up to 0.5 g/L) bicarbonate Ca-Mg, Ca and Ca-Na groundwaters prevail in the area. Bicarbonate-sulphate Ca (0.5–1.0 g/L) and sulphate Na-Mg (up to 1–3 g/L) waters occur locally in the south-east of the Chuya Basin [84].

3. Materials and Methods

3.1. Field Studies and Sampling

Seven travertine occurrences in the south-eastern Gorny Altai were examined and sampled in the course of field expeditions in 2017–2021. The field documentation included the determination of the linear dimensions of travertine bodies, the evaluation of their relationships with the underlying and overlying sediments and rocks, and the mapping of zones of active faults and fracturing. Fifty-five samples of fossil travertines, five samples of modern precipitates, and three samples of spring water were collected. Spring water temperature (T) and pH values were measured in situ with a manual Hanna Instruments PH ORP Combo Meter & Temperature Gauge (HI98121) to a precision of 0.1 °C and 0.1 pH.

3.2. Petrography, Mineralogy and Mineral Chemistry

The petrographic study was assessed by optical petrography on thin sections in both transmitted and reflected light using an Olympus BX51 optical microscope at the Analytical Centre for Multi-Elemental and Isotope Research (Sobolev Institute of Geology and Mineralogy (IGM), Novosibirsk, Russia).

Mineral phases (≥1%) were identified by X-ray diffraction analysis (XRD) in powdered samples. All specimens were analysed on a Shimadzu XRD-600 diffractometer (Shimadzu Corporation, Kyoto, Japan) (Cu$K\alpha$ radiation with a graphite monochromator, λ = 1.54178 Å) at the South Urals Research Centre of Mineralogy and Geoecology (SU FRC MG) Miass, Russia). The scans were recorded from 6 to 60° 2θ at 0.05° 2θ increments with a 5 s scanning time per step.

Scanning electron microscopy (SEM) was applied to characterize the phase distribution and to identify minerals based on back-scattered electron (BSE) images, energy-dispersive spectra (EDS), and elemental maps (EDS system). The polished thin sections were sputter-coated with ~15–25 nm carbon films for SEM examination. The measurements were performed on a Tescan Mira 3MLU scanning electron microscope (Tescan Orsay Holding, Brno, Czech Republic) equipped with an Oxford AZtec Energy Xmax-50 microanalyses system (Oxford Instruments Nanoanalysis, Abingdon, UK), at IGM (Novosibirsk, Russia). An accelerating voltage of 20 kV and 1 nA beam current were used in low- (40–60 Pa) or high-vacuum modes at a 20 s count time.

The chemical composition of minerals was determined by wavelength-dispersive electron microprobe analysis (EMPA) on carbon-coated polished samples (IGM, Novosibirsk). Carbonates were analysed on a Jeol JXA 8100 electron microprobe microanalyzer (Jeol, Tokio, Japan) at an accelerating voltage of 20 keV, a beam current of 15 nA, and a peak counting time of 10 s. The mineral compositions were estimated with reference to natural and synthetic standards: albite (Na), diopside (Mg, Ca), $BaSO_4$ (Ba, S), pyrope (Fe), Sr-Si glass (Sr), Mn-almandine (Mn), fluorapatite (P), $Y_3Al_5O_{12}$ (Y). The detection limits for the elements were (in wt%): 0.02 for P; 0.03 for Na, Fe and Ca; 0.04 for Mn; 0.05 for S and Sr; 0.06 for Y; and 0.08 for Ba (3σ). The matrix correction using the ZAF algorithm (generalized algebraic procedure; assumes a linear relation between concentration and X-ray intensity) was applied to the raw data prior to recalculation into elements. Analytical accuracy was within 2 rel% for [C] > 5 wt% elements, and ca. 5 rel% for [C] < 2 wt% elements [85].

3.3. Rock Chemistry

Minor and trace elements, including REE, in isolated carbonate fractions from travertine samples were determined with inductively coupled plasma mass spectrometry (ICP-MS) on an Agilent 7700× spectrometer (Agilent Technologies, Inc., Santa Clara, CA, USA) at the SU FRC MG (Miass, Russia). Selective dissolution of $CaCO_3$ followed a modified procedure of Feng et al. [86]. Powdered samples (0.2 g) were rinsed in 50 mL of water (Milli-Q) in an ultrasonic bath. The obtained solutions were centrifuged and the supernatants were carefully decanted to remove any water-soluble salts. Then, the insoluble residues were dissolved in 5% HNO_3 in a centrifuge tube for 2–3 h to isolate the carbonate phase. The solutions were centrifuged and the supernatants were stored for analysis. The analyses were ran in triplicate, and the results differed for <5%. Precision and accuracy were estimated to be 10–15 rel% for all elements. The detection limits for trace elements were in the range 0.01–0.5 μg/l. The analytical reproducibility was monitored using a BCR-2 standard (U.S. Geological Survey, Denver, CO, USA), in which the measured values were consistent with the certified ones.

3.4. Stable Isotope Geochemistry of Carbonates

For the measurement of C and O stable isotopes in bulk travertine samples (Table S1), 26 powdered carbonate samples were dissolved in 100% phosphoric acid at 50 °C for over 48 h and then analysed on a Gas Bench II instrument coupled with a MAT 253 mass spectrometer (Thermo Finnigan, Waltham, MA, USA) at IGM (Novosibirsk, Russia). All isotope values are quoted in per mill (‰) using δ-notation relative to the Vienna Pee Dee Belemnite (V-PDB), with accuracy of <0.1‰ for $\delta^{13}C$ and 0.2‰ for $\delta^{18}O$ (1 σ).

Additionally, the 105 powdered carbonate samples were taken from 17 hand specimens (Table S1). For each specimen, multiple samples were taken across the growth zones of mineral aggregates. Powders were reacted with 99% orthophosphoric acid and analysed using a Delta V Plus isotope ratio mass spectrometer equipped with a Gasbench II (Thermo Fisher Scientific, Waltham, MA, USA) at the Institute of Geology, University of Innsbruck. The results were calibrated against international calcite reference materials and reported relative to the VPDB standard. The long-term precision for $\delta^{13}C$ and $\delta^{18}O$ was 0.06 and 0.08‰, respectively (1 σ [87]).

3.5. Water Chemistry and Stable Isotope Analysis

Major- and trace-element concentrations in three water samples were measured at the Institute of Geology, Innsbruck University by inductively coupled plasma–optical emission spectrometry (ICP-OES) on an Activa spectrometer (HORIBA Jobin Yvon SAS, Edison, NJ, USA) and by titration. Major cations and trace elements were analysed in 15 water aliquots which were stored separately and then acidified with 0.5 mL of 15 N distilled nitric acid; anions were determined in another 50 mL aliquot of unacidified water. Oxygen and hydrogen isotope compositions of H_2O were studied at the Institute of Geology, Innsbruck University using a L-2130-i analyser (Picarro, Santa Clara, CA, USA). The results were statistically processed and normalized against the Vienna Standard Mean Ocean Water (V-SMOW). The measurement accuracy was 0.1‰ for $\delta^{18}O$ and 0.4 for $\delta^{2}H$ (1σ). The PHREEQC software can be used to calculate the saturation index (SI) values of the minerals in water [88].

3.6. ^{230}*Th-U Geochronology*

Nineteen samples of travertines were dated by the ^{230}Th-U method. Thirteen samples were prepared and analysed at the Institute for Geosciences, University of Mainz, Germany, using a Neptune Plus multi-collector inductively coupled plasma mass spectrometer (MC-ICPMS) (Thermo Scientific, Bremen, Germany). Six samples were prepared and analysed at the Max Planck Institute for Chemistry, Mainz, Germany, using a Nu Plasma MC-ICPMS (Nu Instruments™; Nu Instrument Ltd., Wrexham, Wales, UK). Typical sample sizes were 0.3 g. Chemical separation of U and Th in both laboratories was performed as described by Yang et al. [89]. Details of the MC-ICPMS analyses in both laboratories are described by Obert et al. [90], a detailed description of the calibration of the mixed U-Th spike used in both laboratories is given by Gibert er al. [91]. To account for the potential effects of detrital contamination, all ages were corrected assuming an average upper continental crust ^{232}Th/^{238}U mass ratio of 3.8 for the detritus and ^{230}Th, ^{234}U and ^{238}U in secular equilibrium. All activity ratios were calculated using the half-lives of Cheng et al. [92]. Ages are reported in thousands of years before present (ka BP); age uncertainties are quoted at 2σ-level.

4. Results

4.1. *Description of the Travertines*

4.1.1. The Cheybek-Kohl 1 Travertines

The Cheybek-Kohl 1 travertine is located at the western flank of the KFZ (Figures 1 and 2), in close proximity to the active Kurai Fault line (Figure 3A). The area of the Kurai Fault in the terrain is expressed by a deep gorge that divides the Kurai and Aygulak ranges. The gorge hosts the Cheybek-Kohl lake, which formed as a result of the Chibitka River being dammed by a debris flow cone. The travertine body is located in the upper reaches of a creek, a left tributary of the Chibitka River, 168 m above the Cheybek-Kohl lake, at 1958 m a.s.l. (50.400556° N, 87.611389° E) (Figure 3A). In the modern relief, only a few remnants of the original large travertine complex have survived. In August 2015, the largest remnant with the size 12 × 6.5 × 2 m was found (Figure 3B). The valley hosts abundant boulders of breccia with travertine cement (Figure 3C). In travertines, calcium carbonate cements grey Holocene colluvium, forming breccias (Figure 3B-E). Carbonate also lines the walls of tensional fissures in underlying shale and/or fill them along with shale clasts (Figure 3F).

4.1.2. The Cheybek-Kohl 2 Travertines

The Cheybek-Kohl 2 travertine crops out 0.5 km west of the Cheybek-Kohl 1 travertines, at the shore of the lake with the same name (Figures 1, 2 and 3A; 50.40144° N, 87.60491° E; 1818 m a.s.l.). Travertine carbonate cements grey colluvium, which predominantly consists of fragments of Devonian shale, forming a breccia (Figure 4A–D). The apparent thickness of the outcrop is 2.5–4.0 m (Figure 4A). The travertine body can be traced for 29 m along the base of the mountain slope. Water seeps are observed in the south-eastern part of the breccia body (Figure 4E). At these discharge sites, fresh carbonate precipitates on the older breccia, forming a white-yellow or brownish laminated travertine crust, not exceeding 1–2 mm in thickness. The surface of the modern precipitates is coated by microbial colonies and algae (Figure 4F).

Figure 3. Location and morphology of the Cheybek-Kohl 1 and Cheybek-Kohl 2 travertines. (**A**) Active faults, location, and U-Th dates of the Cheybek-Kohl 1 (50.400556° N, 87.611389° E) and Cheybek-Kohl 2 (50.40144° N, 87.60491° E) travertines. (**B**) Remnant of the breccia with travertine cement capping the colluvium. (**C**) Boulders of breccia with travertine cement truncated by a creek. (**D**,**E**) White-yellow sparry and micritic calcite cementing Lower Devonian shale clasts; druse calcite aggregates are present in voids and small cavities. (**F**) Tensional fissures, lined with travertine carbonate and hosting shale clasts.

4.1.3. The Kuraika Travertines

The Kuraika travertines are located on the southern slope of the Kurai Range, overthrusted along the KFZ on the deposits of the Kurai Basin (Figures 1, 2 and 5A). The spring is located some 770 m above the bottom of the Kurai Depression (50.28034° N, 88.00408° E, 2029 m a.s.l.). The spring discharges from the debris, formed by the destruction of the Lower-Middle Devonian siltstones and sandstones (Figure 5B). At the spring orifice, the ferruginization of debris surfaces is notable. No signs of modern travertine formation were found (Figure 5C). A scree slope of yellowish colour can be traced from the spring down the slope for 200 m (Figure 5B). The colour of the talus is due to the abundance of travertine debris (Figure 5D). These fragments are formed during the destruction of travertine-cemented breccias. Small outcrops of such breccias can be found locally under the scree.

Figure 4. The Cheybek-Kohl 2 travertines. (**A**) Outcrop of breccia with travertine cement. (**B**,**C**) Close-ups: breccia with travertine cement. (**D**) Travertine crusts lining the inter-clast space. (**E**,**F**) Carbonate crusts covered with algae precipitating in the present seepage zone.

4.1.4. The Meshtuyaryk Travertines

The Meshtuyaryk travertines are located on the Kurai Fault line, immediately northeast of the Chagan Uplift (Figures 1, 2 and 6A). They crop out on the right side of the valley of the river with the same name, at the base of one of the triangular facets (50.13817° N, 88.34074° E, 1860 m a.s.l.). The outcrop of coarse-layered breccias with travertine cement is ca. 5 m thick, 15 m wide, and extends 80 m up the slope. The breccia overlies the ferruginous greenschists and phyllites. In the breccia, white, white-yellow and honey-yellow microsparitic-to-sparitic carbonate crusts, 2 cm in thickness, cement sand- to gravel-sized detrital clasts. The layering of breccias is the result of sorting of the colluvial material (Figure 6B–D). Along with greenschist and phyllite debris dominating in the breccias, fragments of dolerites, limestones, and calcite veins were found. There were no obvious angular unconformities in the body of the breccias that would suggest the presence of travertines of different ages.

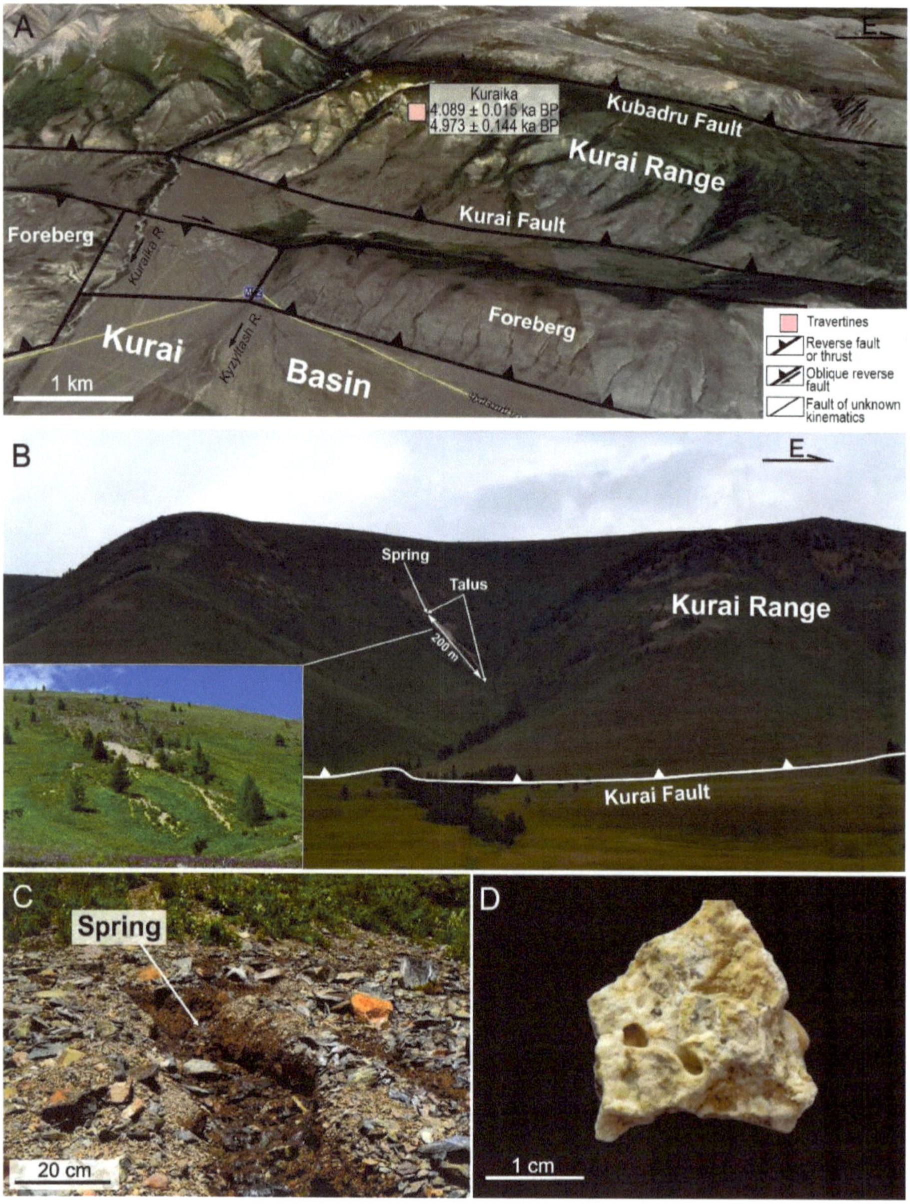

Figure 5. Location and morphology of the Kuraika travertines. (A) Structure of the Kurai Fault Zone and northern part of the Kurai Basin, showing the location and U-Th dates of the Kuraika travertines (50.28034° N, 88.00408° E). (B,C) Emerging spring and scree slope, composed of travertine on the southern slope of the Kurai Range. (D) Travertine fragment.

Figure 6. Location and morphology of the Meshtuyaryk travertines. (**A**) Active faults, location, and U-Th dates of the Meshtuyaryk travertines (50.13817° N, 88.34074° E). (**B**) Fragment of a layered breccia body with travertine cement. (**C**) Layered characteristics of travertine-cemented breccia; layering is due to variations in grain sizes of the clastic material. (**D**) Close-up: breccia with travertine cement.

4.1.5. The Totugem Travertines

The Totugem travertines are located on one of the faults, which is part of the KFZ structure at the junction of the Chuya Basin and the Kurai Range (Figures 1, 2 and 7A; 50.11039° N, 88.60268° E, 2434 m a.s.l.). The thrust crosses the Totugem River valley in its middle part, and forms in its sides a tectonic scarp ca. 300 m in height. During floods, the Totugem River intensively erodes the left side of the valley; as a result, a 6 m high cliff has formed. Its lower part is composed of strongly fissured siltstones (Middle Devonian) with a visible thickness of 3–5 m, overlain by a scree (Figure 7B). The lower part of this talus is cemented by travertine, making a 1–3 m thick and 30 m long breccia body (Figure 7C). The white-yellow carbonate crusts lining the debris surface reaches a thickness of 1.5 cm (Figure 7D).

4.1.6. The Baltyrgan Travertines

The Baltyrgan travertines are exposed within the forberg located north-west of the Chagan Uplift (Figures 1, 2 and 8A; 50.157524° N, 88.023709° E, 1700 m a.s.l.). The asymmetric foreberg is bordered from the south-east by an active thrust and rises above the bottom of the Kurai Depression from the south-west to the north-east from several tens to 100 m. Within the foreberg Middle Pleistocene alluvial fan gravels are involved in the uplift [56]. Late Proterozoic limestones and metamorphosed effusives of the Lower Cambrian age dominate in the debris, ranging in size from 1 cm to 30 cm. Travertine material cements the alluvial fan gravels, transforming them into conglomerate, roughly layered beds which dip towards the Chagan Uplift at 60–70° (Figure 8B). In conglomerates, carbonates line the pore space between clasts, forming crusts up to 0.5 cm thick (Figure 8C).

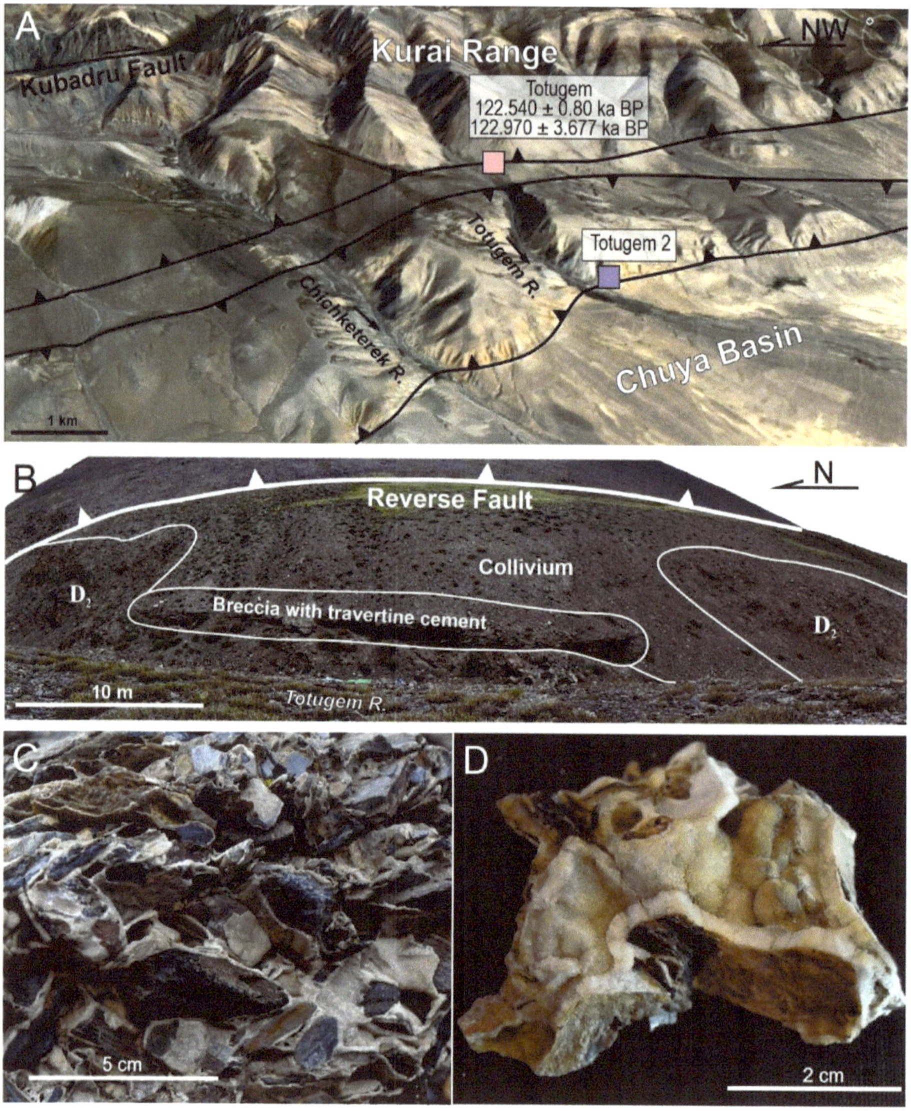

Figure 7. Location and morphology of the Totugem travertines. (**A**) Active faults, location, and U-Th dates of the Totugem travertines (50.11039° N, 88.60268° E). (**B**) Outcrop of the breccia with travertine cement on the left bank of the Totugem River. (**C**) Breccia with travertine cement. (**D**) Close-up: travertine cement of breccia.

4.1.7. The Chibit Travertines

Chibit travertines are confined to the north-western flank of the South Chuya Fault (Figures 1, 2 and 9A; 50.31277° N, 87.49337° E, 1187 m a.s.l.). In the erosional ledge of the left bank of the Chuya River valley, 15 m above the river, travertines cement alluvial sand and gravel, creating a 20 m body. The cementation produces sandstones and conglomerates up to 3 m thick (Figure 9B,C). In conglomerates, travertine crusts (several mm to 3 cm thick) coat the gravel surfaces and fill the free space between them. In the lower part of the outcrop, the seepage of water was observed. At the spring outlet, modern precipitates (crusts up to 0.2 cm thick) have been deposited (Figure 9D). The modern vegetation and soil are partially covered by powdery gypsum (Figure 9E).

Figure 8. Location and morphology of the Baltyrgan travertines. (**A**) Active faults, location, and U-Th dates of the Baltyrgan travertines (50.157524° N, 88.023709° E). (**B**) Inclined thick-bedded conglomerates with travertine cement. (**C**,**D**) Conglomerates with travertine cement.

Summarizing, the most common type of fossil travertine in the south-eastern Gorny Altai is represented by carbonate cements that bind the debris clasts of different genesis: colluvial, alluvial, and alluvial fan. Much less common travertines in the south-eastern Gorny Altai are the crystalline flowstone crusts (up to 2 mm thick), forming near the discharge sites of modern springs and seeps.

Figure 9. Location and morphology of the Chibit travertines. (**A**) Active faults, location, and U-Th date of the Chibit travertines (50.31277° N, 87.49337° E). (**B**,**C**) Sandstones and conglomerates formed by the cementation of alluvial deposits by travertines. (**D**) Travertine crusts, formed at the point of discharge of a modern spring. (**E**) Powdery gypsum aggregates covering modern vegetation and soil.

4.2. Hydrochemical Characteristics of Travertine Spring Water

All waters sampled at the travertine sites were slightly alkaline (pH = 7.9–8.2) (Table 1). The temperature recorded in the Cheybek-Kohl 2 and Kuraika springs were similar (3–4 °C, 3–4 °C, August 2021), water of the Chibit spring was substantially warmer (11 °C, 3–4 °C, August 2021). The TDS (total dissolved solids) values were 1629 mg/L at Chibit, 431 mg/L at Kuraika and 312 mg/L at Cheybek-Kohl 2. The Cheybek-Kohl 2 and Kuraika spring water were of HCO_3-SO_4–Ca-Mg and HCO_3-SO_4–Mg-Ca types, whereas Chibit water was of the SO_4-HCO_3–Mg-Ca type (Figure 10). The SO_4^{2-} content was the highest in the Chibit spring water (733.6 mg/L), whereas 13.6 mg/L and 18.9 mg/L were found at Kuraika and Cheybek-Kohl 2, respectively. Similar to SO_4^{2-}, HCO_3^- was higher in the Chibit spring water (507 mg/L). The HCO_3^- concentrations were 330 mg/L at Kuraika and 224 mg/L at Cheybek-Kohl 2. Ca^{2+} and Mg^{2+} were the dominant cations in all the sampled spring waters (mg/L): 35 and 45 at Kuraika, 34 and 28 at Cheybek-Kohl 2, and 211 and 150 at Chibit. All the sampled spring waters showed low concentrations of Na^+ (3–16 mg/L), K^+ (1–2 mg/L) and Cl^- (0.7–1.6 mg/L). Travertine spring water at the Kuraika, Cheybek-Kohl 2 and Chibit sites were characterized with similar anion and cation compositions in 2015 and 2020 (Figure 10). The saturation index values of calcite and aragonite for all the sampled water samples were greater than 0: $SI_{calcite}$ = 0.45 and $SI_{aragonite}$ = 0.29 for Kuraika; $SI_{calcite}$ = 0.09 and $SI_{aragonite}$ = 0.25 for Cheybek-Kohl 2; and $SI_{calcite}$ = 1.00 and $SI_{aragonite}$ = 0.85 for Chibit (Table 1).

Table 1. Chemistry and stable isotope composition of waters of the travertine springs of the south-eastern Gorny Altai.

Spring	Kuraika	Cheybek-Kohl 2	Chibit
Location	50.280503° N 88.00398 °E 2258 m a.s.l.	50.40144° N 87.60491° E 1818 m a.s.l.	50.31277° N 87.49337° E 1187 m a.s.l.
Sample ID	PN-2021/2.1	PN-2021/4.1	PN-2021/5.1
Year of sampling	2021	2021	2021
T (°C)	3	4	11
pH	8.21	8.14	7.92
TDS (mg/L)	431	312	1629
K^+ (mg/L)	0.9	1.0	2.5
Na^+ (mg/L)	4.9	2.9	15.9
Ca^{2+} (mg/L)	34.6	34.3	211.4
Mg^{2+} (mg/L)	44.8	27.9	150.4
Ba^{2+} (mg/L)	0.009	0.010	0.017
Sr^{2+} (mg/L)	0.25	0.35	5.14
Al^{3+} (mg/L)	0.022	0.021	0.051
Cl^- (mg/L)	0.67	1.59	1.21
SO_4^{2-} (mg/L)	13.56	18.85	733.6
NO_3^- (mg/L)	1.36	0.61	1.66
HCO_3^- (mg/L)	330	224	507
Mg/Ca	1.29	0.82	0.71
$\delta^{18}O$ (‰ VSMOW)	−17.4	−15.9	−16.9
δ^2H (‰ VSMOW)	−126.7	−114.6	−125.5
$SI_{calcite}$	0.45	0.09	1.00
$SI_{aragonite}$	0.29	0.25	0.85

The Chibit and Kuraika spring waters had similar stable isotope compositions: $\delta^{18}O$ = −16.9 and −17.4‰ and δD = −126 and −127‰, respectively. The heaviest isotopic composition was typical of the Cheybek-Kohl 2 spring water ($\delta^{18}O$ = −15.9‰; δD = −145‰) (Figure 11).

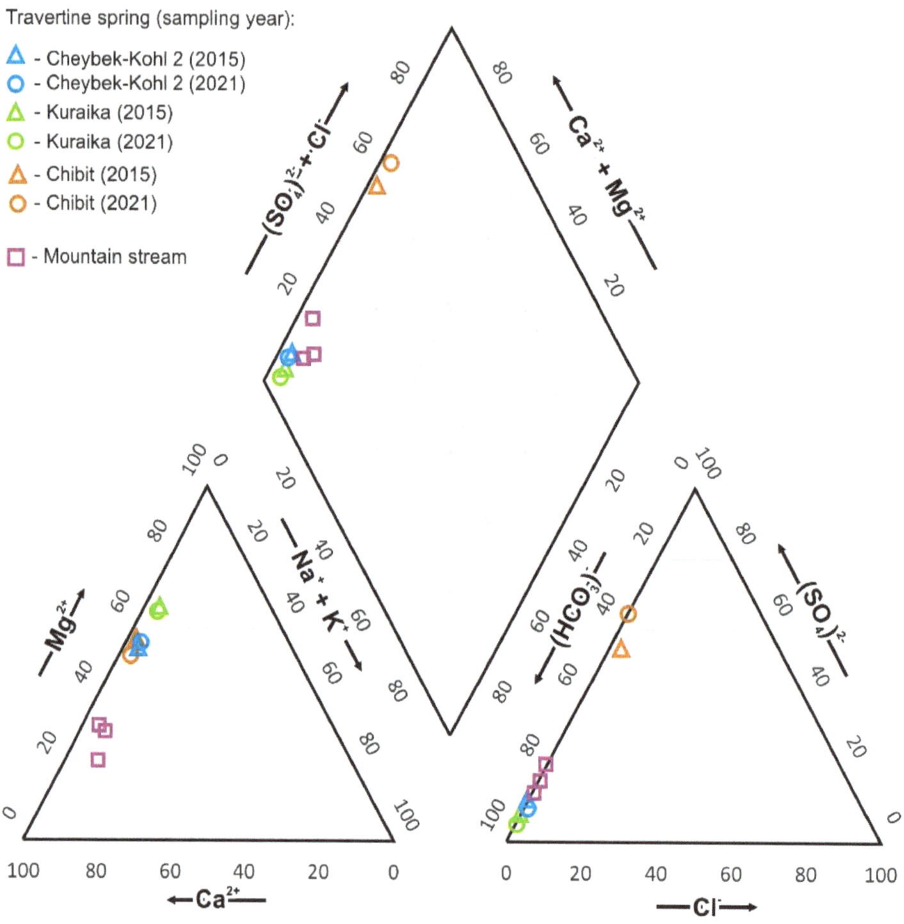

Figure 10. Piper diagram for the travertine spring water (cf. Table 1) and mountain steam water [41] of the south-eastern Gorny Altai.

4.3. Travertine Petrography and Mineralogy

All fossil travertines of Totugem, Baltyrgan, Cheybek-Kohl 1, Cheybek-Kohl 2, Chibit, and most parts of travertines of the Meshtuyaryk were composed of calcite (Table 2, Figures 12 and 13). Sparry and micritic calcite was white-yellow or grey; cementing clasts of different shapes and sizes. Calcite forms multiple layers which rimmed the clasts (Figure 12A–E). The free space of cavities and pores were filled with closely packed elongate crystals, up to 1 mm in size, with rhombohedral terminations. At the base of such aggregates there was a zone usually formed by anhedral calcite. In some cases, a series of druse rhythms alternated. Breaks in calcite precipitation were caused by poisoning by detrital silicate materials (Figure 12E,G–I). Euhedral terminations of calcite crystals displayed evidence of recurrent surface etching and subsequent regeneration (Figure 12F).

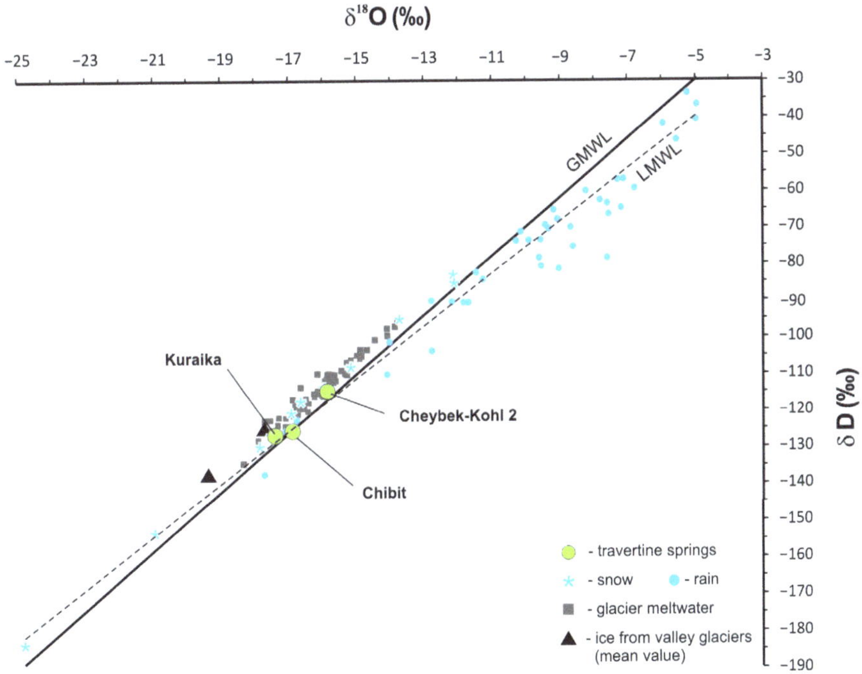

Figure 11. The isotope composition of the travertine water compared with ice and meltwater of the south-eastern Gorny Altai as well as meteoric precipitation. Sources: Glacier meltwater and ice from valley glaciers in Tavan-Bogd and Mongun-Taiga massifs [93,94]. Local meteoric water line (LMWL; $\delta D = 7.1 \cdot \delta^{18}O - 4.3$) [95]. Global meteoric water line (GMWL) [96].

Table 2. Stable isotope composition of $CaCO_3$ and mineralogy of fossil travertines and modern precipitates from the south-eastern Gorny Altai.

Travertine Deposit	Mineralogy	$\delta^{13}C$ (‰VPDB)	$\delta^{18}O$ (‰VPDB)
Fossil travertines			
Cheybek-Kohl 1	calcite	−4.4 to −2.9	−13.8 to −13.0
Baltyrgan	calcite	−2.5 to −2.0	−13.4 to −13.1
Cheybek-Kohl 2	calcite	−2.5 to 1.2	−13.8 to −12.9
Meshtuyaryk	calcite; calcite + aragonite	−1.2 to 1.9	−16.3 to −14.4
Kuraika	aragonite; aragonite + calcite	−0.3 to 0.7	−14.8 to −13.2
Totugem	calcite	0.4 to 2.0	−15.0 to −14.5
Chibit	calcite	1.5 to 4.3	−14.3 to −12.2
Modern precipitates			
Cheybek-Kohl 2	calcite + aragonite	na	na
Chibit	gypsum + calcite + aragonite	na	na

Mineralogy according to XRD and SEM data; na—not analysed.

Kuraika was the only fossil travertine site, where aragonite was the dominant mineral phase (62–98%) and calcite was present as a minor phase. Aragonite formed colourless and white-yellow fan-shaped radiating aggregates up to 2 mm in size, built up of needle-like crystals. Usually such aggregates are formed in the free pore space, where they overgrow with elongated calcite crystals (Figure 13A–D).

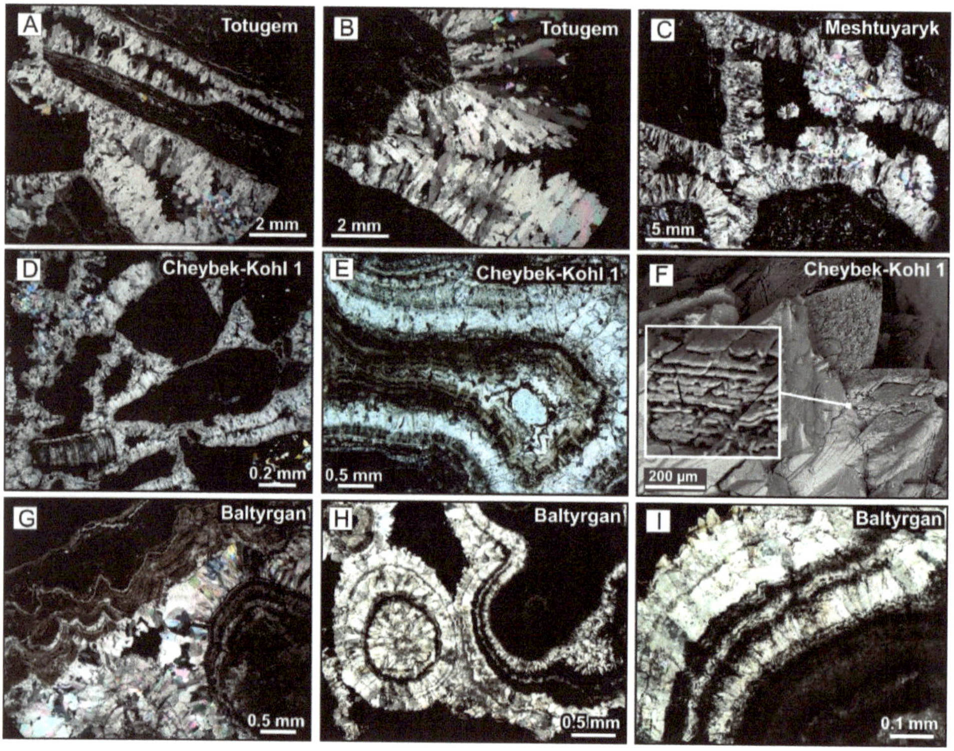

Figure 12. Photomicrographs of the fossil travertine samples composed of calcite. (**A–D**) Closely packed druse aggregates of elongate calcite crystals covering angular detrital clasts. (**E**,**G–I**) Calcite-laminated crusts composed of translucent sparite layers alternating with brown micrite laminae enriched in detrital material. (**F**) Calcite crystals with evidence of repeated surface etching and regeneration. Cross-polarized light (**A–D**,**G**), plane-polarized light (**E**,**H**,**I**), and backscattered electron (BSE) images (**F**).

In the Meshtuyaryk travertines calcite drusy aggregates sometimes contained aragonite (up to 15%), overgrowing anhedral sparry calcite at the base (Figure 13M–P). Aragonite formed layers of acicular crystals (up to 150 µm long) which grew nearly perpendicularly to the substratum. In some samples aragonite formed thicker layers (up to 1 mm), which overgrew drusy aggregates of calcite, and, in turn, were overgrown by calcite. The fan-shaped radiating acicular aggregates of aragonite originating from a nucleation point (detrital particles or calcite grains) are quite rare.

Samples of modern travertine crusts found only in the active seepage zone at the Cheybek-Kohl 2 and the Chibit sites also contained aragonite as well as calcite (Figure 13E–L). The upper surface of the travertine crust was colonized by algae and moss. Aragonite (5 to 40%) was present only in laminated crusts that grew upon oriented aggregates of elongate calcite crystals, up to 300 µm in length. The laminae of aragonite alternated with layers of brown micrite (<150 µm thick) or, less often, colourless anhedral sparite calcite. Aragonite formed colourless fan-shaped radiating acicular aggregates. The carbonate rhythms were locally separated by thin layers of fine-grained detrital material or organic films. The Chibit travertines contained fossilized moss cushion and algae (Figure 13G). Gypsum was found in the Chibit travertines in association with the active water seepage zone. Gypsum occurred as anhedral grains filling pores between calcite and aragonite or forming thin layers overlying laminated calcite–aragonite aggregates (Figure 13H).

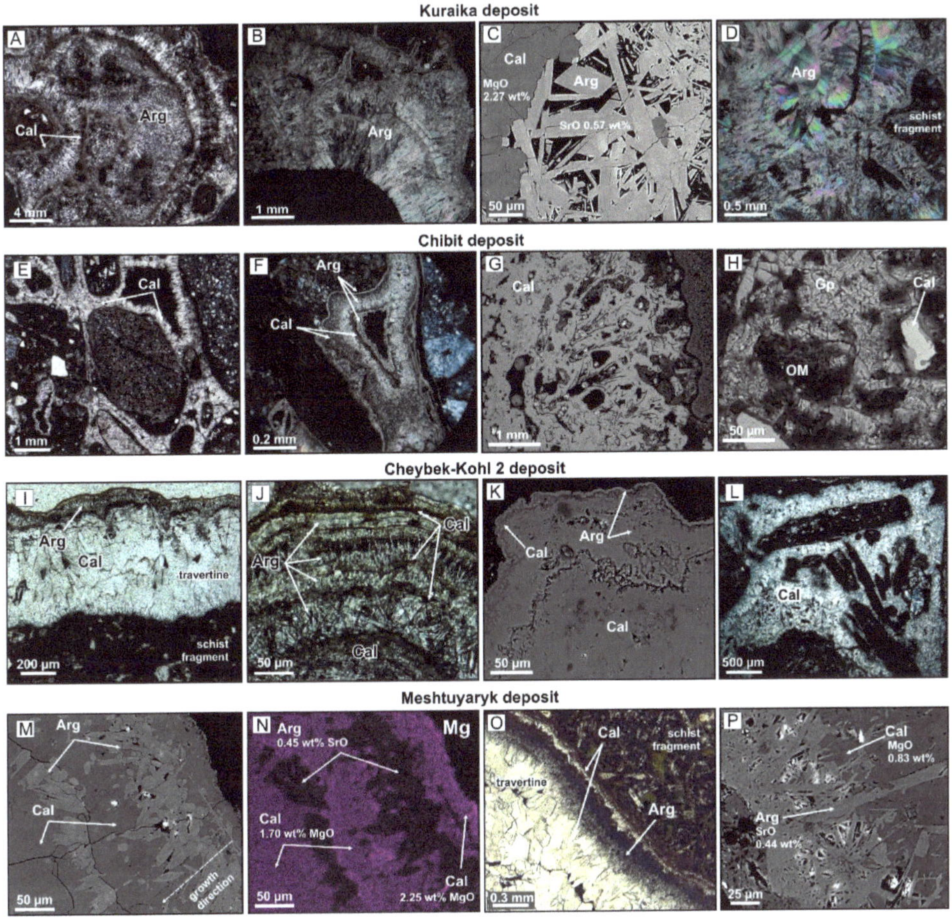

Figure 13. Photomicrographs of the fossil travertine samples and modern precipitates composed of aragonite and calcite. (**A**,**B**,**D**) Fan-shaped radiating and acicular aragonite layers. (**C**) Porous aggregate of elongated aragonite crystals and monolithic calcite aggregate. (**E**) Calcite crystals cementing detrital clasts. (**F**) Calcite–aragonite-laminated crust in void. Present seepage zone. (**G**) Fossilized moss cushion and algae in calcite precipitated in the present seepage zone. (**H**) Aggregate of the anhedral gypsum grains with calcite clasts and organic matter particles. Present seepage zone. (**I**,**J**,**K**) Laminated crust with multiple alternations of micrite calcite and acicular aragonite laminae covering an aggregate of oriented calcite crystals. Present seepage zone. (**L**) Closely packed calcite crystal-cemented sand particles. (**M**,**N**) BSE image and elemental map (Mg) of laminated crust with aragonite layers at the base. (**O**) Calcite–aragonite crust covering a fragment of schist. (**P**) Fan-shaped radiating and acicular aragonite in a Mg–calcite matrix. Cross-polarized light (**A**,**B**,**D**,**E**,**F**), plane-polarized light (**I**,**J**,**L**,**O**), and backscattered electron (BSE) (**C**,**G**,**H**,**K**,**M**,**P**) images; Mg elemental map (**N**). Arg = aragonite, Cal = calcite, Gp = gypsum; OM = organic matter. Fossil travertines (**A**–**E**,**L**–**P**); modern precipitates (**F**–K).

4.4. Mineral Chemistry

Calcites from all travertines contained significant amounts of MgO (0.5–5.7 wt%) while FeO, MnO, SrO, and Y_2O_3 were below detection. Aragonite showed SrO contents of 0.2–2.0 wt%, occasionally containing Y_2O_3 (up to 0.5 wt%) and BaO (up to 0.15 wt%). Other impurities (Fe, Mn, Na) in aragonite were negligible.

4.5. Trace Elements

The calcitic and calcite–aragonitic travertines showed contrasting trace-element compositions (Table 3). The calcite travertines were enriched in Mg (4264–7925 ppm) but had low Sr and Ba abundances (235–1515 ppm and 11–41 ppm, respectively). Travertines composed of aragonite and calcite showed low Mg contents ($\leq$ 594 ppm) but higher Sr (3106–4586 ppm) and Ba (up to 138 ppm) contents. The Fe and Mn contents in travertines were variable (18–465 ppm and 1–36 ppm, respectively) due to commonly observed secondary impregnations of cements by Fe^{3+}-hydroxides. Kuraika and Cheybek-Kohl 2 travertines were enriched in U relative to other travertines of the south-eastern Gorny Altai (13.9–15.8 ppm vs. 0.6–2.0 ppm). Travertines from Cheybek-Kohl 1, Cheybek-Kohl 2, Totugem, and Baltyrgan deposits had the greatest Zn content (4.9–8.2 ppm). All travertines were depleted in Th (<0.13ppm) and Rb (<0.5ppm).

Table 3. Trace-element and REE concentrations (ICP-MS data, in ppm) in carbonate fractions extracted from travertines of the south-eastern Gorny Altai.

Deposit	Cheybek-Kohl 1 *	Baltyrgan	Totugem	Meshtuyaryk	Cheybek-Kohl 2 *	Kuraika	Chibit
Phases	Cal	Cal	Cal	Cal	Cal + Arg	Cal + Arg	Cal + Arg
Li	1.90	3.30	3.35	2.95	1.50	2.86	4.80
Na	93.5	n.a.	146	204	114	141	n.a.
Mg	7049	n.a.	4264	7925	273	594	n.a.
Al	378	107	19.8	25.9	169	45.9	110
P	50.3	n.a.	32.5	n.a.	35.3	8.08	n.a.
K	108	34.7	25.9	13.0	85.7	19.0	237
Mn	35.6	n.a.	1.30	1.60	17.2	2.78	n.a.
Fe	376	89.2	18.4	51.4	465	64.0	183
Co	0.6	0.17	n.d.	0.07	0.50	0.07	0.16
Cu	3.40	0.59	n.d.	5.21	1.00	1.77	4.70
Zn	5.20	4.90	7.47	0.49	8.20	0.69	2.50
Rb	0.50	0.30	n.d.	n.d.	0.40	0.01	0.10
Sr	344	1515	454	235	4586	3817	3106
Y	0.50	3.04	6.90	n.d.	0.40	n.d.	0.12
Ba	11.1	40.8	26.3	20.7	121	138	31.0
La	0.42	2.02	3.41	n.d.	0.25	0.04	0.07
Ce	0.86	2.76	3.14	0.03	0.48	0.07	0.15
Pr	0.11	0.44	0.73	n.d.	0.06	n.d.	0.02
Nd	0.48	1.90	3.20	0.02	0.25	0.02	0.09
Sm	0.11	0.34	0.71	0.01	0.08	n.d.	0.02
Eu	0.02	0.08	0.16	n.d.	0.01	n.d.	0.01
Gd	0.10	0.39	0.96	0.02	0.10	n.d.	0.02
Tb	0.02	0.07	0.13	n.d.	0.01	n.d.	n.d.
Dy	0.13	0.39	0.68	n.d.	0.06	n.d.	0.02
Ho	0.02	0.09	0.15	n.d.	0.02	n.d.	n.d.
Er	0.06	0.27	0.50	n.d.	0.04	n.d.	0.02
Tm	0.01	0.04	0.08	n.d.	0.01	n.d.	n.d.
Yb	0.06	0.22	0.54	n.d.	0.05	n.d.	0.01
Lu	0.01	0.03	0.09	n.d.	0.01	n.d.	n.d.
ΣREE	2.41	9.05	14.49	0.08	1.43	0.13	0.43
Th	0.13	0.12	n.d.	n.d.	0.09	n.d.	n.d.
U	0.60	1.76	3.1	1.96	15.8	13.9	0.79

n.a. = not analysed, n.d. = not detected. Arg = aragonite; Cal = calcite. Phase composition according to XRD data.
* Composition according to [41].

All travertines were markedly depleted in ΣREE (0.08–14.49 ppm) relative to the PAAS (post-Archean Australian shale) value of 83 ppm [97]. The PAAS-normalized REE + Y patterns of all travertines were similar, with a slight enrichment in heavy REEs (Figure 14). Weak negative anomalies for redox-sensitive REEs were observed for travertines Baltyrgan and Totugem (Ce anomaly) and Cheybek-Kohl 2. Travertine Chibit showed a positive Eu anomaly.

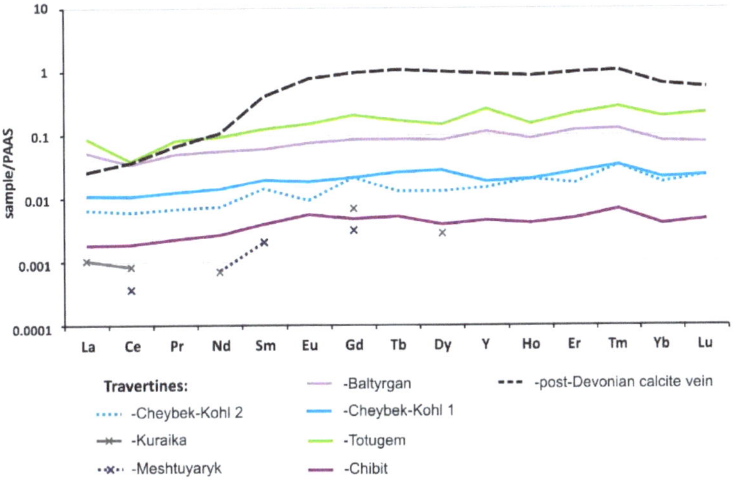

Figure 14. REE + Y patterns of $CaCO_3$ fractions extracted from travertines of the south-eastern Gorny Altai (cf. Table 3) compared to the post-Devonian calcite vein [41]. Normalized to PAAS [97].

4.6. Stable Isotope Compositions of $CaCO_3$

The studied travertines slightly differed in their $\delta^{18}O$ values (−15.8 to −12.2‰) (Table 2, Table S1, Figure 15). The $\delta^{13}C$ values showed a rather wide range (−4.4 to 4.3‰), with 70% of the data falling within the −1.0 to 2.0 ‰ range. The $CaCO_3$ from Cheybek-Kohl 1 and Baltyrgan travertines had a $\delta^{18}O$ varying between −13.8 and −13.0‰, and a $\delta^{13}C$ varying between −4.4 and −2.0‰. Both positive and negative values of the $\delta^{13}C$ (−2.5 to 1.9‰) were measured in the Cheybek-Kohl 2, Kuraika and Meshtuyaryk travertines. Calcites from Totugem site were markedly enriched in ^{13}C (0.4 to 2.0‰). The $CaCO_3$ from Meshtuyaryk and Totugem had the lowest $\delta^{18}O$ values (−16.3 to −14.4‰). The highest $\delta^{13}C$ (1.5 to 4.3‰) and $\delta^{18}O$ −4.3 to −12.2‰) values were observed in $CaCO_3$ from the Chibit deposits.

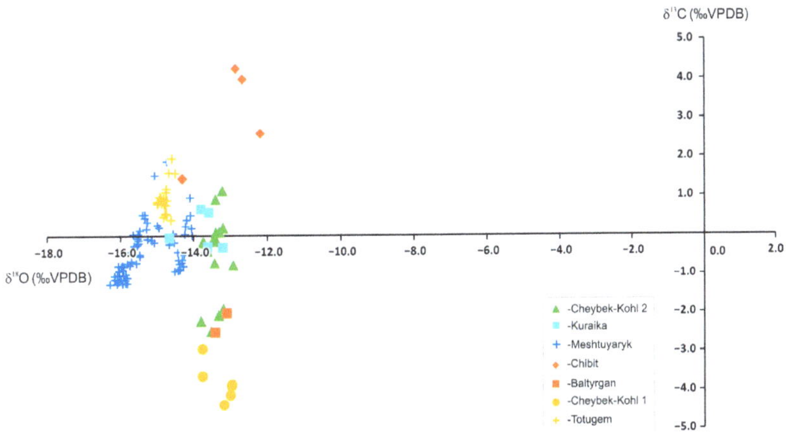

Figure 15. Stable isotope properties of the $CaCO_3$ from travertines of the south-eastern Gorny Altai (see Table S1).

4.7. Ages of Travertines

We obtained nineteen of ^{230}Th-U age dates [48,49] of travertines from the south-eastern Gorny Altai (Table 4 and Table S2, Figures 3A, 5A and 6A). The oldest Baltyrgan travertine (401,455 +41,000/−28,700 ka BP) is located within the foreberg west of the Chagan Uplift (Figure 8A). The Totugem travertines (ca. 123 ka BP) are located on the eastern flank of the KFZ (Figure 7A). Further to the west, along the active faults of the zone, there are only Holocene travertines, ranging in ages from 11.0 to 3.5 ka BP. The youngest Chibit travertine (1.243 ± 0.266 ka BP) is spatially related to the South Chuya Fault.

Table 4. ^{230}Th-U ages of travertines from the south-eastern Gorny Altai, data from [48,49].

Sample	^{238}U (mg/g)	±	^{232}Th (ng/g)	±	[^{234}U/^{238}U]$_A$	±	[^{230}Th/^{238}U]$_A$	±
Cheybek-Kohl 1								
20.1-2 $^{+}$	0.441	0.003	3.0930	0.0245	2.6888	0.0002	0.1715	0.0013
RYALT-15-17-1 $^{++}$	0.623	0.005	2.452	0.026	2.687	0.012	0.151	0.001
Cheybek-Kohl 2								
5.12-3 $^{+}$	36.670	0.246	0.6821	0.0057	2.8820	0.0008	0.2383	0.0011
RYALT-15-16-11 $^{++}$	0.886	0.006	33.247	0.339	2.840	0.010	0.278	0.004
RYALT-15-16-8-1 $^{+}$	0.518	0.015	2.166	0.018	2.846	0.002	0.275	0.002
Kuraika								
15-14-1 $^{+}$	8.276	0.055	0.2274	0.0019	3.4099	0.0010	0.1258	0.0004
RYALT-13-18-2 $^{++}$	12.229	0.238	0.651	0.007	3.385	0.070	0.151	0.003
Meshtuyaryk								
PN 2019/1.1_2 $^{+}$	1.404	0.008	0.4626	0.0031	1.1687	0.0003	0.0980	0.0005
PN 2019/1.5_1 $^{+}$	1.561	0.010	0.0170	0.0002	1.1321	0.0004	0.0371	0.0003
PN 2019/1.5_3a $^{+}$	1.661	0.010	0.0349	0.0003	1.1390	0.0004	0.0367	0.0004
PN 2019/1.5_3b $^{+}$	2.012	0.013	0.0363	0.0003	1.1391	0.0003	0.0390	0.0004
PN 2019/1.6 $^{+}$	1.077	0.007	0.7849	0.0055	1.1537	0.0004	0.0802	0.0008
PN 2019/1.10 $^{+}$	2.160	0.014	0.3897	0.0026	1.1653	0.0003	0.0810	0.0007
PN 2019/1.11 $^{+}$	1.596	0.010	0.0327	0.0002	1.1587	0.0004	0.0758	0.0007
PN-2019/1.14 $^{+}$	0.969	0.006	0.0765	0.0011	1.1310	0.0005	0.0760	0.0004
Totugem								
RYALT-13-8.1 $^{++}$	2.852	0.037	5.069	0.060	2.345	0.028	1.728	0.023
RYALT-15-12-13-2 $^{+}$	2.260	0.015	12.689	0.097	2.3237	0.0011	1.7078	0.0067
Baltyrgan								
RYALT-13-7.1 $^{++}$	1.519	0.011	12.064	0.138	1.045	0.004	1.034	0.007
Chibit								
RYALT-15-32-1 $^{++}$	0.476	0.004	12.746	0.169	1.490	0.003	0.017	0.004

Sample	Age Uncorrected (ka BP)	±	Age Corrected (ka BP)	±	[^{234}U/^{238}U]$_A$ Initial	±
Cheybek-Kohl 1						
20.1-2 $^{+}$	7.226	0.045	7.152	0.056	2.7233	0.0020
RYALT-15-17-1 $^{++}$	6.337	0.056	6.296	0.059	2.7171	0.0124
Cheybek-Kohl 2						
5.12-3 $^{+}$	9.340	0.042	9.340	0.043	2.9324	0.0008
RYALT-15-16-11 $^{++}$	11.484	0.053	11.109	0.168	2.8992	0.0098
RYALT-15-16-8-1 $^{+}$	11.022	0.061	10.981	0.060	2.9038	0.0016
Kuraika						
15-14-1 $^{+}$	4.090	0.015	4.089	0.015	3.43788	0.0010
RYALT-13-18-2 $^{++}$	4.973	0.140	4.973	0.144	3.418	0.072

Table 4. *Cont.*

Meshtuyaryk						
PN 2019/1.1_2 +	9.553	0.048	9.545	0.049	1.1733	0.0003
PN 2019/1.5_1 +	3.639	0.031	3.638	0.031	1.1335	0.0004
PN 2019/1.5_3a +	3.575	0.036	3.574	0.037	1.1404	0.0004
PN 2019/1.5_3b +	3.804	0.041	3.804	0.041	1.1406	0.0003
PN 2019/1.6 +	7.870	0.079	7.852	0.078	1.1572	0.0004
PN 2019/1.10 +	7.859	0.066	7.855	0.067	1.1690	0.0003
PN 2019/1.11 +	7.379	0.075	7.379	0.074	1.1620	0.0005
PN-2019/1.14 +	7.589	0.040	7.587	0.040	1.1339	0.0006
Totugem						
RYALT-13-8.1 ++	122.987	3.652	122.970	3.677	2.904	0.030
RYALT-15-12-13-2 +	122.60	0.80	122.540	0.80	2.8716	0.0006
Baltyrgan						
RYALT-13-7.1 ++	401.656	+39.900/ −29.000	401.455	+41.000/ −28.700	1.139	0.011
Chibit						
RYALT-15-32-1 ++	1.761	0.067	1.243	0.266	1.4922	0.0025

All measurements are reported with ±2 σ absolute uncertainties. Subscript A denotes activity ratio. U decay constants: λ_{238} = 1.55125 × 10^{-10} [98] and λ_{234} = 2.82206 × 10^{-6} [92]. Th decay constant: λ_{230} = 9.1705 × 10^{-6} [92]. Corrected ^{230}Th ages assume the initial ^{230}Th/^{232}Th atomic ratio of 4.4 ± 2.2 × 10^{-6}. Those are the values for a material at secular equilibrium, with the bulk Earth ^{232}Th/^{238}U value of 3.8. The errors are arbitrarily assumed to be 50%. BP stands for "before present" where the "present" is defined as the year 1950 CE. +—Measured on the Neptune Plus at the Institute for Geosciences, ++—Measured on the Nu Plasma at the Max Planck Institute for Chemistry, Mainz.

5. Discussion

5.1. Paleofluid Source Rocks

A combination of isotope and elemental analyses can be used to constrain the origin of CO_2 and paleofluid provenance fossil travertine systems [14–16,99]. The carbon and oxygen isotope data of fossil travertines from the south-eastern Gorny Altai plot was within the hypogean travertine (according to Teboul et al. [16]), with CO_2 derived from carbonates or igneous (except carbonatites and ultramafics) source rocks (Figure 16A). The high strontium and low barium contents in the Gorny Altai travertines indicate that the fluids had hypogean origin and were in contact with source rocks consisting of mixed limestones, evaporites and dolostone [16] (Figure 16B). The relatively elevated concentrations of Mg can be related to the interaction of the fluids with the dolomitic limestones or dolostones. Thus, the isotope and trace-element data jointly indicate hypogean origin of the paleofluid from the fossil Gorny Altai travertine system and the dolomitic limestones or dolostones as fluid source rocks. The paleofluid originated from groundwaters subjected to prolonged water–rock interactions with aquifer carbonate rocks, Upper Neoproterozoic to Devonian limestone and dolostone, widespread in the area.

5.2. Calcite–Aragonite Precipitation

Fossil travertine cements in the south-eastern Gorny Altai comprise calcite and aragonite. At different sites, mineral compositions varied from exclusively calcitic (Totugem, Baltyrgan, Cheybek-Kohl 1, Cheybek-Kohl 2, and Chibit) to predominantly calcitic with minor aragonite (Meshtuyaryk) to predominantly aragonitic (Kuraika). The Mg-calcite and Sr-aragonite showed no signs of diagenetic alteration and retained their typical morphological features. This indicates that both $CaCO_3$ polymorphs were in primary phases, and that calcite was not a replacement of precursor aragonite. As it was shown previously, [1,2,100–104] the precipitation of aragonite and calcite in spring systems is controlled by a number of factors, such as: (1) temperature of the spring water, (2) microbial activity, (3) pCO_2 and the rate of CO_2 degassing, (4) mineral precipitation rates, and (5) Mg/Ca ratio in water. There is not a single or universal control applicable in all situations [102].

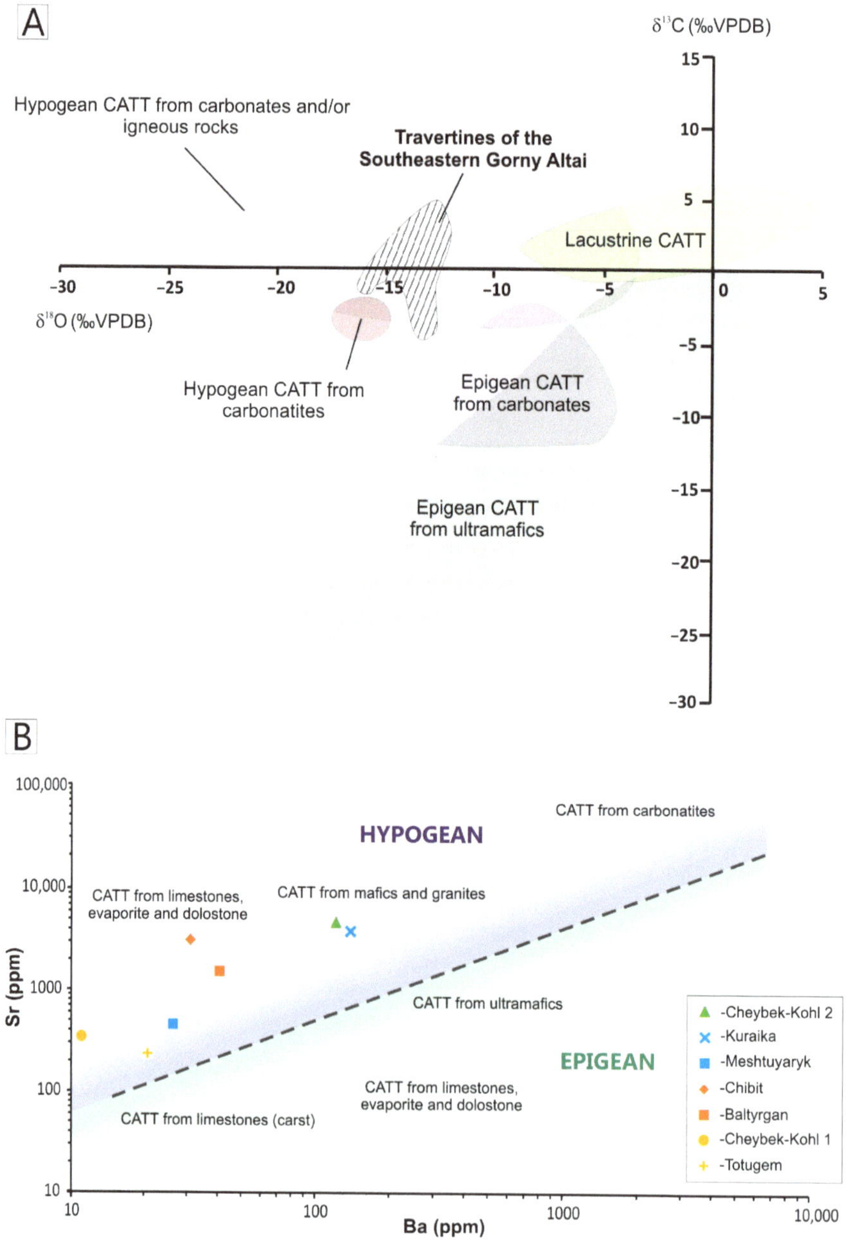

Figure 16. $\delta^{13}C$-$\delta^{18}O$ (**A**) and Sr-Ba (**B**) cross-plots of travertines from the south-eastern Gorny Altai overlain on clusters of Teboul et al. [16].

Aragonite precipitation commonly takes place in thermal springs with temperatures of 30–80 °C [1,100,105–108]. Meanwhile, some studies have suggested that temperature does not play any significant role in the precipitation of aragonite or calcite [109–113]. The studied travertine springs in the south-eastern Gorny Altai had temperatures ranging from 3 to 11 °C; yet both calcite and aragonite were deposited in them (Table 2). Stable isotope studies suggested that fossil travertines were precipitated from ascending ambient-temperature bicarbonate waters [41,48,49,65,66]. Thus, the spring water temperature does not seem to be a controlling factor in aragonite precipitation in the case of the Altai travertines.

It is known that biofilms can affect the precipitation of $CaCO_3$ polymorphs [1,106,107, 113–116]. The Cheybek-Kohl 2 travertine complex provided evidence that microbial activity was involved in the modern calcite–aragonite precipitation, but was suppressed during the formation of fossil travertines [41]. In the case of the Chibit travertine site, limited aragonite precipitation occurred only in present seepage zone featuring abundant microbial mats around the spring orifice; modern precipitates were enriched in organic compounds (Figure S1). Fossil travertines at this site were composed of calcite only. Evidently, at the Chibit site the microbial activity and high $SI_{aragonite}$ in water may favour aragonite formation in modern precipitates.

Neither optical microscopy nor SEM analysis on fossil aragonite-rich Kuraika and aragonite-bearing Meshtuyaryk travertines yielded evidence of biological influence on their precipitation. In Particular, weak UV-luminescence of Meshtuyaryk travertines [49] was consistent with the lack of involvement of biological processes in travertine precipitation. In the Kuraika travertines aragonite occurred as fan-shaped radiating aggregates associated with calcite layers in the inner parts of the fossil travertine body. At the fossil Meshtuyaryk travertines aragonite formed thin layers at the base of calcite crusts. The absence of organic compounds corroborates the hypothesis about abiotic precipitation of calcite and aragonite.

Many researchers note, that the precipitation of aragonite is favoured by high Mg/Ca ratios of the spring water (>1), because the Mg inhibits the growth of calcite [117–121]. In water of the south-eastern Gorny Altai travertine springs the Mg content was typically lower than the Ca content (Table 1). Only one water sample from the Kuraika spring had a Mg/Ca = 1.29. According to Folk [100], such Mg/Ca ratio at low spring temperatures (3 °C) should promote the precipitation of aragonite. All calcites from Meshtuyaryk travertines contained appreciable amounts of Mg. The anhedral sparry calcite that coexisted with aragonite at the base of druse had the greatest CaO content (up to 6.0 mol% $MgCO_3$). There, fine layers of Mg-enriched calcite overgrew the aragonite layers. The Mg/Ca ratio in carbonates of abiotic origin depend on the Mg/Ca ratio in solution [119] and temperature, which affects the distribution coefficient [122]. The Mg/Ca ratio of waters is controlled by the lithology of host rock in zones, where most of water–rock interactions occur. For Meshtuyaryk travertine-forming water, the Paleozoic limestone and dolostone served as the aquifer. It is likely that the water was characterized by a slightly elevated Mg/Ca ratio, favouring aragonite precipitation.

Reviewing calcium carbonate polymorph precipitations, Jones [102] reported abundant evidence of aragonite (± calcite) precipitation from spring water with a high CO_2 content and rapid CO_2 degassing. Chafetz et al. [123] also suggested that aragonite precipitates from water with high $CaCO_3$ supersaturation, whereas low-supersaturation waters tend to precipitate calcite. The formation of travertines in the south-eastern Gorny Altai was triggered by seismic events, which induced the activation of faults and the rapid rise of bicarbonate groundwaters previously sealed in carbonate aquifers. The rapid rise of groundwaters to the surface was accompanied by rapid CO_2 degassing and increased $CaCO_3$ supersaturation levels, which led to aragonite precipitation. Decreasing overpressure led to the rise in subsequent portions of groundwaters in a quieter mode. The lowering of $CaCO_3$ supersaturation favoured the precipitation of calcite.

It is likely, therefore, that rapid CO_2 degassing along with elevated Mg/Ca ratios of the travertine-forming water were the primary controls of calcite and aragonite precipitation in the fossil travertines of the south-eastern Gorny Altai.

5.3. Stable Isotopes and Geochemistry of Present-Day Spring Water

The linear distribution of the δD and $\delta^{18}O$ in spring water along global and local meteoric water lines (Figure 11) strongly suggests its meteoric origin [96]. The stable isotope values of the travertine spring water were lower than that of the summer rainwater in the south-eastern Gorny Altai [95]. On the contrary, the δD and $\delta^{18}O$ in travertine spring water were plotted close to the meteoric precipitation of cold and transitional seasons (snow) and water from melted glaciers/ice. Bantsev et al. [93,94] showed that precipitation of the transitional seasons (spring-autumn) played the most important role in the recharge of glaciers in the arid parts of the Altai Mountains. Thus, the travertine spring water appeared to be recharged by glaciers meltwater and/or winter precipitation, with isotopically depleted compositions.

The chemistry of groundwater is controlled by many factors, including aquifer rock type, residence time, and geochemical processes along the groundwater flow paths (evaporation–crystallization, flow regime, etc.) [124,125]. Waters from Cheybek-Kohl 2 and Kuraika travertine springs were dominated by HCO_3^-, Ca^{2+}, and Mg^{2+} (Table 1). Such composition is indicative of carbonate aquifers. Mg/Ca ratios depend on the proportion of calcite and dolomite present in the aquifer rock and/or chemical kinetics [126]. The Upper Neoproterozoic fractured limestone and dolostone are the main aquifer rocks for travertine-forming solution within the south-eastern Gorny Altai (Figure 2) [84]. Hence, the Mg/Ca ratios of travertine-forming solutions (and, by extension, of travertine carbonates) was controlled by reactions of water with dolostone-poor or dolostone-rich rocks.

The sulphate, Ag and In contents were low for almost all the travertine spring water samples, except for the Chibit sample, showing the highest values of 733.6 mg/L for SO_4^{2-}, 5 μg/L for Ag, and 10 μg/L for In (Table 1). Gypsum dissolution or oxidation of sulphides is a common source of sulphate in groundwaters [127,128]. The Mg/Ca ratio was similar in the sulphate-rich Chibit and sulphate-poor Cheybek-Kohl 2 waters. Therefore, gypsum should not be considered as the main source of SO_4^{2-} in the travertine-forming solutions, because it dissolution would increase the Ca^{2+} content and lower the Mg/Ca ratios. Sulphate, In and Ag in Chibit spring water can be sourced from the oxidation of sulphide minerals, which are abundant in Devonian Chibit Cu-sulphide-quartz-carbonate ore deposits and numerous Cu-Au, Ag-Cu-Pb, Cu-Ag, Au-Cu-Ag sulphide ore occurrences [84].

Thus, the studied spring waters were similar to groundwater in the area in their HCO_3-SO_4–Ca-Mg water chemistry. Their isotopic composition suggests the dominant contribution of winter precipitation; chemical characteristics suggest minor water–rock interactions. These waters rarely deposit carbonates. The saturation index values of calcite ($SI_{calcite}$) and aragonite ($SI_{aragonite}$) in Cheybek-Kohl 2 water were 0.09 and 0.25, respectively. In Kuraika water $SI_{calcite}$ (0.45) and $SI_{aragonite}$ (0.29) values were higher. According to Pentecost [1], in freshwater environments the minimum critical $SI_{calcite}$ for calcite precipitation is perhaps around 0.48. In many travertine systems calcite and aragonite precipitation starts significantly after $SI_{calcite}$ exceeds 0.8 (or around one) [104,107,129,130]. In the studied spring waters $SI_{calcite}$ (1.00) and $SI_{aragonite}$ (0.85) values only exceeded 0.8 in the Chibit deposits, where minor amounts of $CaCO_3$ precipitated along with gypsum.

5.4. Comparison of Travertine Ages with Paleoearthquake Ages

The age of the Chibit travertines (1.243 ± 0.266 ka BP) agrees well with the age of the paleoearthquake (Figure 17A) with Mw = 6.6–6.9 and intensity IX-X on the ESI-2007 intensity scale, associated with the KFZ. Its surface ruptures can be traced along the zone for 40 km; in the north of the Kurai Basin, they are dated to 1.210 ± 0.07 cal ka BP ($IGAN_{AMS}$7749; 0.954 probability; this study), 1.240 ± 0.110 cal ka BP (IGAN −3205), and 1.150 ± 0.105 cal ka BP [131], in the north-western part of the Chuya Basin to 1.305 ± 0.015 cal ka BP [57,79]. The western end of the surface ruptures is 30 km east of the Chibit travertine outcrop.

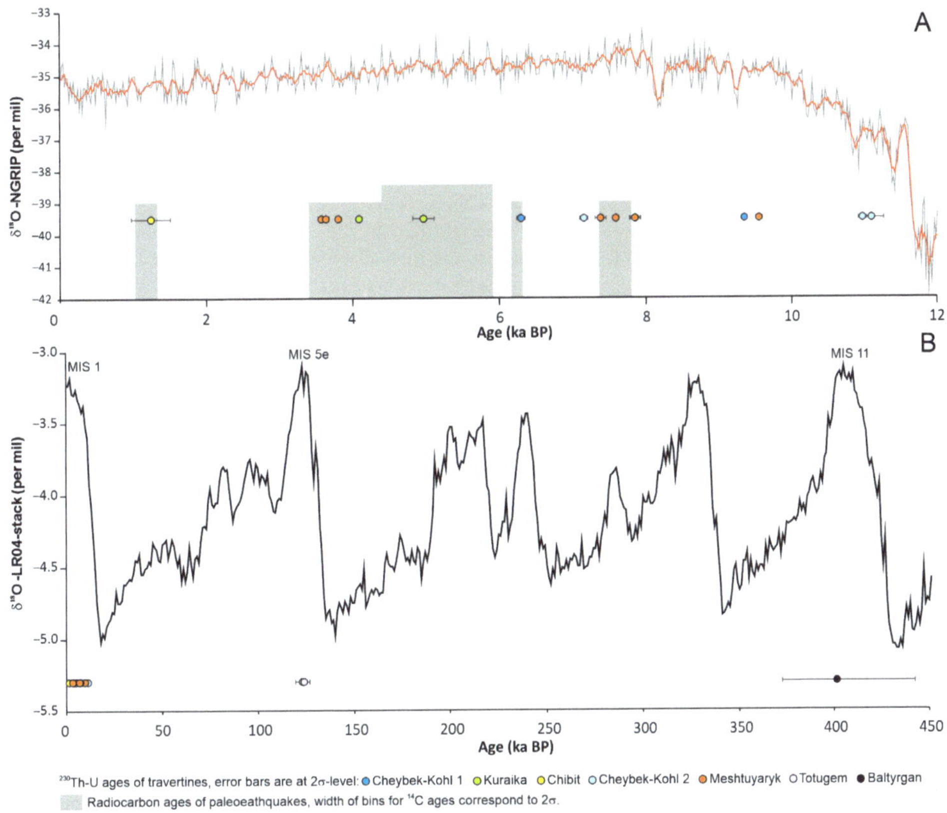

Figure 17. Age dates of the travertines (U-Th) and large paleoearthquakes (^{14}C) in the south-eastern Gorny Altai shown in the context of the paleoclimate proxy records: (**A**) North Greenland Ice Core Project (NGRIP) $\delta^{18}O$ [132,133]; grey line—complete data set; red line—5-point moving average. (**B**) LR04-stack $\delta^{18}O$ [134].

The youngest age cluster of the Meshtuyaryk travertines (3.574 $\pm$ 0.037, 3.638 $\pm$ 0.031, and 3.804 $\pm$ 0.041 ka BP), as well as the age of the youngest Kuraika travertines (4.089 $\pm$ 0.015 ka BP) fit into the time frame (Figure 17A) of the surface rupture of the paleoearthquakes: from 4.630 $\pm$ 0.120 cal ka BP to 3.420 $\pm$ 0.030 cal ka BP [49]. The synchronicity of travertine formation at different locations may indicate that travertine springs began to discharge in the north of the Chuya and Kurai basins as a result of a single seismic event. Consequently, the zones of open fracturing along the faults of the Kurai zone opened up simultaneously at a distance of ca. 30 km. This distance can be used as an analogue of the "surface rupture length" parameter. If so, according to the empirical relationships [135] the paleoearthquakes had a magnitude of Mw = 6.8. This value is close to the magnitude estimates (6.5–6.7) obtained from trenching studies.

A series of different-age surface ruptures of paleoearthquakes were identified and dated by us in the north of the Kurai Basin. Thus, the age of the oldest Kuraika travertines (4973 $\pm$ 144 yr BP) falls within a relatively wide age range of one of the primary paleoseismic deformations (Figure 17A): from 5.910 $\pm$ 0.050 cal ka BP ($IGAN_{AMS}$7731; 0.954 probability; this study) to 4.550 $\pm$ 0.170 cal ka BP ($IGAN_{AMS}$7715; 0.954 probability; this study). At the same time, the Cheybek-Kohl 1 travertines of 6.296 $\pm$ 0.059 ka BP match the age of the surface rupture (Figure 17A), which occurred in the northern part of the Kurai Basin during the 6.3 ka paleoearthquake (Mw = 6.5–6.7, shaking intensity VIII–IX) [57,79].

The age dates of the Meshtuyaryk travertines (7.379 ± 0.074, 7.587 ± 0.040, 7.852 ± 0.078, and 7.855 ± 0.067 ka BP) agree well with the age of the surface ruptures caused by one paleoearthquake (Figure 17A) documented only 3.7 km south-east (7.850–7.500 cal ka BP [49]). A seismic rupture with a similar age at 7.460 ± 0.060 cal ka BP (IGAN 7752; 0.916 probability; this study), was also found in the north of the Kurai Basin. The Cheybek-Kohl 1 travertine 30 km north-west of it was dated to 7.152 ± 0.056 ka BP.

In recent years, there has been much evidence that the removal of glacial loading on continents after the Last Glacial Maximum (LGM) has increased the rate of displacement along active faults and resulted in the overall increase in the seismicity of these areas [136,137]. For example, the glacial deloading during the degradation of the last (Vistulian) glaciation in Fennoscandia between 15 and 9 ka BP was accompanied by fault movements and large earthquakes with Mw ≈ 7–8 [138–142].

Similar processes apparently occurred during the degradation of the LGM glaciers in the south-western Gorny Altai. According to Butvilovsky [63] and Blyakharchuk et al. [143], the degradation occurred between 16 and 8 ka BP. Seismic movements along the KFZ at this time led to the formation of surface ruptures at ca. 8.5 ka BP [131], as well as large landslides and rockslides dated to around 8.3, 9.7, and 16 ka BP [81,131]. Our ^{230}Th-U age dates for Cheybek-Kohl 2 (9.340 ± 0.043, 10.981 ± 0.060, and 11.109 ± 0.168 ka BP) and Meshtuyaryk (9.545 ± 0.049 ka BP) correlate well with both the onset of Holocene warming (Figure 17A) and increased seismicity.

The ^{230}Th/U date of the Totugem travertines (ca. 123 ka BP) corresponds the MIS 5e (Eemian) (Figure 17B). At this time, the most widespread sheet glaciation in the Gorny Altai was degrading [62]. The intensification of neotectonic movements along the KFZ at this time could have been caused by an isostatic response of the Earth's crust to the removal of the glacial load. An additional isostatic effect could have been caused by the catastrophic release of water from the Kurai–Chuya limnosystem during the destruction of glacial dams at the end of the late glacial–early interglacial period [58].

The age of the oldest Baltyrgan travertines, ca. 400 ka BP, corresponds to the accumulation of the upper part of the Early Middle Pleistocene molasse in the intermontane basins. At the peak of Cenozoic orogenesis in the region, forebergs formed in the marginal parts of the Kurai Basin under submeridional compression [54]. Within the foreberg the travertine material cements the Middle Pleistocene alluvial fan gravels. Deposition of travertine among the gravels occurred during the degassing of carbonaceous groundwater along the fracture zone at that time. Obviously, the Middle Pleistocene fault movements were also accompanied by earthquakes. The high seismicity of the region at that time is indicated by numerous seismites associated with the liquefaction and fluidization of soils, which were recorded in the Lower-Middle Pleistocene sediments in the west of the Chuya Basin.

Summarizing, we posit that the ages of the large paleoearthquakes and ages of the episodes of travertine formation in the rim of the Chuya and Kurai basins during the last 8000 years exhibit a close match. The stages of travertine formation between 11and 9 ka BP fit into the period of strong paleoseismicity in the region (16 to 8 ka BP). Travertines dated to 123 and 400 ka BP mark seismic movements, which could be induced by the isostatic rebound of the Earth's crust following the removal of the glacial load and related to the most intense phase of the Cenozoic mountain formation, respectively.

5.5. Paleoclimate and Travertine Deposition

The age of the Chibit travertine, 1.243 ± 0.266 ka BP, falls into the Medieval Climate Optimum (1.2–1.3 ka), during which time the average summer temperatures in the alpine zone of the Gorny Altai were higher than today by 0.7–0.9 °C [144]. However, Ganyushkin et al. [145] showed that the average temperatures and annual precipitation between 1.3 and 1.0 ka ago at the Mongun-Taiga Uplift (Figure 1) were close to the present-day temperatures.

Five ^{230}U-Th ages of the Meshtuyaryk and the Kuraika travertines were dated to 3.6–5.0 ka BP, which is within the time frame of the Late Holocene Akkem glacial stage

(3.5–5.3 ka BP [145]). At the same time, in the Mongun-Taiga Uplift area, the annual precipitation exceeded the modern one by 10%, and the temperature could have been 1 °C colder relative to modern values [145]. Galakhov et al. [144], however, placed the Akkem glacial stage into a narrower time range of 4.3–4.0 ka BP. At 4.0 ka BP, the mean summer temperature was 1 °C lower than the present-day values, while for the 3.5 and 5.0 ka BP this parameter was 0.5 and 0.9 °C lower than the present-day value, respectively.

Cheybek-Kohl 1 travertines (ca. 6.3 ka BP) formed under wetter and warmer conditions. The mean annual temperatures could have exceed modern ones by 1.5–2.5 °C, whereas annual precipitation exceeded the modern norm twice [144,145].

Similar climatic conditions were present in the south-eastern part of the Gorny Altai during the formation of the Cheybek-Kohl 1 and Meshtuyaryk travertines at 7.2 to 7.9 ka BP, Cheybek-Kohl 2 and Meshtuyaryk travertines at 9.3 and 9.5 ka BP, and Cheybek-Kohl 2 travertines at 11.0 ka BP [143,145].

The ages of the Totugem travertines (ca. 123 ka BP) and the Baltyrgan travertines (ca. 400 ka BP) are poorly characterized in climatic terms. One can only state that the formation of the former coincides with the warm interglacial MIS 5e (Eemian), while the latter corresponds to the MIS 11 interglacial (Figure 17B).

As it was shown previously [22,33,146,147] warm and humid (interglacial) periods are favourable for travertine formation. All travertine-forming events at the south-western Gorny Altai fall within warm (Figure 17) and wet climatic phases.

5.6. Travertine Formation Model

All Quaternary travertine deposits of the south-eastern Gorny Altai are spatially associated with the active faults bordering the Chuya and Kurai basins. Carbonate cements of the travertines have similar isotopic–geochemical characteristics, and their ages match the ages of the paleoearthquakes and periods of increased tectonic activity of the region in the Quaternary [41,48,49,65,66]. This suggests a similar mechanism of their formation and similar compositional sources. According to stable isotope and trace-element compositions, travertines precipitated from cold groundwaters remained in prolonged contact with Upper Neoproterozoic–Paleozoic carbonate aquifers (Figure 18). The rapid rise of these waters towards the surface was triggered by earthquakes. During seismic shifts, zones of open fracturing were formed and/or renewed along active faults, serving as pathways for groundwater migration to the Earth's surface. Near and on the surface the rapid rise of water was accompanied by rapid CO_2 degassing and $CaCO_3$ precipitation. One of the important arguments supporting the seismogenic characteristic of travertines is the observed existence of three periods of travertine precipitation in the Meshtuyaryk deposit, separated by long periods of no carbonate formation (Table 4). All three groups of ^{230}Th-U dates obtained for travertines of the Meshtuyaryk deposit correspond to the ages of large paleoearthquakes whose epicentral zones were located in the immediate vicinity of this deposit [49]. According to the ^{230}Th-U dating, travertine deposition occurred during warm (Figure 17) and humid periods (interglacials). Warming, ice rebound, and permafrost degradation may result in seismicity and fracturing, which channelled groundwaters previously sealed in carbonate aquifers. During interglacials, the overall volume and dynamics of groundwater are expected to increase.

5.7. Implications for Paleoseismology and Long-Term Fault History

The possibilities of the paleoseismological method in determining the ages of paleoearthquakes are limited, to a significant extent, by the preservation in the relief of fault scarps and other varieties of primary seismic dislocations, as well as seismogravitational deformations. These are destroyed by erosion processes, as a rule, within several thousand years, which limits their use for paleoseismological analysis. For example, in the Gorny Altai and in climatically similar regions of the northern Tien Shan, there are known fault scarps with ages up to 13–14 ka [61,148] and landslides up to 16 ka [81]. Therefore, for the reconstruction of more ancient seismic events, various secondary seismodislocations in unconsolidated soils, united under the term "seismites", have recently become significant. The analysis of these in the Gorny Altai has allowed us to extend its seismic record to the Middle Pleistocene [58–60,62,149,150]. The main disadvantage of the analysis of seismo-gravity structures and seismites is that in most cases they allow neither identification of the focal zones of paleoearthquakes nor specific active faults associated with seismic events.

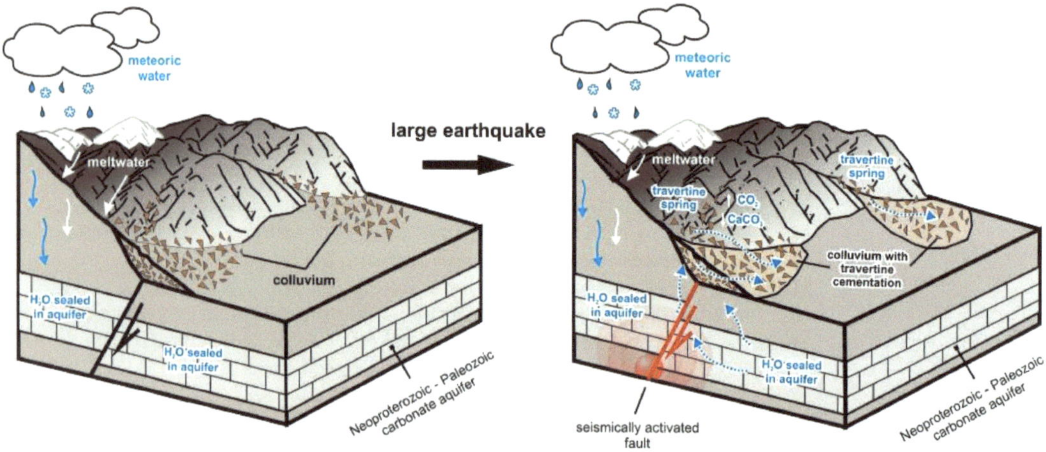

Figure 18. Generalized model of the formation of the south-eastern Gorny Altai travertines.

According to our proposed model, travertines located along active faults and possessing specific carbon isotope properties, mark the fracture zones included in the systems of surface ruptures of paleoearthquakes. The found match between the times of travertine formation in the south-eastern Gorny Altai and the ages of paleoearthquakes over the last 8 ka BP (Figure 17A) allow to consider travertines as an analogue of colluvial wedges, the age of which determines the age of paleoearthquakes. Consequently, the radiometric ages of travertine carbonates can be used as a tool to rebuild the sequence of ancient and historical earthquakes in the region. In contrast to surface earthquake ruptures and seismo-gravitational structures, travertine bodies are more resistant to weathering. This allows us, by determining their radiometric ages, to identify episodes of seismic activity on a much longer temporal scale. The results of our ^{230}Th-U dating demonstrated the possibility of detecting travertine with ages up to 400 ka BP (Figure 17B). Unlike seismites, large landslides and rockslides, the travertines we studied clearly indicate the fault on which the seismogenic displacement occurred. Furthermore, their resistance to weathering allows using coeval travertines to assess the length of the surface rupture system of a particular earthquake; in other words, the distance between such travertines can be used as an analogue of the "surface rupture length" parameter when estimating the moment magnitudes of paleoearthquakes. Consequently, seismogenic travertines become an important tool in the analysis of seismically active regions.

6. Conclusions

This study describes seven travertine deposits of the south-western Gorny Altai. The analysis of the morphology of travertine bodies, their petrography and mineralogy, mineral and rock chemistry, stable isotope geochemistry of carbonates, and ^{230}Th-U geochronology of carbonates in comparison with data on neotectonics, active tectonics, paleoseismicity and paleoclimate of the region allow the following main conclusions.

(1) The travertine deposits have been found to be spatially associated with the active faults bordering the Chuya and Kurai basins. Travertines predominantly cement Middle

Pleistocene-Holocene colluvial, alluvial, and alluvial fan deposits, transforming them into breccias, conglomerates, and sandstones.

(2) Ca-carbonates of fossil travertines are primary precipitates and show no indications of diagenetic alteration. Precipitation of calcite and aragonite in fossil travertines was probably controlled by rapid CO_2 degassing and the Mg/Ca ratios of the spring water.

(3) The isotope and trace-element data jointly indicate a hypogean origin of the paleofluid from the fossil Gorny Altai travertine system and the Upper Neoproterozoic–Paleozoic limestone-dolostone as fluid source rocks.

(4) The times of travertine formation match the ages of paleoearthquakes over the last 8 ka BP (Figure 17A), and episodes of travertine deposition at 9.0–11.0 ka BP correspond to the period of elevated paleoseismicity of the region at 8.0–16.0 ka BP. The 123 ka BP travertine resulted from a slip triggered by the Middle Pleistocene deglaciation, while that of 400 ka BP represents seismic motions during the main Cenozoic orogenic phase. Episodes of travertines formation were triggered by large seismic events, which induced the rapid rise of bicarbonate groundwaters previously sealed in carbonate aquifers along seismically activated ruptures. All travertine-forming events at the south-western Gorny Altai fall within warm and wet climatic phases (interglacials), when ice rebound and permafrost degradation occurred.

(5) The obtained results allow using travertines with similar isotopic characteristics and structural position as indicators of fault activity both in the south-western Siberian Altai and in other seismically active regions of the world. Radiometric ages of carbonates make it possible to determine the age of ancient and historical earthquakes, detailing and significantly extending the data of instrumental and historical seismology and paleoseismology.

Supplementary Materials: The following supporting information can be downloaded at: https://www.mdpi.com/article/10.3390/min13020259/s1, Table S1: Carbon and oxygen isotope compositions of $CaCO_3$ from travertines of the south-eastern Gorny Altai. Table S2: Description of travertine samples, dated by the ^{230}Th-U method. Figure S1: Biota in the seepage zone of a modern spring at the Chibit deposit.

Author Contributions: Project idea, E.V.D., S.N.K., Y.D. and E.V.S.; conceptualization, E.V.D., S.N.K. and Y.D.; field work, E.V.D., G.G.R. and S.N.K.; paleoseismological study, E.V.D.; ^{230}Th-U dating, Y.D. and D.S.; SEM and EPMA analyses, S.N.K. and E.V.S.; isotope analysis, Y.D. and V.N.R.; interpretation of analytical data: E.V.D., S.N.K. and Y.D.; writing—original draft preparation, E.V.D., S.N.K. and Y.D.; visualization: E.V.D. and S.N.K.; funding acquisition, E.V.D. All authors have read and agreed to the published version of the manuscript.

Funding: The work was supported by the Russian Science Foundation [Project 21-17-00058].

Data Availability Statement: Not applicable.

Acknowledgments: D. Scholz thanks the DFG (German Research Foundation) for funding (INST 247/889-1 FUGG) and M. Großkopf, M. Weber, and V. Blumrich for assistance with U-series dating at the University of Mainz. D. Scholz is also thankful to K.P. Jochum, M.O. Andreae and G.H. Haug from the Max Planck Institute for Chemistry, Mainz, for long-lasting support. We thank the anonymous reviewers for their valuable comments, which we accepted with gratitude.

Conflicts of Interest: The authors declare no conflict of interest.

References

1. Pentecost, A. *Travertine*; Springer: Berlin, Germany, 2005; p. 446.
2. Jones, B.; Renaut, R.W. Calcareous spring deposits in continental settings. In *Carbonates in Continental Settings: Facies, Environments, and Processes*; Alonso Zarza, A.M., Taner, L.H., Eds.; Elsevier: Amsterdam, The Netherlands, 2010; Volume 61, pp. 177–224.
3. Capezzuoli, E.; Gandin, A.; Pedley, M. Decoding tufa and travertine (freshwater carbonates) in the sedimentary record: The state of the art. *Sedimentology* **2014**, *61*, 1–21. [CrossRef]
4. Pedley, H.M. Classification and environmental models of cool freshwater tufas. *Sediment. Geol.* **1990**, *68*, 143–154. [CrossRef]
5. Ford, T.D.; Pedley, H.M. A review of tufa and travertine deposits of the world. *Earth Sci. Rev.* **1996**, *41*, 117–175. [CrossRef]
6. Pentecost, A. British travertines: A review. *Proc. Geol. Assoc.* **1993**, *104*, 23–39. [CrossRef]

7. Pentecost, A.; Viles, H. A review and reassessment of travertine classification. *Géographie Phys. Quat.* **1994**, *48*, 305–314. [CrossRef]
8. Zentmyer, R.; Myrow, P.M.; Newell, D.L. Travertine deposits from along the South Tibetan Fault System near Nyalam, Tibet. *Geol. Mag.* **2008**, *145*, 753–765. [CrossRef]
9. Ruszkiczay-Rüdiger, Z.; Csillag, G.; Fodor, L.; Braucher, R.; Novothny, Á.; Thamó-Bozsó, E.; Virág, A.; Pazonyi, P.; Timár, G.; Aumaître, G.; et al. Integration of new and revised chronological data to constrain the terrace evolution of the Danube River (Gerecse Hills, Pannonian Basin). *Quat. Geochronol.* **2018**, *48*, 148–170. [CrossRef]
10. Barnes, I.; O'Neil, J.R. Calcium-magnesium carbonate solid solutions from Holocene conglomerate cements and travertines in the Coast Range of California. *Geochim. Cosmochim. Acta* **1971**, *35*, 699–718. [CrossRef]
11. Boch, R.; Spötl, C.; Reitner, J.M.; Kramers, J. A Lateglacial travertine deposit in Eastern Tyrol (Austria). *Austrian J. Earth Sci.* **2005**, *98*, 78–91.
12. Lugli, S.; Tang, Y.; Reghizzi, M.; Qiao, X.; Schreiber, B.C.; Deng, G. Seasonal pattern in the high-elevation fluvial travertine from the Jiuzhaigou national nature reserve, Sichuan, southwestern China. *J. Sediment. Res.* **2017**, *87*, 253–271. [CrossRef]
13. Claes, H.; Soete, J.; Van Noten, K.; El Desouky, H.; Marques Erthal, M.; Vanhaecke, F.; Özkul, M.; Swennen, R. Sedimentology, three-dimensional geobody reconstruction and carbon dioxide origin of Pleistocene travertine deposits in the Ballık area (south-west Turkey). *Sedimentology* **2015**, *62*, 1408–1445. [CrossRef]
14. Claes, H.; Huysmans, M.; Soete, J.; Dirix, K.; Vassilieva, E.; Marques Erthal, M.; Vandewijngaerde, W.; Hamaekers, H.; Aratman, C.; Özkul, M.; et al. Elemental geochemistry to complement stable isotope data of fossil travertine: Importance of digestion method and statistics. *Sediment. Geol.* **2019**, *386*, 118–131. [CrossRef]
15. Crossey, L.J.; Fischer, T.P.; Patchett, P.J.; Karlstrom, K.E.; Hilton, D.R.; Newell, D.L.; Huntoon, P.; Reynolds, A.C.; de Leeuw, G.A.M. Dissected hydrologic system at the Grand Canyon: Interaction between deeply derived fluids and plateau aquifer waters in modern springs and travertine. *Geology* **2006**, *34*, 25–28. [CrossRef]
16. Teboult, P.-A.; Durlet, C.; Gaucher, E.C.; Virgone, A.; Girard, J.-P.; Curie, J.; Lopez, B.; Camoin, G.F. Origins of elements building travertine and tufa: New perspectives provided by isotopic and geochemical tracers. *Sediment. Geol.* **2016**, *334*, 97–114. [CrossRef]
17. Pazdur, A.; Pazdur, M.F.; Starkel, L.; Szulc, J. Stable isotopes of Holocene calcareous tufa in southern Poland as palaeoclimatic indicators. *Quat. Res.* **1988**, *30*, 177–189. [CrossRef]
18. Frank, N.; Braum, M.; Hambach, U.; Mangini, A.; Wagner, G. Warm period growth of travertine during the last Interglaciation in southern Germany. *Quat. Res.* **2000**, *54*, 38–48. [CrossRef]
19. Minissale, A.; Kerrick, D.M.; Magro, G.; Murrell, M.T.; Paladini, M.; Rihs, S.; Sturchio, N.C.; Tassi, F.; Vaselli, O. Geochemistry of Quaternary travertines in the region north of Rome (Italy): Structural, hydrologic, and paleoclimatic implications. *Earth Planet. Sci. Lett.* **2002**, *203*, 709–728. [CrossRef]
20. Martin-Algarra, A.; Martin-Martin, M.; Andreo, B.; Julia, R.; Gonzalez-Gomez, C. Sedimentary patterns in perched spring travertines near Granada (Spain) as indicators of the paleohydrological and palaeoclimatical evolution of a karst massif. *Sediment. Geol.* **2003**, *161*, 217–228. [CrossRef]
21. Liu, Z.; Zhang, M.; Li, Q.; You, S. Hydrochemical and isotope characteristics of spring water and travertine in the Baishuitai area (SW China) and their meaning for paleoenvironmental reconstruction. *Environ. Geol.* **2003**, *44*, 698–704. [CrossRef]
22. Faccenna, C.; Soligo, M.; Billi, A.; De Filippis, L.; Funiciello, R.; Rossetti, C.; Tuccimei, P. Late Pleistocene depositional cycles of the Lapis Tiburtinus travertine (Tivoli, central Italy): Possible influence of climate and fault activity. *Glob. Planet. Change* **2008**, *63*, 299–308. [CrossRef]
23. Altunel, E.; Hancock, P.L. Active fissuring, faulting and travertine deposition at Pamukkale, western Turkey. In *Neotectonics and Active Faulting*; Stewart, I.S., Vita-Finzi, C., Owen, L.A., Eds.; Zeitschrift für Geomorphologie: Stuttgart, Germany, 1993; Volume 94, pp. 285–302.
24. Çakır, Z. Along-strike discontinuity of active normal faults and its influence on quaternary travertine deposition; examples from western Turkey. *Turk. J. Earth Sci.* **1999**, *8*, 67–80.
25. Hancock, P.L.; Chalmers, R.M.L.; Altunel, E.; Çakir, Z. Travitonics: Using travertine in active fault studies. *J. Struct. Geol.* **1999**, *21*, 903–916. [CrossRef]
26. Altunel, E.; Karabacak, V. Determination of horizontal extension from fissure-ridge travertines: A case study from the Denizli Basin, southwestern Turkey. *Geodin. Acta* **2005**, *18*, 333–342. [CrossRef]
27. Mesci, B.L.; Gürsoy, H.; Tatar, O. The evolution of travertine masses in the Sivas Area (Central Turkey) and their relationships to active tectonics. *Turk. J. Earth Sci.* **2008**, *17*, 219–240.
28. Brogi, A.; Capezzuoli, E. Travertine deposition and faulting: The fault-related travertine fissure-ridge at Terme S. Giovanni, Rapolano Terme (Italy). *Int. J. Earth Sci.* **2009**, *98*, 931–947. [CrossRef]
29. Brogi, A.; Capezzuoli, E.; Aqué, R.; Branca, M.; Voltaggio, M. Studying travertines for neotectonics investigations: Middle–Late Pleistocene syn-tectonic travertine deposition at Serre di Rapolano (Northern Apennines, Italy). *Int. J. Earth Sci.* **2010**, *99*, 1383–1398. [CrossRef]
30. Nishikawa, O.; Furuhashi, K.; Masuyama, M.; Ogata, T.; Shiraishi, T.; Shen, C. Radiocarbon dating of residual organic matter in travertine formed along the Yumoto Fault in Oga Peninsula, northeast Japan: Implications for long-term hot spring activity under the influence of earthquakes. *Sediment. Geol.* **2012**, *243–244*, 181–190. [CrossRef]
31. Temiz, U. Age of fault reactivation by using U/Th dating method of carbonate precipitates: An example from the active Manisa Fault (West Turkey). *Geodin. Acta* **2012**, *1–2*, 1–11. [CrossRef]

32. Temiz, U.; Gökten, Y.E.; Eikenberg, J. Strike-slip deformation and U/Th dating of travertine deposition: Examples from North Anatolian Fault Zone, Bolu and Yeniçağ Basins, Turkey. *Quat. Int.* **2013**, *312*, 132–140. [CrossRef]
33. De Filippis, L.; Faccenna, C.; Billi, A.; Anzalone, E.; Brilli, M.; Soligo, M.; Tuccimei, P. Plateau versus fissure ridge travertines from Quaternary geothermal springs of Italy and Turkey: Interactions and feedbacks between fluid discharge, paleoclimate, and tectonics. *Earth-Sci. Rev.* **2013**, *123*, 35–52. [CrossRef]
34. De Filippis, L.; Anzalone, E.; Billi, A.; Faccenna, C.; Poncia, P.P.; Sella, P. The origin and growth of a recently-active fissure ridge travertine over a seismic fault, Tivoli, Italy. *Geomorphology* **2013**, *195*, 13–26. [CrossRef]
35. Ascione, A.; Iannace, A.; Imbriale, P.; Santangelo, N.; Santo, A. Tufa and travertines of southern Italy: Deep-seated, fault-related CO_2 as the key control in precipitation. *Terra Nova* **2014**, *26*, 1–13. [CrossRef]
36. Mohajjel, M.; Taghipour, K. Quaternary travertine ridges in the Lake Urmia area: Active extension in NW Iran. *Turk. J. Earth Sci.* **2014**, *23*, 602–614. [CrossRef]
37. Erol, S.Ç.; Özkul, M.; Aksoy, E.; Kele, S.; Ghaleb, B. Travertine occurrences along major strike-slip fault zones: Structural, depositional and geochemical constraints from the Eastern Anatolian Fault System (EAFS), Turkey. *Geodin. Acta* **2015**, *27*, 155–174. [CrossRef]
38. Selim, H.H.; Yavuz, O.; Gürer, Ö.F.; Karakaş, A.; Taş, K.Ö. Age determination for segments of the North Anatolian Fault (NAF) northern branch by $^{234}U/^{230}Th$ dating of Soğucak (Yalova) range-front travertines, south Marmara, Turkey. *Quat. Int.* **2016**, *425*, 416–424. [CrossRef]
39. Sinisi, R.; Petrullo, A.V.; Agosta, F.; Paternoster, M.; Belviso, C.; Grassa, F. Fault and fluid interaction in the Bradano Trough, southern Italy. *Tectonophysics* **2016**, *690*, 206–218. [CrossRef]
40. Frery, E.; Gratier, J.P.; Ellouz-Zimmerman, N.; Deschamps, P.; Blamart, D.; Hamelin, B.; Swennen, R. Geochemical transect through a travertine mount: A detailed record of CO_2-enriched fluid leakage from Late Pleistocene to present-day e Little Grand Wash fault (Utah, USA). *Quat. Int.* **2017**, *437*, 98–106. [CrossRef]
41. Kokh, S.N.; Sokol, E.V.; Deev, E.V.; Ryapolova, Y.M.; Rusanov, G.G.; Tomilenko, A.A.; Bul'bak, T.A. Post-Late Glacial calcareous tufas from the Kurai fault zone (Southeastern Gorny Altai, Russia). *Sediment. Geol.* **2017**, *355*, 1–19. [CrossRef]
42. Martínez-Díaz, J.J.; Hernández-Enrile, J.L. Using travertine deformations to characterize paleoseismic activity along an active oblique-slip fault: The Alhama de Murcia fault (Betic Cordillera, Spain). *Acta Geol. Hisp.* **2001**, *36*, 297–313.
43. Uysal, I.T.; Feng, Y.; Zhao, J.; Isik, V.; Nuriel, P.; Golding, S.D. Hydrothermal CO_2 degassing in seismically active zones during the late Quaternary. *Chem. Geol.* **2009**, *265*, 442–454. [CrossRef]
44. Uysal, I.T.; Feng, Y.; Zhao, J.; Altunel, E.; Weatherley, D.; Karabacak, V.; Cengiz, O.; Golding, S.D.; Lawrens, M.G.; Collerson, K.D. U-series dating and geochemical tracing of Late Quaternary travertines in co-seismic fissures. *Earth Planet. Sci. Lett.* **2007**, *257*, 450–462. [CrossRef]
45. Brogi, A.; Capezzuoli, E. Earthquake impact on fissure-ridge type travertine deposition. *Geol. Mag.* **2014**, *151*, 1135–1143. [CrossRef]
46. Gradziński, M.; Wróblewski, W.; Duliński, M.; Hercman, H. Earthquake-affected development of a travertine ridge. *Sedimentology* **2014**, *61*, 238–263. [CrossRef]
47. Karabacak, V.; Uysal, I.T.; Mutlu, H.; Ünal-İmer, E.; Dirik, R.K.; Feng, Y.-X.; Akıska, S.; Aydoğdu, İ.; Zhao, J.-X. Are U-Th dates correlated with historical records of earthquakes? Constraints from co-seismic carbonate veins within the North Anatolian Fault Zone. *Tectonics* **2019**, *38*, 2431–2448. [CrossRef]
48. Deev, E.V.; Dublyansky, Y.V.; Pozdnyakova, N.I.; Scholz, D.; Kokh, S.N.; Sokol, E.V.; Rusanov, G.G. $^{230}Th/U$ dating of travertines related to paleoearthquakes in Gorny Altai: First results. *Dokl. Earth Sci.* **2021**, *500*, 820–825. [CrossRef]
49. Deev, E.; Dublyansky, Y.; Kokh, S.; Scholz, D.; Rusanov, G.; Sokol, E.; Khvorov, P.; Reutsky, V.; Panin, A. Large Holocene paleoseismic events and synchronized travertine formation: A case study of the Kurai Fault Zone (Gorny Altai, Russia). *Int. Geol. Rev.* 2022. [CrossRef]
50. Mancini, A.; Frondini, F.; Capezzuoli, E.; Mejia, E.G.; Lezzi, G.; Matarazzi, D.; Brogi, A.; Swennen, R. Evaluating the geogenic CO2 flux from geothermal areas by analysing quaternary travertine masses. New data from western central Italy and review of previous CO2 flux data. *Quat. Sci. Rev.* **2019**, *215*, 132–143. [CrossRef]
51. Molnar, P.; Tapponnier, P. Cenozoic tectonics of Asia: Effects of a continental collision. *Science* **1975**, *189*, 419–426. [CrossRef] [PubMed]
52. Delvaux, D.; Cloetingh, S.; Beekman, F.; Sokoutis, D.; Burov, E.; Buslov, M.M.; Abdrakhmatov, K.E. Basin evolution in a folding lithosphere: Altai–Sayan and Tien Shan belts in Central Asia. *Tectonophysics* **2013**, *602*, 194–222. [CrossRef]
53. Deev, E.V.; Nevedrova, N.N.; Zol'nikov, I.D.; Rusanov, G.G.; Ponomarev, P.V. Geoelectrical studies of the Chuya basin sedimentary fill (Gorny Altai). *Russ. Geol. Geophys.* **2012**, *53*, 92–107. [CrossRef]
54. Nevedrova, N.N.; Deev, E.V.; Sanchaa, A.M. Deep structure and margins of the Kurai Basin (Gorny Altai), from controlled-source resistivity data. *Russ. Geol. Geophys.* **2014**, *55*, 98–107. [CrossRef]
55. Rusanov, G.G.; Deev, E.V.; Zolnikov, I.D.; Khazin, L.B.; Khazina, I.V.; Kuz'mina, O.B. Reference section of Neogene–Quarternary deposits in the Uimon Basin (Gorny Altai). *Russ. Geol. Geophys.* **2017**, *58*, 973–983. [CrossRef]
56. Deviatkin, E.V. *Cenozoic Deposits and Neotectonics in South-Eastern Altai*; Nauka: Moscow, Russia, 1965; p. 244.
57. Deev, E.V. Localization Zones of Ancient and Historical Earthquakes in Gorny Altai. *Izv. Phys. Solid Earth* **2019**, *55*, 451–470. [CrossRef]

58. Deev, E.V.; Zolnikov, I.D.; Gus'kov, S.A. Seismites in Quaternary sediments of southeastern Altai. *Russ. Geol. Geophys.* **2009**, *50*, 546–561. [CrossRef]
59. Deev, E.V.; Zolnikov, I.D.; Bortodovsky, A.P.; Goltsova, S.V. Neotectonics and paleoseismicity of the lower Katun' valley (Gorny Altai). *Russ. Geol. Geophys.* **2012**, *53*, 883–894. [CrossRef]
60. Deev, E.V.; Zolnikov, I.D.; Goltsova, S.V.; Rusanov, G.G.; Emanov, A.A. Traces of paleoearthquakes in the Quaternary deposits of intermontane basins in central Gorny Altai. *Russ. Geol. Geophys.* **2013**, *54*, 312–323. [CrossRef]
61. Deev, E.V.; Zolnikov, I.D.; Turova, I.V.; Rusanov, G.G.; Ryapolova, Y.M.; Nevedrova, N.N.; Kotler, S.A. Paleoearthquakes in the Uimon basin (Gorny Altai). *Russ. Geol. Geophys.* **2018**, *59*, 437–452. [CrossRef]
62. Deev, E.; Turova, I.; Borodovskiy, A.; Zolnikov, I.; Pozdnyakova, N.; Molodkov, A. Large earthquakes in the Katun Fault zone (Gorny Altai): Paleoseismological and archaeoseismological evidence. *Quat. Sci. Rev.* **2019**, *203*, 68–89. [CrossRef]
63. Butvilovsky, V.V. *Paleogeography of the Latest Glaciation and Holocene of Altai: An Event–Catastrophic Model*; Tomsk University: Tomsk, Russia, 1993; p. 253. (In Russian)
64. Rusanov, G.G.; Deev, E.V.; Ryapolova, Y.M.; Zolnikov, I.D. Paleohydrothermal activity of faults in the Gorny Altai based on travertine dating. *Geol. Miner. Resour. Sib.* **2013**, *4*, 53–64.
65. Deev, E.V.; Sokol, E.V.; Ryapolova, Y.M.; Kokh, S.N.; Rusanov, G.G. Quaternary travertine of the Kurai fault zone (Gorny Altai). *Dokl. Earth Sci.* **2017**, *473*, 261–265. [CrossRef]
66. Kokh, S.N.; Sokol, E.V.; Deev, E.V.; Ryapolova, Y.M.; Rusanov, G.G. Modern and paleotravertines from Gorny Altai: Indicative features. *Dokl. Earth Sci.* **2018**, *483*, 1399–1404. [CrossRef]
67. GeoMapApp, Version 3.6.15. Available online: http://www.geomapapp.org (accessed on 5 December 2022).
68. Turkin, Y.A.; Fedak, S.I. *Geology and Structural-and-Compositional Complexes of Gorny Altai*; STT: Tomsk, Russia, 2008; p. 460. (In Russian)
69. Nevedrova, N.N.; Epov, M.I.; Antonov, E.Y.; Dashevskii, Y.A.; Duchkov, A.D. Deep structure of the Chuya basin (Gomy Altai) as imaged by TEM soundings. *Russ. Geol. Geophys.* **2001**, *42*, 1399–1416.
70. Delvaux, D.; Theunissen, K.; Van Der Meer, R.; Berzin, N. Dynamics and paleostress of the Cenozoic Kurai-Chuya depression of Gorny Altai (South Siberia): Tectonic and climatic control. *Russ. Geol. Geophys.* **1995**, *36*, 26–45.
71. Deev, E.V.; Krzhivoblotskaya, V.E.; Borodovskiy, A.P.; Entin, A.L. Active faults and Late Holocene surface rupturing earthquakes in the Kokorya Basin (Gorny Altai, Russia). *Dokl. Earth Sci.* **2022**, *506*, 666–670. [CrossRef]
72. Zol'nikov, I.D.; Deev, E.V.; Lyamina, V.A. New data on Quaternary morpholithogenesis in the Chuya basin (Gorny Altai). *Russ. Geol. Geophys.* **2010**, *51*, 339–348. [CrossRef]
73. Shokalsky, S.P.; Turkin, Y.A.; Fedak, S.I. *State Geological Map of the Russian Federation. Scale 1: 1 000 000, 3rd. Series Altai-Sayan. Sheet M-45–Gorno-Altaisk*, 3rd ed.; VSEGEI: St. Petersburg, Russia, 2011.
74. Deev, E.V.; Turova, I.V.; Borodovskiy, A.P.; Zolnikov, I.D.; Oleszczak, L. Unknown large ancient earthquakes along the Kurai fault zone (Gorny Altai): New results of paleoseismological and archaeoseismological studies. *Int. Geol. Rev.* **2017**, *59*, 293–310. [CrossRef]
75. Bachmanov, D.M.; Kozhurin, A.I.; Trifonov, V.G. The active faults of Eurasia database. *Geodyn. Tectonophys.* **2017**, *8*, 711–736. [CrossRef]
76. Leskova, E.V.; Emanov, A.A. Hierarchical properties of the tectonic stress field in the source region of the 2003 Chuya earthquake. *Russ. Geol. Geophys.* **2013**, *54*, 87–95. [CrossRef]
77. Rogozhin, E.A.; Ovsyuchenko, A.N.; Marakhanov, A.V.; Ushanova, E.A. Tectonic setting and geological manifestations of the 2003 Altai earthquake. *Geotectonics* **2007**, *41*, 87–104. [CrossRef]
78. Lunina, O.V.; Gladkov, A.S.; Novikov, I.S.; Agatova, A.R.; Vysotskii, E.M.; Emanov, A.A. Geometry of the fault zone of the 2003 Ms=7.5 Chuya earthquake and associated stress fields, Gorny Altai. *Tectonophysics* **2008**, *453*, 276–294. [CrossRef]
79. Turova, I.; Deev, E.; Pozdnyakova, N.; Entin, A.; Nevedrova, N.; Shaparenko, I.; Bricheva, S.; Korzhenkov, A.; Kurbanov, R.; Panin, A. Surface-rupturing paleoearthquakes in the Kurai Fault Zone (Gorny Altai, Russia): Trenching and geophysical evidence. *J. Asian Earth Sci.* **2020**, *197*, 104399. [CrossRef]
80. Korzhenkov, A.M.; Deev, E.V.; Pozdnyakova, N.I. Young landforms created by earthquakes in the foothills of the Kurai Range (Gorny Altai). *Seism. Instrum.* **2022**, *58*, 424–441. [CrossRef]
81. Deev, E.V.; Zolnikov, I.D.; Kurbanov, R.N.; Panin, A.V.; Murray, A.; Korzhenkov, A.M.; Turova, I.V.; Pozdnyakova, N.I.; Vasiliev, A.V. OSL dating of the Sukor earthquake-induced rockslide in Gorny Altai: Paleoseismological and paleogeographic implications. *Russ. Geol. Geophys.* **2022**, *63*, 743–754. [CrossRef]
82. Ershov, E.D. (Ed.) *Geocryology of the USSR. Vol. V. Mountain Regions of the Southern USSR*; Nedra: Moscow, Russia, 1989; p. 359. (In Russian)
83. Olenchenko, V.V.; Kozhevnikov, N.O.; Antonov, E.Y.; Pospeeva, E.V.; Potapov, V.V.; Shein, A.N.; Stefanenko, S.M. Distribution of permafrost in Chuiskaya Basin (Gorny Altai) according to transient electromagnetic soundings data. *Earth's Cryosphere* **2011**, *15*, 15–22.
84. Fedak, S.I.; Turkin, Y.A.; Gusev, A.I.; Shokalsky, S.P.; Rusanov, G.G.; Borisov, B.A.; Belyaev, G.M.; Leontyeva, E.M. *State Geological Map of the Russian Federation. Scale 1: 1 000 000. Third Generation. Series: Altai-Sayan. Sheet M-45—Gorno-Altaisk. Explanatory Note*; VSEGEI: St. Petersburg, Russia, 2011; p. 567. (In Russian)

85. Lavrent'ev, Y.G.; Korolyuk, V.N.; Usova, L.V.; Nigmatulina, E.N. Electron probe microanalysis of rock-forming minerals with a JXA-8100 electron probe microanalyzer. *Russ. Geol. Geophys.* **2015**, *56*, 1428–1436. [CrossRef]
86. Feng, D.; Chen, D.; Peckmann, J. Rare earth elements in seep carbonate as tracers of variable redox conditions at ancient hydrocarbon seeps. *Terra Nova* **2009**, *21*, 49–56. [CrossRef]
87. Spötl, C. Long-term performance of the Gasbench isotope ratio mass spectrometry system for the stable isotope analysis of carbonate microsamples. *Rapid Commun. Mass Spectrom.* **2011**, *25*, 1683–1685. [CrossRef]
88. Parkhurst, D.L.; Appelo, C.A.J. User's Guide to PHREEQC (Version 2) a Computer Program for Speciation, Batch-Reaction, One-Dimensional Transport, and Inverse Geochemical Calculations: U.S Geological Survey Water-Resources Investigations Report 99-4259. 1999. Available online: https://pubs.er.usgs.gov/publication/wri994259 (accessed on 1 December 2022).
89. Yang, Q.; Scholz, D.; Jochum, K.P.; Hoffmann, D.L.; Stoll, B.; Weis, U.; Schwager, B.; Andreae, M.O. Lead isotope variability in speleothems—A promising new proxy for hydrological change? First results from a stalagmite from western Germany. *Chem. Geol.* **2015**, *396*, 143–151. [CrossRef]
90. Obert, J.C.; Scholz, D.; Felis, T.; Brocas, W.M.; Jochum, K.P.; Andreae, M.O. $^{230}Th/U$ dating of Last Interglacial brain corals from Bonaire (southern Caribbean) using bulk and theca wall material. *Geochim. Cosmochim. Acta* **2016**, *178*, 20–40. [CrossRef]
91. Gibert, L.; Scott, G.R.; Scholz, D.; Budsky, A.; Ferrandez, C.; Martin, R.A.; Ribot, F.; Leria, M. Chronology for the Cueva Victoria fossil site (SE Spain): Evidence for Early Pleistocene Afro-Iberian dispersals. *J. Hum. Evol.* **2016**, *90*, 183–197. [CrossRef]
92. Cheng, H.; Edwards, R.L.; Shen, C.C.; Polyak, V.J.; Asmerom, Y.; Woodhead, J.; Hellstrom, J.; Wang, Y.; Kong, X.; Spötl, C.; et al. Improvements in ^{230}Th dating, ^{230}Th and ^{234}U half-life values, and U–Th isotopic measurements by multi-collector inductively coupled plasma mass spectrometry. *Earth Planet. Sci. Lett.* **2013**, *371*, 82–91. [CrossRef]
93. Bantsev, D.V.; Ganyushkin, D.A.; Chistyakov, K.V.; Ekaykin, A.A.; Tokarev, I.V.; Volkov, I.V. Formation of glacier runoff on the northern slope of Tavan Bogd mountain massif based on stable isotopes data. *Ice Snow* **2018**, *58*, 333–342. [CrossRef]
94. Bantcev, D.; Ganyushkin, D.; Terekhov, A.; Ekaykin, A.; Tokarev, I.; Chistyakov, K. Isotopic composition of glacier ice and meltwater in the arid parts of the Altai Mountains (Central Asia). *Water* **2022**, *14*, 252. [CrossRef]
95. Bantsev, D.V. Isotopic composition of glacio-nival systems components as an indicator of their runoff-forming features in South-Eastern Altai. Ph.D. Thesis, Saint-Petersburg State University, Saint-Petersburg, Russia, 2020.
96. Craig, H. Isotopic variations in meteoric waters. *Science* **1961**, *133*, 1702–1703. [CrossRef]
97. Taylor, S.R.; McLennan, S.M. *The Continental Crust: Its Evolution and Composition*; Blackwell Scientific Publications: Oxford, UK, 1985; p. 312.
98. Jaffey, A.H.; Flynn, K.F.; Glendenin, L.E.; Bentley, W.C.; Essling, A.M. Precision measurement of the half-lives and specific activities of ^{235}U and ^{238}U. *Phys. Rev. C* **1971**, *4*, 1889–1906. [CrossRef]
99. Janssens, N.; Capezzuoli, E.; Claes, H.; Muchez, P.; Yu, T.-L.; Shen, C.-C.; Ellam, R.M.; Swennen, R. Fossil travertine system and its palaeofluid provenance, migration and evolution through time: Example from the geothermal area of Acquasanta Terme (Central Italy). *Sediment. Geol.* **2020**, *398*, 105580. [CrossRef]
100. Folk, R.L. Interaction between bacteria, nannobacteria, and mineral precipitation in hot springs of central Italy. *Géographie Phys. Et Quat.* **1994**, *48*, 233–246. [CrossRef]
101. Loste, E.; Wilson, R.M.; Seshadri, R.; Meldrum, F.C. The role of magnesium in stabilizing amorphous calcium carbonate and controlling calcite morphologies. *J. Cryst. Growth* **2003**, *354*, 206–218. [CrossRef]
102. Jones, B. Review of calcium carbonate polymorph precipitation in spring systems. *Sediment. Geol.* **2017**, *353*, 64–75. [CrossRef]
103. Boch, R. Carbonates in natural and geotechnical settings–chemical sediments as environmental archives. *Jahrb. Der Geol. Bundesanst.* **2019**, *159*, 67–130.
104. Luo, L.; Capezzuoli, E.; Rogersonb, M.; Vaselli, O.; Wen, H.; Lu, Z. Precipitation of carbonate minerals in travertine-depositing hot springs: Driving forces, microenvironments, and mechanisms. *Sediment. Geol.* **2022**, *438*, 106207. [CrossRef]
105. Friedman, I. Some investigations of the deposition of travertine from Hot Springs—I. The isotopic chemistry of a travertine-depositing spring. *Geochim. Et Cosmochim. Acta* **1970**, *34*, 1303–1315. [CrossRef]
106. Fouke, B.W.; Farmer, J.D.; Des Marais, D.J.; Pratt, L.; Sturchio, N.C.; Burns, P.C.; Discipulo, M.K. Depositional facies and aqueous-solid geochemistry of travertine depositing hot springs (Angel Terrace, Mammoth hot springs, Yellowstone national park, USA). *J. Sediment. Res.* **2000**, *70*, 565–585. [CrossRef] [PubMed]
107. Okumura, T.; Takashima, C.; Shiraishi, F.; Akmaluddin; Kano, A. Textural transition in an aragonite travertine formed under various flow conditions at Pancuran Pitu, Central Java, Indonesia. *Sediment. Geol.* **2012**, *265–266*, 195–209. [CrossRef]
108. Kanellopoulos, C. Distribution, lithotypes and mineralogical study of newly formed thermogenic travertines in Northern Euboea and Eastern Central Greece. *Cent. Eur. J. Geosci.* **2012**, *4*, 545–560. [CrossRef]
109. Gonzalez, L.A.; Lohmann, K.C. Controls on mineralogy and composition of spelean carbonates: Carlsbad Caverns, New Mexico. In *Paleokarst*; James, N.P., Choquette, P.W., Eds.; Springer: Berlin, Germany, 1987; pp. 81–101.
110. Hill, C.A.; Forti, P. *Cave Minerals of the World*; National Speleological Society Inc.: Huntsville, AL, USA, 1997; p. 463.
111. Frisia, S.; Borsato, A.; Fairchild, I.J.; McDermott, F.; Selmo, E.M. Aragonite-calcite relationships in speleothems (Grotto de Clamouse, France): Environment, fabrics and carbonate geochemistry. *J. Sediment. Res.* **2002**, *72*, 687–699. [CrossRef]
112. Rowling, J. Studies on aragonite and its occurrence in caves, including New South Wales caves. *J. Proc. R. Soc. N. S. W.* **2004**, *137*, 123–149.

113. Okumura, T.; Takashima, C.; Kano, A. Textures and processes of laminated travertines formed by unicelluar cyanobacteria in Myokwn hot spring, southwestern Japan. *Isl. Arc* **2013**, *22*, 410–426. [CrossRef]
114. Dupraz, C.; Reid, P.R.; Braissant, O.; Decho, A.W.; Norman, R.S.; Visscher, P.T. Processes of carbonate precipitation in modern microbial mats. *Earth-Sci. Rev.* **2009**, *96*, 141–162. [CrossRef]
115. Rainey, D.K.; Jones, B. Abiotic versus biotic controls on the development of the Fairmont Hot Springs carbonate deposit, British Columbia, Canada. *Sedimentology* **2009**, *56*, 1832–1857. [CrossRef]
116. Peng, X.; Jones, B. Patterns of biomediated $CaCO_3$ crystal bushes in hot spring deposits. *Sediment. Geol.* **2013**, *294*, 105–117. [CrossRef]
117. Fernández-Díaz, L.; Putnis, A.; Prieto, M.; Putnis, C.V. The role of magnesium in the crystallization of calcite and aragonite in porousmedium. *J. Sediment. Res.* **1996**, *66*, 482–491. [CrossRef]
118. Davis, K.J. The role of Mg^{2+} as an impurity in calcite growth. *Science* **2000**, *290*, 1134–1137. [CrossRef] [PubMed]
119. De Choudens-Sánchez, V.; González, L.A. Calcite and aragonite precipitation under controlled instantaneous supersaturation: Elucidating the role of $CaCO_3$ saturation state and Mg/Ca ratio on calcium carbonate polymorphism. *J. Sediment. Res.* **2009**, *79*, 363–376. [CrossRef]
120. Rossi, C.; Lozano, R.P. Hydrochemical controls on aragonite versus calcite precipitation in cave dripwaters. *Geochim. Et Cosmochim. Acta* **2016**, *192*, 70–96. [CrossRef]
121. Fermani, S.; Njegić Džakula, B.; Reggi, M.; Falini, G.; Kral, D. Effects of magnesium and temperature control on aragonite crystal aggregation and morphology. *Crystengcomm* **2017**, *19*, 2451–2455. [CrossRef]
122. Oomori, T.; Kaneshima, H.; Maezato, Y.; Kitano, Y. Distribution coefficient of Mg^{2+} ions between calcite and solution at 10–50 °C. *Mar. Chem.* **1987**, *20*, 327–336. [CrossRef]
123. Chafetz, H.S.; Folk, R.L. Travertines: Depositional morphology and the bacterially constructed constituents. *J. Sediment. Res.* **1984**, *54*, 289–316. [CrossRef]
124. Gibbs, R.J. Mechanisms controlling world water chemistry. *Science* **1970**, *170*, 1088–1090. [CrossRef]
125. Moral, F.; Cruz-Sanjulián, J.J.; Olías, M. Geochemical evolution of groundwater in the carbonate aquifers of Sierra de Segura (Betic Cordillera, southern Spain). *J. Hydrol.* **2008**, *360*, 281–296. [CrossRef]
126. Fairchild, I.J.; Borsato, A.; Tooth, A.F.; Frisia, S.; Hawkesworth, C.J.; Huang, Y.; McDermott, F.; Spiro, B. Controls on trace element (Sr–Mg) compositions of carbonate cave waters: Implications for speleothem climatic records. *Chem. Geol.* **2000**, *166*, 255–269. [CrossRef]
127. Spence, J.; Telmer, K. The role of sulfur in chemical weathering and atmospheric CO_2 fluxes: Evidence from major ions, $\delta^{13}C_{DIC}$, and $\delta^{34}S_{SO4}$ in rivers of the Canadian Cordillera. *Geochim. Et Cosmochim. Acta* **2005**, *69*, 5441–5458. [CrossRef]
128. Calmels, D.; Gaillardet, J.; Brenot, A.; France-Lanord, C. Sustained sulfide oxidation by physical erosion processes in the Mackenzie River basin: Climatic perspectives. *Geology* **2007**, *35*, 1003–1006. [CrossRef]
129. Hammer, Ø.; Jamtveit, B.; Benning, L.G.; Dysthe, D.K. Evolution of fluid chemistry during travertine formation in the Troll thermal springs, Svalbard, Norway. *Geofluids* **2005**, *5*, 140–150. [CrossRef]
130. Arenas, C.; Vázquez-Urbez, M.; Auqué, L.; Sancho, C.; Osacár, C.; Pardo, G. Intrinsic and extrinsic controls of spatial and temporal variations in modern fluvial tufa sedimentation: A thirteen-year record from a semi-arid environment. *Sedimentology* **2014**, *61*, 90–132. [CrossRef]
131. Rogozhin, E.A.; Ovsyuchenko, A.N.; Marakhanov, A.V. Major earthquakes of the southern Gornyi Altai in the Holocene. *Izv. Phys. Solid Earth* **2008**, *44*, 469–486. [CrossRef]
132. Rasmussen, S.O.; Andersen, K.K.; Svensson, A.M.; Steffensen, J.P.; Vinther, B.M.; Clausen, H.B.; Siggaard-Andersen, M.-L.; Johnsen, S.J.; Larsen, L.B.; Dahl-Jensen, D.; et al. A new Greenland ice core chronology for the last glacial termination. *J. Geophys. Res.* **2006**, *111*, D06102. [CrossRef]
133. Vinther, B.M.; Clausen, H.B.; Johnsen, S.J.; Rasmussen, S.O.; Andersen, K.K.; Buchardt, S.L.; Dahl-Jensen, D.; Seierstad, I.K.; Siggaard-Andersen, M.-L.; Steffensen, J.P.; et al. A synchronized dating of three Greenland ice cores throughout the Holocene. *Journ. Geophys. Res.* **2006**, *111*, D13102. [CrossRef]
134. Lisiecki, L.E.; Raymo, M.E. A Pliocene-Pleistocene stack of 57 globally distributed benthic $\delta^{18}O$ records. *Paleoceanography* **2005**, *20*, PA1003. [CrossRef]
135. Wells, D.L.; Coppersmith, K.J. New empirical relationship among magnitude, rupture length, rupture width, rupture area and surface displacement. *Bull. Seismol. Soc. Am.* **1994**, *84*, 974–1002. [CrossRef]
136. Hampel, A.; Hetzel, R.; Maniatis, G. Response of faults to climate-driven changes in ice and water volumes on Earth's surface. *Philos. Trans. R. Soc. A* **2010**, *368*, 2501–2517. [CrossRef] [PubMed]
137. Hampel, A.; Karow, T.; Maniatis, G.; Hetzel, R. Slip rate variations on faults during glacial loading and post-glacial unloading: Implications for the viscosity structure of the lithosphere. *J. Geol. Soc. Lond.* **2010**, *167*, 385–399. [CrossRef]
138. Muir Wood, R. Extraordinary deglaciation reverse faulting in northern Fennoscandia. In *Earthquakes at North-Atlantic Passive Margins: Neotectonics and Postglacial Rebound*; Gregersen, S., Basham, P.W., Eds.; Kluwer Academic Publishers: Dordrecht, The Netherlands, 1989; pp. 141–173.
139. Arvidsson, R. Fennoscandian earthquakes: Whole crustal rupturing related to postglacial rebound. *Science* **1996**, *274*, 744–746. [CrossRef]
140. Gregersen, S. Earthquakes and change of stress since the ice age in Scandinavia. *Bull. Geol. Soc. Den.* **2002**, *49*, 73–78.

141. Olesen, O.; Blikra, L.H.; Braathen, A.; Dehls, J.F.; Olsen, L.; Rise, L.; Roberts, D.; Riis, F.; Faleide, J.I.; Anda, E. Neotectonic deformation in Norway and its implications: A review. *Nor. J. Geol.* **2004**, *84*, 3–34.
142. Kukkonen, I.T.; Olesen, O.; Ask, M.V.S. The PFDP Working Group Postglacial faults in Fennoscandia: Targets for scientific drilling. *GFF* **2010**, *132*, 71–81. [CrossRef]
143. Blyakharchuk, T.A.; Wright, H.E.; Borodavko, P.S.; van der Knaap, W.O.; Ammann, B. Late-glacial and Holocene vegetational changes on the Ulagan high-mountain plateau, Altai Mountains, southern Siberia. *Palaeogeogr. Palaeoclimatol. Palaeoecol.* **2004**, *209*, 259–279. [CrossRef]
144. Galakhov, V.P.; Chernykh, D.V.; Mardasova, E.V. Mid-Late Holocene mean temperature fluctuations of a warm period in high mountains of the central Russian Altai. *Eurasian J. Math. Comput. Appl.* **2017**, *5*, 14–26. [CrossRef]
145. Ganyushkin, D.; Chistyakov, K.; Volkov, I.; Bantcev, D.; Kunaeva, E.; Brandová, D.; Raab, G.; Christl, M.; Egli, M. Palaeoclimate, glacier and treeline reconstruction based on geomorphic evidences in the Mongun-Taiga massif (south-eastern Russian Altai) during the Late Pleistocene and Holocene. *Quat. Int.* **2018**, *470*, 26–37. [CrossRef]
146. Croci, A.; Della Porta, G.; Capezzuoli, E. Depositional architecture of a mixed travertine-terrigenous system in a fault-controlled continental extensional basin (Messinian, Southern Tuscany, Central Italy). *Sediment. Geol.* **2016**, *332*, 13–39. [CrossRef]
147. Priewisch, A.; Crossey, L.J.; Karlstrom, K.E.; Polyak, V.J.; Asmerom, Y.; Nereson, A.; Ricketts, J.W. U-series geochronology of large-volume Quaternary travertine deposits of the southeastern Colorado Plateau: Evaluating episodicity and tectonic and paleohydrologic controls. *Geosphere* **2014**, *10*, 401–423. [CrossRef]
148. Landgraf, A.; Dzhumabaeva, A.; Abdrakhmatov, K.E.; Strecker, M.R.; Macaulay, E.A.; Arrowsmith, J.R.; Sudhaus, H.; Preusser, F.; Rugel, G.; Merchel, S. Repeated large-magnitude earthquakes in a tectonically active, low-strain continental interior: The northern Tien Shan, Kyrgyzstan. *J. Geophys. Res. Solid Earth* **2016**, *121*, 3888–3910. [CrossRef]
149. Deev, E.; Korzhenkov, A.; Turova, I.; Pavlis, T.L.; Luzhanskii, D.; Mažeika, J.; Abdieva, S.; Yudakhin, A. Large ancient earthquakes in the western Issyk-Kul basin (Kyrgyzstan, northern Tien Shan). *J. Asian Earth Sci.* **2018**, *166*, 48–65. [CrossRef]
150. Deev, E.V.; Zolnikov, I.D.; Lobova, E.Y. Late PleistoceneeHolocene coseismic deformations in the Malyi Yaloman River valley (Gorny Altai). *Russ. Geol. Geophys.* **2015**, *56*, 1256–1272. [CrossRef]

minerals

MDPI

Article

Uranium Geochemistry of Italian Travertines and Calcareous Tufas: Exploring the Relationship between Carbonate Deposition, Groundwater Circulation and Subsurface Geology

Francesca Giustini * and Mauro Brilli *

Istituto di Geologia Ambientale e Geoingegneria (IGAG), Consiglio Nazionale delle Ricerche (CNR), Area della Ricerca di Roma 1, Via Salaria km 29,300, Monterotondo Street, 00015 Rome, Italy
* Correspondence: francesca.giustini@igag.cnr.it (F.G.); mauro.brilli@cnr.it (M.B.)

Abstract: A database consisting of 163 data on the uranium content and $^{234}U/^{238}U$ initial activity ratio of 15 Italian travertine and calcareous tufa sites was created using data from the relevant literature. Using a graphical method, data were interpreted considering the U geochemistry in natural environments as well as the geological, hydrogeological and hydrogeochemical settings of each site. The U content and $^{234}U/^{238}U$ initial activity ratio in travertine and tufa appear to be affected by different factors, such as the availability of U in the aquifer rocks, the redox state of the waters, and the alpha-active radionuclide recoil phenomenon. The data allow the identification of four groups of travertines/tufas: (i) those precipitated from circulating groundwater, with a short/fast flow path, in volcanic rocks with a high radionuclide content; (ii) those precipitated from circulating groundwater, with a long, deep flow path in carbonate/evaporite formations with a relatively low radionuclide content; and (iii) those precipitated from cold waters associated with riverine systems, which are characterized by oxidizing conditions and fed by high-discharge springs recharged by carbonate aquifers. The fourth group represents the intermediate situations frequently occurring due to the mixing of waters from different aquifers. The results suggest an interpretative model that might contribute to the paleo-environmental reconstruction of fossil travertine and calcareous tufa depositing systems.

Keywords: uranium geochemistry; travertine; calcareous tufa; Italy

Citation: Giustini, F.; Brilli, M. Uranium Geochemistry of Italian Travertines and Calcareous Tufas: Exploring the Relationship between Carbonate Deposition, Groundwater Circulation and Subsurface Geology. *Minerals* **2023**, *13*, 782. https://doi.org/10.3390/min13060782

Academic Editor: Javier Sánchez-España

Received: 29 April 2023
Revised: 1 June 2023
Accepted: 6 June 2023
Published: 8 June 2023

1. Introduction

Travertine and calcareous tufa are lithological terms generally used to describe continental carbonates that form in subaerial environments through the precipitation of calcite/aragonite from waters ranging in temperature from ambient to boiling, around groundwater seepages, springs, and along streams and rivers. These carbonate deposits are commonly found in the Quaternary deposits in central and southern Italy [1,2]. They are particularly abundant along the western sector of the Italian peninsula, where they can form large deposits of remarkable thickness. The most notable example of this type is represented by the travertine of Tivoli (a town near Rome), a continental tabular carbonatic plateau that reaches in its depocenter a thickness of 85 m [3]. It has been exploited as a building material since the Roman period; even the word 'travertine' probably originated from *Lapis Tiburtinus*, i.e., the stone of *Tibur*, from the Latin name of Tivoli.

Travertines and calcareous tufas are important terrestrial archives of past climate and environmental changes [4,5], tectonic activity [6], and paleohydrological–hydrothermal circulations [7,8]. They may also indicate the possible location of geothermal resources [9,10] and contribute to defining the carbon cycle and budget [11,12]. Therefore, during the last twenty years, many studies have been conducted on these deposits. Many of these studies rely on U-series geochronology, because they can be reliably dated through the U-Th method, which allows the establishment of a precise chronology of up to 600 thousand years (ka) [13].

Radiometric dating techniques require the measurement of uranium content (U) and the determination of the $^{234}U/^{238}U$ initial activity ratio (i.e., at time of deposition) in the precipitated carbonate; these data are generally published in scientific papers and are used to calculate the age, but they are rarely discussed in relation to travertine/tufa formation and are not considered for their geochemical significance. In this study, we created a dataset of 163 published data on the U content and $^{234}U/^{238}U$ initial activity ratio (herein referred to as ($^{234}U/^{238}U$)i), of travertine and calcareous tufa samples from the relevant literature, in an effort to investigate the relationship between travertine/tufa deposition, groundwater circulation and mixing, and subsurface geology. The dataset includes fifteen Italian travertine and tufa deposits originating from cold and thermal waters. Data were interpreted using a graphical method. The data interpretation also made use of the related information on the groundwaters, the aquifers and the geological settings of the various deposits.

Introduction to Uranium Geochemistry and Rationale of This Study

The behavior of uranium in the environment has been studied by many authors ([14,15], and reference therein). In nature, U mainly exists in two oxidation states, U(IV) (tetravalent) and U(VI) (hexavalent). In surface environments, U(VI) is the dominant form; in reducing environments and in the mantle, tetravalent U is dominant and insoluble in water, and is generally less mobile than hexavalent U.

In aqueous solutions, the uranium concentration depends on several factors, including pH, Eh, temperature, ionic strength, and the ability to complex other ions (fluorides, chlorides, phosphates, and carbonates) and organic ligands [16,17].

Under oxidizing conditions, hexavalent uranium can be highly mobile, as it forms the uranyl ion $(UO_2)^{2+}$ which forms compounds that are soluble in water in a wide range of pH conditions. The ability to form complexes with inorganic (carbonate, bicarbonate and phosphate) and organic ligands further enhances the uranium's mobility, because these complexes do not participate in sorption processes. Under near-neutral conditions, uranium forms soluble complexes with carbonate and phosphates, while at lower pH values and saline waters, it forms soluble complexes with fluorides, chlorides and sulphates [15].

Under reducing conditions, tetravalent uranium tends to be insoluble and precipitates as insoluble uraninite (UO_2). At a low pH, uranium (IV)'s solubility tends to increase due to the formation of complexes with fluoride, and at pH > 7–8 it tends to increase due to complexation with hydroxyl ions [15].

In continental carbonates, such as travertine/tufa, traces of U, Ra and other decayed products of the ^{238}U series are common [18]. Uranium is generally present in carbonate because of its incorporation in the crystal structure of calcite [19], but the way in which uranium is incorporated into the calcite lattice is poorly understood [20]. Generally, uranium is not present in the carbonate as a separate phase, as evidenced by the experimental study of [21]. Uranium can be adsorbed on mineral surfaces and particulate impurities and associated with organic matter [22].

Naturally occurring uranium is an isotopic mixture of three long-lived isotopes: ^{238}U ($t_{1/2}$ = 4.5 billion years), ^{235}U ($t_{1/2}$ = 0.7 billion years), and ^{234}U ($t_{1/2}$ = 245,000 years). In a closed system, ^{238}U and ^{234}U are in secular radioactive equilibrium, which corresponds to the steady state reached by the whole decay series (i.e., their activity is equal) [23]. However, physical and chemical processes occurring during water–rock interactions can preferentially remove different uranium isotopes from the system, altering the secular equilibrium and changing the activity ratios of uranium isotopes. The preferential dissolution of ^{234}U is mainly caused by leaching owing to crystal lattice instability after alpha emission during ^{238}U decay, and by the direct recoil ejection of the ^{234}Th nucleus into the water via the alpha recoil effect or via the etching of alpha recoil tracks [24]. Other reasons for isotope fractionation could be associated with adsorption, changes in U speciation or redox chemistry, including microbially mediated uranium reduction [25]. As a consequence, natural waters are generally enriched in ^{234}U, and their $^{234}U/^{238}U$ activity ratios can deviate from the radioactive equilibrium by up to more than 10% [26].

Based on the studies of [27,28], it is possible to argue that the total U content and $^{234}U/^{238}U$ initial activity ratio, generally measured on continental carbonates for U-Th dating determinations, can provide information about the groundwaters from which they precipitated and the respective geological nature of the aquifer. Three basic assumptions are needed: (i) the U concentration in continental carbonates reflects the U content in the groundwater at the time of deposition, (ii) the $^{234}U/^{238}U$ initial activity ratio in continental carbonates is the same occurring in the groundwater, and (iii) the carbonates maintain a closed-system behavior (i.e., uranium and the products of its decay used in the dating are neither added nor removed from the carbonate system after deposition).

The mobilization of U and fractionation of $^{234}U/^{238}U$ take place in different geological contexts through different processes. Groundwater dissolution preferentially mobilizes U in a carbonate aquifer over a silicate aquifer; however, its content in the former is generally lower than in the latter, so it is more frequently possible to find more U in groundwaters flowing through silicate aquifers. $^{234}U/^{238}U$ fractionation is a process due to the alpha-active radionuclide recoil phenomenon that preferentially leaches ^{234}U into the circulating waters. This phenomenon is directly related to the groundwater residence time, which in hydrogeological Italian contexts is generally longer in the carbonate aquifers that constitute the Meso-Cenozoic bedrock of the Plio-Quaternary volcanic and alluvial formations.

Therefore, it can be stated that a relatively high $^{234}U/^{238}U$ initial activity ratio measured in continental carbonate may be indicative of precipitating groundwaters that circulated through a Meso-Cenozoic carbonate aquifer, in which the recoil enrichment of ^{234}U was more pronounced; on the contrary, a low $^{234}U/^{238}U$ initial activity ratio may be indicative of waters that circulated through a silicate (volcanic) aquifer. In summary, U content, used in conjunction with the $^{234}U/^{238}U$ ratio, contributes to distinguishing groundwaters that flowed deep into the crust *versus* those that resided at shallower levels.

2. Materials and Methods

2.1. Travertine and Calcareous Tufa Deposits

Fifteen travertine and calcareous tufa deposits are considered in this study (Figure 1). Most of these deposits are located in the western sector of central Italy: in the southern Tuscany (sites 4—Bagno Vignoni, 11—Rapolano, 13—Sarteano, 14—Semproiano and 15—Saturnia) and Lazio (sites 2—Acque Albule, 3—Aniene Valley, 5—Bolsena lake, 6—Canino, 9—Middle Velino valley and 10—Prima Porta) regions. Two deposits (sites 1—Acquasanta T. and 12—Rocchetta al Volturno) are located in the eastern regions of central Italy (Marche and Molise, respectively). The northern deposit is located in the Veneto region (site 8—Euganean Hill), whereas the southern one is in Sicily (site 7—Mt. Etna). As a consequence of the different geological contexts and genetic processes, these deposits show different morphological and geochemical features, which are synthetized in Table 1.

According to a widely used classification which considers the genetic environment and the presence of biota [5,29,30], the term "travertine" indicates the hard and crystalline lithotype, related to abiotic processes in waters which are typically hydrothermal in origin, whereas the term "calcareous tufa" indicate a material more porous than travertine, rich in microphytes, macrophytes, invertebrate and bacterial remains, generally formed in waters of an ambient temperature. This classification implies some uncertainties, especially when carbonate deposition occurs from cooled thermal waters, where tufa-like deposits may be precipitated when the water temperature decreases with distance from the spring orifice. In Table 1, we use the terms "travertine" and "calcareous tufa" for the deposits based on the description and terminology used in the reference papers.

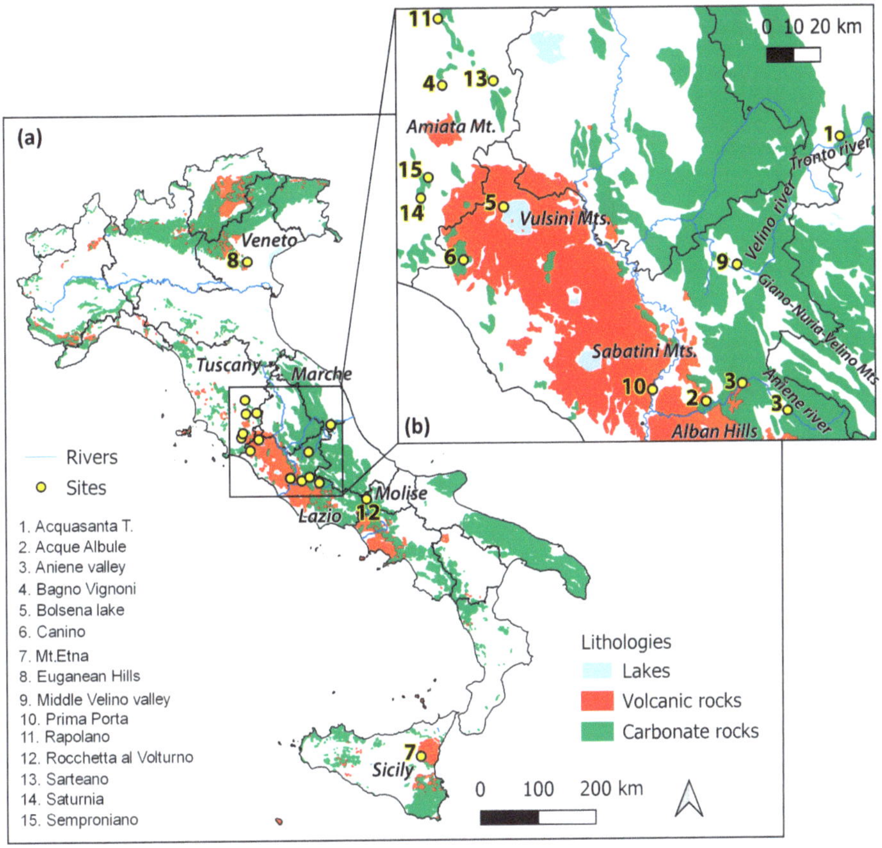

Figure 1. (**a**) Location of travertine and calcareous tufa sites and simplified lithological map: carbonate rocks include carbonate platform and pelagic formations, limestone-marls (Mesozoic-Cenozoic), volcanic rocks include extrusive volcanites (Miocene, Plio-Quaternary, and Holocene), (**b**) location of deposits in central Italy. The site 3—Aniene valley includes two outcrops, the first at Subiaco (the eastern site) and the second at Vicovaro Mandela (the western site).

The travertine and tufa deposition in the sites included in this study occurred during the last 600 ka [3,12,27,28,31–46]. At present, carbonate deposition is still active (at sites 1, 2, 3, 4, 6, 7, 11, 14), but on a very small scale [2]; for example, in the Sarteano site, the deposition is restricted to the discharge channel of the local thermal spa [44]. Some exceptions are the Rapolano and Saturnia sites, where the carbonate still precipitates abundantly.

Several morphologies were recognized: mounds, fissure ridges, plateaus, terraces, cascades, and dams. Travertine/tufa deposits are complex systems, and therefore such morphologies coexist frequently in the same deposits. Different geometries depend on multiple factors—firstly the substrate geometry and topographic gradient, but also the spring orifice position and its change during time, as well as the water geochemistry. Mounds and fissure ridges are depositional types typical of vent environments, i.e., associated with spring orifices. Circular mounds, consisting of domes situated at the spring orifices, can be identified at sites 1, 4 and 8 [32,35,39]. Linear-to-arcuate fissure ridges are elongated mounds with a central fissure along the long axis, developing along fractures; this morphology is recognizable at sites 4 and 15 [12,35,45]. Moving away from the spring orifice, different depositional systems can occur, varying from gentle inclines to steep slopes, depending on the underlying morphology. Variably inclined lobate bodies characterized by smooth to well-developed terraced slopes in their frontal part can be formed by waters flowing away from the vent. Terraced pools, aprons, and self-building channels are distinct forms developing on steep and smooth slopes, identifiable at sites 1, 2, 4, 11, 12 and 14 [3,28,31,32,35,43,45]. At sites 1, 3, 9 and 12, deposits are developed along the course of the fluvial valley, i.e., the Tronto, Aniene, Velino and Volturno rivers, respectively. In these cases, dams, terraces and slope systems, grading from a proximal feeding system into flat marginal deposits, are recognized [31,32,34,40,43]. Finally, in a low relief topography, a tabular plateau can be formed, as in the case of sites 2 and 15 [3,12,33,45]. With the exception of mounds and fissure ridges, spring orifices are rarely recognizable in fossil deposits, hence the difficulty of reconstructing the temporal and spatial evolution of travertine/tufa deposits.

Groundwaters from which the travertine/tufa precipitated vary significantly in temperature and chemical composition [47–67]. At sites 3, 5, 7, 9 and 12, the travertine/tufa precipitated from cold waters ($T < 15$ °C) and Ca-HCO_3 and Na-HCO_3 chemical composition, typical of cold karstic springs [49,53,55,57,63]. In the other cases, the travertine precipitated from thermal ($T > 20$ °C) or intermediate ($15 < T < 20$ °C) waters, whose chemical compositions include Ca-HCO_3 (sites 4, 10, 11 [51,59,61]) and are sometimes enriched in Mg (site 2 [48]) or Na (site 7 [55]), Ca-Cl (sites 14 and 15 [66,67]), Ca-SO_4 (site 13 [64,65]), Ca-SO_4 and Na-Cl (sites 1, 6 and 8 [47,54,56]) hydrochemical facies. Their chemistry and temperature are due to the hydrothermal origins of the groundwaters. In several sites (1, 2, 5, 6, 9, 10, 11, 13), the hydrochemistry of groundwater and hydrogeological data suggest a mixing process of groundwater from deep and shallow aquifers [27,47,48,54,58,60,61,64,65]. In general, Meso-Cenozoic carbonate formations act as deep, sometimes confined, aquifers that can be hydraulically connected with shallow, unconfined aquifers through faults and fractures [27,47,48,60].

Indeed, most of the deposits are located in active hydrothermal settings, in tectonically active areas, where active faults and fractures offer possible pathways for the circulation and mixing of deep and shallow fluids and lead to diffuse hydrothermal manifestations, such as thermal springs and gas vents, next to continental carbonate deposition.

In southern Tuscany, several travertine deposits (sites 4, 11, 13, 14 and 15) occur at the periphery of the Mt. Amiata volcanic center [12,28,35,42,44–46]. Its magmatism is characterized by intrusive acidic and volcanic products with associated high-temperature metamorphism [68]. The location of travertine/tufa deposits and parental thermal springs is tectonically controlled by regional and local faults; for example, the fluid upwelling connected to the deposition of Sarteano was favored by the intersection of transcurrent faults with pre-existing regional normal faults [44], while in the case of the Bagno Vignoni site, travertine deposition was tectonically controlled by WSW–ENE striking, oblique and normal faults, associated with a main fault (named as the Violante Fault) [28].

In the Lazio region, magmatism generated several volcanic centers, the activity of which started approximately 600 ka, with the emplacement of large volumes of potassic and ultrapotassic lavas and pyroclastics [69]. Several travertine/tufa deposits are associated with the volcanic centers of the Roman Magmatic Province, such as the Vulsini Mts. (sites 5 and 6 [27]), the Sabatini Mts. (site 10 [41,59,60]) and Alban Hills (site 2 [3,33]) volcanic complexes. Additionally, in these cases, tectonics controlled hydrothermal setting and travertine deposits: examples are the transverse faults (NE–SW and N–S) identified in the Acque Albule basin [3], or the normal faulting that controls the rising of water from the deep aquifer in the Canino area [70].

In the Veneto region, a travertine deposit (site 8) is located in the Euganean Hills district, an isolated volcanic body within the Veneto alluvial plain and the most important geothermal field of northern Italy [56]. This area is within the complex Lessini-Euganei structural domain, which is part of the transcurrent western margin of the Adria Indenter and, at the same time, the western boundary of the foreland of the South Alpine thrust and fold belt [71]. Travertine crops out in the spa town of Abano Terme, and is closely associated with a fault buried under the Quaternary alluvial deposits [39].

Finally, in eastern Sicily, a travertine deposit (site 7) crops out on the lower southwestern flank of Mt. Etna., the largest strato-volcano in Europe and one of the major emitters of magmatic CO_2 on the Earth's surface [37]. This volcano grew in proximity to the collision boundary of the African and Eurasian continental plates due to repeated eruptions of alkali basalts–hawaiites over the last 200 ka [69].

In the inner and eastern sectors of central Italy, the travertine/tufa deposits crop out far from the main volcanic centers of the peninsula. Some deposits occur within the Apennine chain, the Neogene fold and thrust belt formed by the convergence of the European and Adria-African continental blocks during the late Cretaceous. These deposits (sites 1, 3, 9 and 12) developed along the course of fluvial valleys, and formed depositional terraces [31,32,34,40,43]; their formation in some cases was directly connected to hydrothermal systems and tectonically active areas (sites 1 and 9), and not in other cases (sites 3 and 12). In the inner sector of central Apennine, several deposits occur along the Middle Velino valley (site 9); they crop out along the northern boundary of the San Vittorino plain, an intramontane depression that is the result of extensional and/or transtensive tectonics, displacement along major fault planes and regional uplift [40]. The plain is characterized by the occurrence of mineralized springs, gas emissions and sinkholes, as well as recent and historical seismicity. The travertine here likely formed due to the emergence of mineralized water [72] following intense tectonic activity at the end of the Middle Pleistocene. Other travertine/tufa deposits, in the inner sector of central Apennine, crop out along the course of Aniene river (site 3); their formation was due to fluvial barrage that developed in correspondence with a morphological step, possibly of tectonic origin [34].

In the eastern part of central Apennine, deposit of Acquasanta T. is located (site 1) [31,32,47,73,74]. This deposit is associated with an important compressional feature, the Acquasanta Terme anticline, which was cross-cut by a local transtensional tectonic regime resulting in dextral strike-slip and minor normal faulting during the Neogene-Quaternary, that may have controlled the hydrothermal circulation system [73]. Additionally, the Acquasanta Terme system is sustained by a low temperature geothermal field probably due to a normal geothermal gradient [74], a consequence of long and deep circulation of meteoric fluids infiltrating nearby mountain ranges, such as the Sibillini mountains to the east [74] or the Laga mountains to the south [47].

Finally, in the southeastern sector of central Apennine, the Rocchetta al Volturno tufa (site 12) crops out along the Volturno river valley. This deposit was formed by waterfall that, aggrading and prograding, acted as barrier behind which small accumulation basins developed [43].

2.2. The Dataset

The dataset is listed in Appendix A. It consists of 163 data on the uranium content (in ppb) and $(^{234}U/^{238}U)i$ at 15 Italian travertine and calcareous tufa sites, reviewed from the following studies: [3,12,27,28,31–46]. The dataset includes only uranium data from travertine and tufa samples, and not from groundwater, because such data are not available.

The uranium content data from travertines and calcareous tufas are, in principle, not directly comparable, since the respective depositional contexts differ mainly in terms of carbonate accumulation rate, temperature and ionic strength of the precipitating waters. The uranium concentrations in the carbonate precipitates may be influenced by these variables, and reflect the uranium concentrations in the parent waters. It should also be noted that there is a considerable uncertainty in the correct classification of tufas or travertines in the sites reviewed, and travertines can often belong to depositional facies that recall those typical of tufa, especially if deposited far from the spring orifice. It is therefore difficult to identify depositional features in each sample. For these reasons, in this study, it is assumed that the uranium concentration in carbonate precipitates mainly reflects the concentration in groundwater, irrespective of their attribution to travertines or tufas.

Table 1. Overview of the Italian travertine and calcareous tufa deposits examined for this study, with ages, description of sites, hydrogeochemistry of waters associated with carbonate precipitation and hydrogeological settings of the sites.

Site Code	Site Name	Lithotype	Age (ka)	Deposit Description	Hydrogeochemistry	Hydrogeology	Reference
1	Acquasanta T.	Travertine	221 ± 5.8 to 26 ± 3.0	Slope system grading from a proximal feeding system into flat marginal deposits bordered by waterfall travertine through smoothly sloping deposits and eventually a more distal system. Active.	Acquasanta Terme spring: Na-Cl to Ca-SO_4 thermal waters (T = 51 to 73 °C), associated with CO_2 emission; pH = 6.8	A deep circulation within Triassic $CaSO_4$-rich evaporitic formations; close to surface, dilution with HCO_3 waters. Recharge area of deep aquifer: Laga Mts. (30 km far).	[31,32,47]
2	Acque Albule	Travertine	219 ± 26 to 28 ± 16	Cycles of travertine aggradational deposition in sub-horizontal pools and terraces occurring over the inclined erosional surface. Active.	Lago Regina spring: Ca-Mg-HCO_3 intermediate waters (T = 23–24 °C), associated with CO_2 emission; pH = 6.24; Eh = −9 mV	A deep confined carbonate aquifer, and a shallow travertine aquifer, connected through tectonic discontinuities. Recharge area of deep aquifer: Cornicolani Mts. (10 km far).	[3,33,48]
3	Aniene valley	Calcareous tufa	112 ± 7 to 5.5 ± 0.3	From palustrine to fluvio-lacustrine environment. Three-order depositional terraces. Active.	Agosta spring: Ca-HCO_3 cold water (T = 13 °C); pH = 7.30. Pertuso spring: Ca-HCO_3 cold water (T = 8–9.5 °C); pH = 6.9–10.8; Eh = +110 to +481 mV.	Recharge area: carbonate unit of Simbruini Mts.	[34,49,50]
4	Bagno Vignoni	Travertine	160.6 ± 6.6 to 95 ± 1.3	Fissure ridges; cone-shape mounds; slope depositional system that passed to a distal zone characterized by sub-horizontal morphology and low-energy environment. Active.	Bagno Grande pool: Ca-HCO_3 thermal water (T = 33 °C); pH = 7.06; Eh = +180 mV. Borehole: Ca-SO_4 thermal water (T = 48.4 °C), Eh = −106 mV.	Deep circulation in a Mesozoic evaporite carbonate reservoir. Residence time of deep carbonate aquifer > 40 and up to 70 years.	[35,51,52]
5	Bolsena Lake	Travertine	218 ± 39 to 3.6 ± 2	Travertine plates, from the bottom of the lake and along the Fiora river. Fossil.	Bolsena lake water: Na-HCO_3 cold water (T = 9–15 °C); pH = 8.5	One aquifer connected to deep carbonatic formations, and another to the surface volcanic products.	[27,53]
6	Canino	Travertine	57 ± 5.5 to 1.2 ± 1.8	Depressional environments representing the distal portion the thermal springs, related to normal faulting of Mesozoic sediments on Mt. Canino. Active.	Canino-Torlonia springs: Ca-SO_4 and Na-Cl thermal waters (T up to 50 °C), associated with CO_2 emission; pH = 6.30	Hydrothermal system (located within buried carbonate structures) is mede up of shallower zone (Ca-SO_4 water type) and deeper zone (Na-Cl brines). Recharge area: partly by carbonate and partly by regional volcanic aquifer.	[27,36,54]

Table 1. *Cont.*

Site Code	Site Name	Lithotype	Age (ka)	Deposit Description	Hydrogeochemistry	Hydrogeology	Reference
7	Mt. Etna	Travertine	24 ± 4 to 4 ± 0.7	Stratified and massive in the lower part, porous and with clay lens in the upper part. Active.	Adrano spring-borehole: Ca-Na-HCO_3 cold water (T = 7–14 °C); pH = 7.98–8.48	Volcanic aquifer. Recharge area: Western flank of Mount Etna volcano.	[37,38,55]
8	Euganean Hills	Travertine	30 ± 4 to 25 ± 5	Sub-circular mound, up to 15 m high. Fossil.	Abano Terme springs: Na-Cl thermal waters (T = 53.5 to 86.5 °C), associated with CO_2 emission; pH = 6.3–7.3	Recharge area: carbonatic complex of Prealps (70 km far).	[39,56]
9	Middle Velino Valley	Travertine–Calcareous tufa	113.6 ± 18 to 5.3 ± 1.1	Dams along the course of fluvial valley. Fossil.	Cotilia spa spring: Ca-HCO_3 cold water (T = 14.2 °C), associated with CO_2 emission; pH = 6.2; Eh = −198 mV	Multi-layered aquifer of the plain is recharged by inflows from the carbonate aquifers on the two sides of the valley and affected by input of deep CO_2-rich fluids. Recharge area: carbonate units of the Giano-Nuria-Velino Mts. hydrogeological system and in part of the Mt. Paterno-Canetra hydrogeological unit.	[40,57,58]
10	Prima Porta	Travertine	53.5 ± 10 to 24.2 ± 2.7	Buried body, probably pool. Fossil.	Prima Porta well: Ca-HCO_3, intermediate water (T = 18 °C), associated with CO_2 emission; pH = 6.15; Eh = −7.4 mV	Probably a mixing between volcanic and alluvial aquifers. Recharge area: Sabatini volcanic complex and alluvial Tiber Valley.	[41,59,60]
11	Rapolano	Travertine	157 ± 15 to 6.7 ± 1.4	Fan-like high spring-water flow depositional systems (N–NW) and flat, low spring-water flow depositional system (E–SE). Active.	Terme Querciolaie spring: Ca-HCO_3 thermal water (T = 35 °C), associated with CO_2 emission; pH = 6.43	Mesozoic limestone and associated travertine hosts shallow cold Ca-HCO_3 waters and deep thermal saline Ca-$SO_4(HCO_3)$ waters. Cold saline Na-Cl waters are hosted in confined aquifer in Neogene sediments. Recharge area of deep aquifer: pre-Apenninic belt.	[28,42,61]
12	Rocchetta al Volturno	Calcareous tufa	75 ± 8 to 4.4 ± 0.5	Slope system. Fossil.	Capo Volturno spring: Ca-HCO_3 cold water (T = 11 °C); pH = 7.2; Eh = +200 to +400 mV	Recharge area: Mt. Genzana and Mt. Greco.	[43,62,63]

Table 1. *Cont.*

Site Code	Site Name	Lithotype	Age (ka)	Deposit Description	Hydrogeochemistry	Hydrogeology	Reference
13	Sarteano	Travertine–Calcareous tufa	268 ± 22 to 16 ± 3	Tectonically controlled, mature perched-spring system. Active.	Le Canalette springs: Ca-SO_4 thermal water (T = 24 °C); pH = 6.7	Groundwater circulates in a highly fractured, quasi-continuous reservoir constituted by the Mesozoic limestone and the underlying Burano anhydrite formation. Mixing between deep and shallow fluids. Recharge area: Mt. Cetona.	[44,64,65]
14	Saturnia	Travertine	142 ± 23 to 40 ± 7	Cascades, pools, and terraced slopes. Active.	Saturnia Terme spring: Ca-Cl thermal waters (T = 34 °C), associated with CO_2 emission; pH = 6.3; Eh = −212 mV	A deep carbonate aquifer, residence time c.a. 30 years. Recharge area of deep aquifer: south of Roccalbegna (10 km far).	[45,66,67]
15	Semproniano	Travertine	613 ± 200 to 33 ± 4	Fissure ridge; travertine plateau. Fossil.	Saturnia Terme spring: Ca-Cl thermal waters (T = 34 °C), associated with CO_2 emission; pH = 6.3; Eh = −212 mV	A deep carbonate aquifer, residence time c.a. 30 years. Recharge area of deep aquifer: south of Roccalbegna (10 km far).	[12,45,46,66,67]

3. Results

The U content of the sites of this study ranges from 3 to 6390 ppb (Appendix A), with a median value of 77 ppb. Among the sites, U varies significantly (Figure 2a). Very low median values (<15 ppb) are calculated for the sites 1, 4, and 8, while very high values are calculated (>500 ppb) for the sites 5 and 7. Intra-site variability is sometimes recognized, e.g., sites 2, 5, 6, and 7. Occasionally, outliers are detected (sites 1, 2, 3, 11, 15). Figure 2b shows the variation of the $(^{234}U/^{238}U)i$ ratio: it varies between 0.91 and 2.879 (Appendix A), with a median value of 1.106. In most of the sites, the isotopic ratio is between 1 and 1.5, except for sites 4, 8, and 11, where the ratio is higher, and in general, it is correlated with a marked intra-site variability. Outliers are statistically detected in sites 2 and 9.

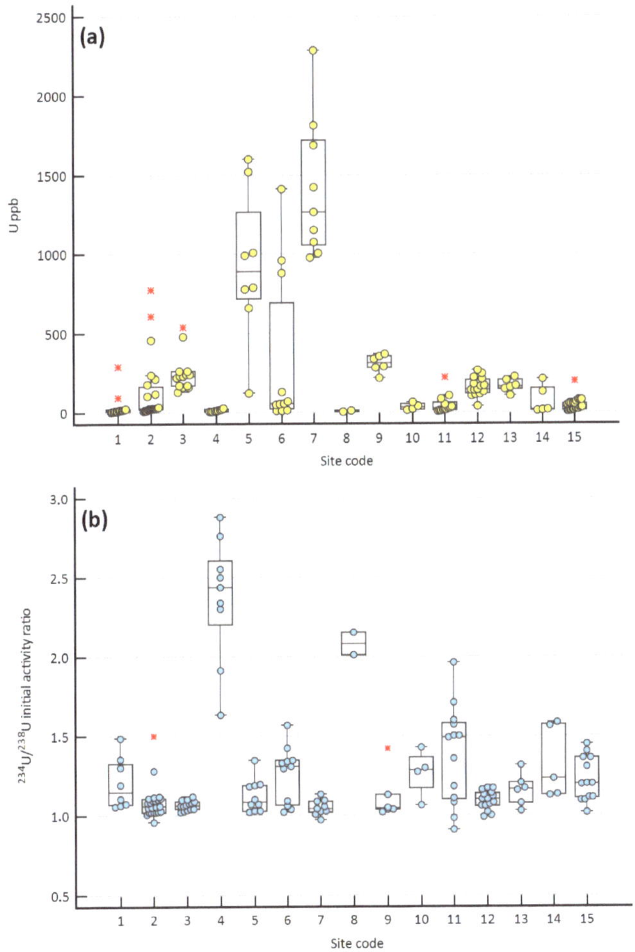

Figure 2. (**a**) Box plot of the uranium content measured in travertine and calcareous tufa samples; (**b**) box plot of the $(^{234}U/^{238}U)i$ ratio. The middle line of the plots represents the median; red asterisks are the outlier values. Site code: 1—Acquasanta T., 2—Acque Albule, 3—Aniene valley, 4—Bagno Vignoni, 5—Bolsena Lake, 6—Canino, 7—Mt. Etna, 8—Euganean Hills, 9—Middle Velino Valley, 10—Prima Porta, 11—Rapolano, 12—Rocchetta al Volturno, 13—Sarteano, 14—Saturnia, 15—Semproniano. See also Figure 1 for the site location.

Figure 3 shows the plots of the ($^{234}U/^{238}U$)i ratio *versus* the inverse of total U content of the reviewed sites. In this type of plot, it is assumed that the ($^{234}U/^{238}U$)i ratio of travertine and tufa is the same of the parental groundwater, and also that U content would reflect the U content in the groundwater. In the reviewed sites, different patterns are recognized, and related to their hydrogeological setting (summarized also in Table 1):

- Site 1—Acquasanta T. The plot shows intermediate to low values of U and a ($^{234}U/^{238}U$)i ratio < 1.5; the U-ratio tends to increase as the U content decreases. According to [47], travertines of Acquasanta are deposited by thermal waters that rise from an aquifer hosted in a carbonate sequence. Chemical and isotope data suggest the co-existence of a deep circuit, involving Triassic evaporitic formations, and a variable amount of cold water coming from fast surface circuits. The recharge area of the deep aquifer could be identified in the Laga Mts., about 30 km away from the site.
- Site 2—Acque Albule. The plot shows intermediate to low values of U and a ($^{234}U/^{238}U$)i ratio < 1.5. According to [48], in the Acque Albule area, a deep confined carbonate aquifer is connected through tectonic discontinuities to a shallow travertine aquifer. The recharge area of this deep aquifer could be identified in the Cornicolani Mts., about 10 km away from the site.
- Site 3—Aniene valley. The plot shows intermediate values of U and a ($^{234}U/^{238}U$)i ratio < 1.2. The calcareous tufa of the Aniene valley is deposited by the cold waters of the Aniene river, fed by many tributaries and the Agosta spring that, with a discharge rate of 5400 l/s, has a recharge area corresponding to the carbonate unit of Simbruini Mts. [49].
- Site 4—Bagno Vignoni. The plot shows low values of U and a ($^{234}U/^{238}U$)i ratio > 1.5; the U-ratio tends to increase as the U content decreases. Groundwater chemistry suggests a deep circulation in a Mesozoic evaporite carbonate reservoir [51]. Tritium data point to a residence time for the deep carbonate aquifer of between 40 and 70 years [52].
- Site 5—Bolsena lake. The plot shows a high U concentration (>1000 ppb) and a ($^{234}U/^{238}U$)i ratio < 1.2. According to [27], the groundwaters are hosted in part in an aquifer connected to deep carbonatic formations, and in part in a surficial aquifer consisting of volcanic products, from the Vulsini volcanic complex, particularly enriched in uranium (up to 22.9 ppm; [69]).
- Site 6—Canino. The plot shows a wide range of U values, from low to high, and a ($^{234}U/^{238}U$)i ratio < 1.5; the U-ratio tends to increase as the U content decreases. According to [54], the hydrothermal system, located within buried carbonate structures, is made up of a shallower zone (where Ca-SO_4 waters are prevalent) and a deeper zone (where Na-Cl brines are present). The recharge area can be identified partly in carbonate structures and partly in the regional volcanic aquifer. As the small carbonate outcrop of Mt. Canino is not sufficient to explain the flow rate of thermal waters, the authors of [54] hypothesize that the carbonate structures are locally in contact with the regional volcanic aquifer that recharges the deep aquifer.
- Site 7—Mt. Etna. The plot shows a high U concentration (>1000 ppb) and a ($^{234}U/^{238}U$)i ratio < 1.2. The recharge area of the spring related to carbonate precipitation is identified in the western flank of Etna Mt volcano [55]. The U content of volcanic products is up to 4.9 ppm [69].
- Site 8—Euganean Hills. The plot shows a very low U concentration (<20 ppb) and a ($^{234}U/^{238}U$)i ratio >2. According to [56], the recharge area of the thermal fluids can be identified in the Venetian Prealps, situated about 70 km north of the Euganean Hills. A magmatic contribution, possibly related to vestiges of the volcanic activity that occurred in the Abano area during the Tertiary age, can be ruled out, on the basis of the chemical and isotope compositions of gases associated with the thermal manifestations [56].

- Site 9—Middle Velino valley. The plot shows intermediate values of U content and a ($^{234}U/^{238}U$)i ratio < 1.5. According to [58], the springs of the middle valley of Velino river and of the San Vittorino plain are recharged by the carbonate units of Giano-Nuria-Velino Mts. hydrogeological system and by the Mt. Paterno-Canetra hydrogeological unit. Mineralized fluids of deep origin affect groundwater, which have particular hydrochemical characteristics (sulphurous water, ferruginous water with slightly hydrothermal character) in some cases [57].
- Site 10—Prima Porta. The plot shows low values of U content and a ($^{234}U/^{238}U$)i ratio < 1.5. The fossil travertine system of Prima Porta is located at the border of the Paleo-Tiber Graben, a tectonic depression filled by an aquifer, comprising a thin layer of volcanic rocks, overlying sandy, gravelly terrains and a thin and discontinuous sandy, gravelly layer at its base [41]. The chemistry of present-day groundwater suggests the circulation in a volcanic aquifer, and probably mixing with the alluvial plain aquifer [59,60].
- Site 11—Rapolano. The plot shows intermediate to low values of U content and a ($^{234}U/^{238}U$)i ratio increasing up to 2 as the U content decreases. According to [61], in the Rapolano area, Mesozoic limestones of the pre-Apenninic belt and associated travertines are the recharge areas of two well-mixed water types, shallow cold Ca-HCO_3 waters and deep thermal saline Ca-SO_4(HCO_3) waters; further aquifers confined in Neogene sediments of the Siena basin host cold saline Na-Cl waters, that appear relatively isolated and are not mixed with the thermal waters.
- Site 12—Rocchetta al V. The plot shows intermediate values of U content and a ($^{234}U/^{238}U$)i ratio < 1.2. The carbonates of Rocchetta al V. are deposited by the cold waters of the Volturno river, fed by the Capo Volturno spring that, with a discharge rate of 6600 L/s, has a recharge area corresponding to the carbonate unit of Mt. Genzana and Mt. Greco [62,63].
- Site 13—Sarteano. The plot shows intermediate values of U content and a ($^{234}U/^{238}U$)i ratio < 1.5. In the Sarteano area, mixing between deep and shallow fluids, initially of meteoric origin, occurs. Groundwater circulation occurs in a highly fractured, quasi-continuous reservoir constituted by the Mesozoic limestone and the underlying Burano anhydrite formation; the recharge area is in Cetona Mt ridge [44].
- Site 14—Saturnia. The plot shows intermediate to low values of U content and a ($^{234}U/^{238}U$)i ratio just over 1.5. Thermal waters currently discharged at Saturnia are hosted in a deep aquifer occurring within the Mesozoic carbonate units of the Tuscan nappe. The residence time, calculated on the basis of tritium data, is about 30 years. The recharge area is located south of Roccalbegna, 10 km away from the site [67].
- Site 15—Semproniano. The plot shows low values of U content and a ($^{234}U/^{238}U$)i ratio < 1.5; the U-ratio tends to increase as the [U] decreases. The fossil system of Semproniano is genetically connected to the hydrothermal system of Saturnia [45], and therefore the hydrogeological setting can be considered not different.

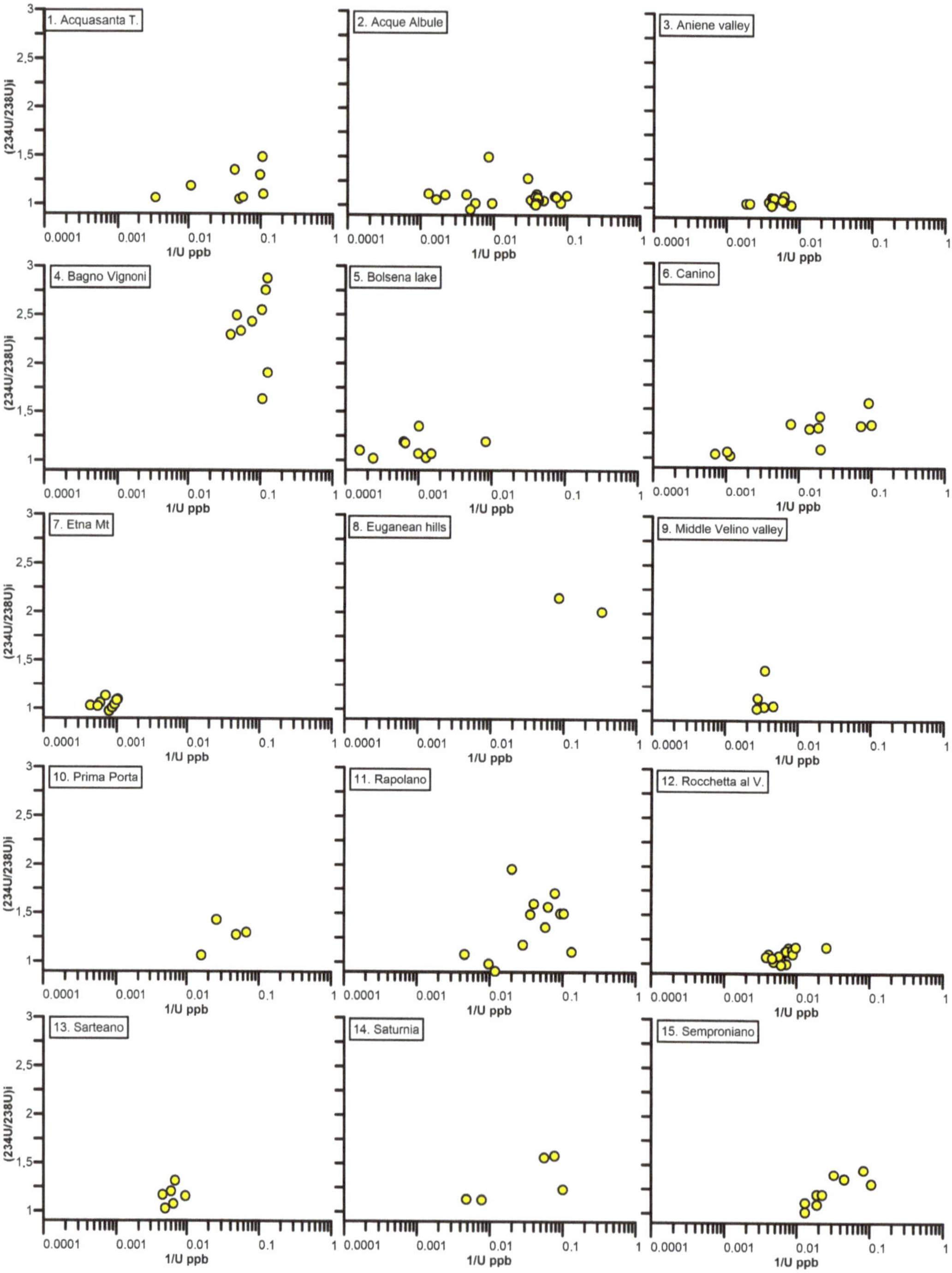

Figure 3. For each site, the $^{234}U/^{238}U$ initial activity ratio is plotted against the uranium content, expressed as 1/U in logarithmic scale.

4. Discussion

On the basis of data collected and plots of Figure 3, the U content and ($^{234}U/^{238}U$)i ratio in the travertine/tufa were interpreted to be the results of the not-trivial interaction between different factors: (a) the U content of the aquifers where groundwater circulated; (b) the redox conditions of the groundwaters; and (c) the alpha-active radionuclide recoil phenomenon.

The main factor affecting the uranium in the travertine/tufa seems to be U enrichment in groundwater through mineral dissolution, which depends firstly on the availability of U in the aquifer rocks. Volcanic aquifers are generally characterized by rocks with high radionuclide contents, up to 22.9 ppm of U in the Vulsini volcanic complex (which hosts the surficial aquifer for site 5—Bolsena lake) and up to 4.9 ppm in Mt. Etna (which is the recharge area of the groundwater precipitating travertine in site 7) [69]. On the contrary, the carbonate rocks are characterized by radionuclide contents generally lower than the volcanic rocks, although data on the U content in carbonate formations are quite scarce. A value of 2.19 ppm of U is reported for limestones, the Meso-Cenozoic carbonate bedrocks of central Italy [75], which represent the aquifer of several sites (1—Acquasanta, 2—Acque Albule, 3—Aniene valley, 6—Canino and 9—Middle Velino valley).

Secondly, the redox state of the waters controls the speciation of uranium and therefore the presence of uranyl compound which is the U mobile phase. For example, uranyl-carbonates are the most common complexes in equilibrium with Mt. Etna oxidizing groundwaters (site 7) [76]. Waters characterized by oxidizing conditions include those of the sites 3—Aniene valley (Eh = +110 to +481 mV [50]) and 12—Rocchetta al V. (Eh = +200 to +400 mV [63]) (Table 1). This could explain the relatively high content of U. Under more reduced conditions, U is very insoluble, and this behavior could explain the lowest contents of U in the sites 14—Saturnia (Eh = −212 mV [66]) and 4—Bagno Vignoni (Eh = −106 mV [51]) (Table 1). The residence time of groundwater could also influence the redox conditions; travertines/tufas will show a lower or higher U content depending on whether groundwaters are deep with long residence times, and thus reducing, or shallow with short residence times, and thus oxidizing.

The recoil phenomenon in the aquifer constitutes the factor controlling the ($^{234}U/^{238}U$)i in travertine and tufa deposits. This ratio is significantly influenced by water–rock interactions and by the alpha recoil phenomenon. A long residence time of groundwater in the aquifer, due to deep circulation and/or remote recharge area, may determine intense water–rock interactions and a selective extraction of ^{234}U, increasing the ($^{234}U/^{238}U$)i ratio. ^{234}U enrichment is especially evident in the reducing groundwaters that precipitated the travertine deposits of site 1—Acquasanta T., 4—Bagno Vignoni, 8—Euganean Hills, 14—Saturnia and 15—Semproniano (Table 1); correspondingly, they show low values of U and a ($^{234}U/^{238}U$)i ratio that increases as the U content decreases. Their long residence times are also confirmed by the related hydrogeological studies. Conversely, the low ($^{234}U/^{238}U$)i ratio could suggest a short residence time (as generally occurs in the shallow, generally oxidizing, groundwaters), determining a scarce ^{234}U recoil enrichment. The mixing process between different groundwaters can occur, determining intermediate values of U and a ($^{234}U/^{238}U$)i ratio in the precipitated carbonates. These characteristics can be observed in some samples of sites 1—Acquasanta, 2—Acque Albule, 5—Bolsena lake, 6—Canino, 9—Middle Velino valley, 10—Prima Porta, 11—Rapolano and 13—Sarteano. These sites demonstrate mixing between deep-circulating fluids from carbonate/evaporite aquifers and shallow-cold groundwaters, that circulate through volcanic (e.g., sites 10, 5 and 6), alluvial (e.g., site 10) and travertine (e.g., site 2) aquifers. In these sites, the mixing processes of groundwater from deep and shallow aquifers are confirmed by geochemical and hydrogeological studies [27,47,48,60].

Figure 4a shows an idealized cross section that summarizes the occurrence of different travertine/tufa deposits and the interaction of the factors that influence the U behavior. In Figure 4b, the samples of travertines/tufas are grouped into four clusters that correspond to the various occurrences of the travertine/tufa deposits in Figure 4a.

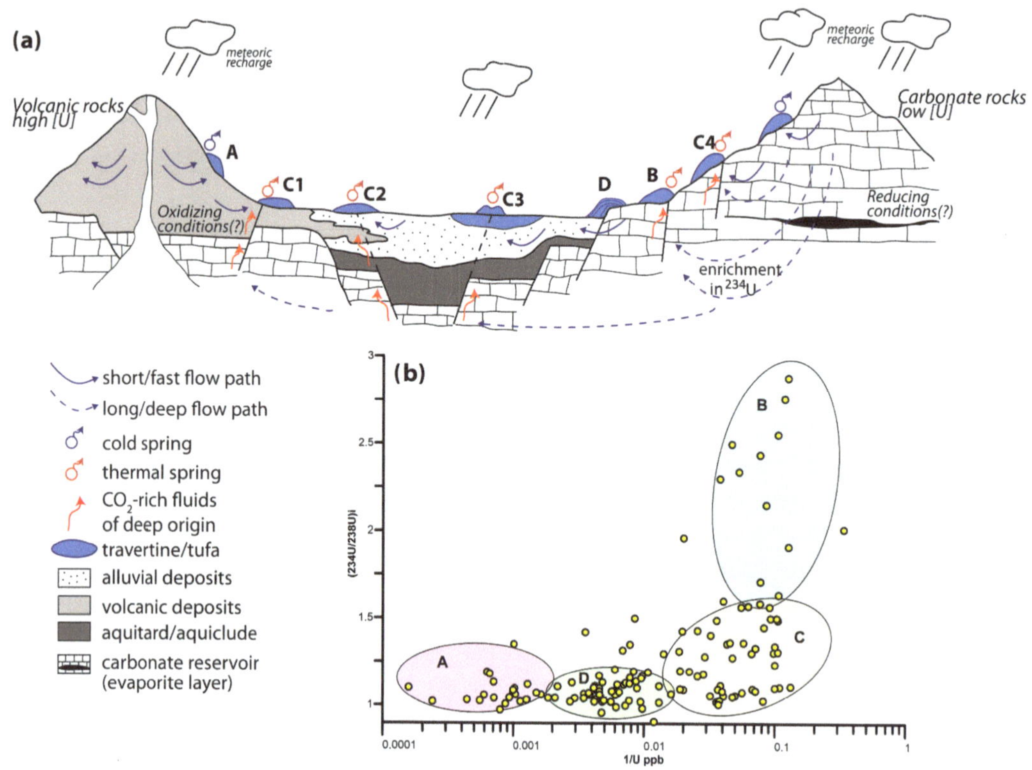

Figure 4. (**a**) Idealized section in which the occurrence of different travertine and calcareous tufa deposits is summarized; (**b**) the $^{234}U/^{238}U$ initial activity ratio *versus* 1/U values with all samples reviewed from literature. Samples are grouped into four clusters (A, B, C and D) which correspond to the different occurrences of travertine/tufa in the idealized section. See text for explanation.

Cluster A groups travertine samples from sites 5 and 7, and partly from 6. The data show very high U concentration (>1000 ppb) and a ($^{234}U/^{238}U$)i ratio of ~1.2. According to our interpretation, groundwater, circulating in volcanic rocks with high radionuclide content, deposited travertines with the highest U contents. The residence time of groundwaters can be short in the volcanic aquifers, and therefore the ^{234}U recoil enrichment is scarce, explaining the relatively low ($^{234}U/^{238}U$)i ratio. Oxidizing groundwater conditions may have favored the U mobility; data on Eh are generally not available to fully confirm this condition.

Cluster B includes travertine samples from sites 4, 8 and partly from 11, 14 and 15. These data show a low U concentration (<50 ppb) and an increasing ($^{234}U/^{238}U$)i ratio from 1.5 to 3. In this case, groundwaters circulating in carbonate/evaporite formations are not particularly enriched in U due to the relatively low content of U in the reservoir rocks. The long residence time of the groundwaters in these aquifers, related to deep circulation, allows a strong ^{234}U recoil enrichment and consequently an increase in the ($^{234}U/^{238}U$)i ratio. In addition, the reducing groundwater conditions, due to a slow and presumably anoxic deep circulation, may have hindered the U mobility.

Cluster C groups travertine/tufa samples from several sites (1, 2, 6, 10, 11, 13, 14, 15). They are characterized by intermediate to low U concentration and a ($^{234}U/^{238}U$)i ratio lower than 1.5, implying mixing processes of the parent groundwaters; mixing can occur between groundwaters from deep/carbonate aquifers and groundwaters from the other types of aquifers as displayed in Figure 4a: a shallow/volcanic aquifer (C1), a shallow/alluvial aquifer (C2), and a shallow/travertine aquifer (C3). C4 represents the mixing between groundwaters from a deep/carbonate aquifer and surface waters from streams and rivers (C4).

Cluster D exclusively groups tufa samples from sites 3, 9 and 12, which are characterized by intermediate values of U content and a ($^{234}U/^{238}U$)i ratio < 1.2. In these cases, carbonate is precipitated from cold waters associated with river systems, which are characterized by oxidizing conditions and fed by high-discharge springs recharged by carbonate aquifers. The oxidizing conditions for these waters are confirmed by the data for redox potential.

In summary, the residence time of groundwater has been found to be crucial for every factor affecting the U content and ($^{234}U/^{238}U$)i ratio: (i) a long residence time, due to deep circulation and/or a remote recharge area, may determine intense water–rock interactions that may increase the uranium content and ($^{234}U/^{238}U$)i ratio; meanwhile, the slow, deep circulation that characterized such conditions may cause anoxia, hindering the uranium mobility and decreasing its concentration; the natural cases observed in this study (Figure 4b) demonstrate that the ^{234}U shows a non-significant depletion due to uranium reduction, thus not showing an evident effect on the fractionation of the ($^{234}U/^{238}U$)i ratio; (ii) a short residence time, due to a shallow and fast circulation, may determine a decrease in the ($^{234}U/^{238}U$)i ratio, and possibly oxidizing conditions that favor the U mobility.

5. Conclusions

This paper reviews the available data on the uranium content and $^{234}U/^{238}U$ initial activity ratio from 15 Italian travertine and calcareous tufa deposits reported in the relevant literature. Using the graphical method previously proposed by [27,28], data have been interpreted considering the U geochemistry in natural environments as well as the geological, hydrogeological and hydrogeochemical setting of each site.

The U content and ($^{234}U/^{238}U$)i ratio in travertines and tufas appear to be affected by different factors, such as the availability of U in the aquifer rocks, the redox state of the groundwaters, and the alpha-active radionuclide recoil phenomenon. These factors interact with each other and determine the U content and isotope ratio of waters responsible for carbonate precipitation. Four groups of travertines/tufas were identified: (i) those that precipitated from groundwater circulating, with a short/fast flow path, in volcanic rocks with a high radionuclide content; these carbonates have a very high U concentration (>1000 ppb) and a ($^{234}U/^{238}U$)i ratio of ~1.2; (ii) those that precipitated from groundwater circulating, with a long and deep flow path, in carbonate/evaporite formations with a relatively low radionuclide content; these continental carbonates have a low U concentration (<50 ppb) and an increasing ($^{234}U/^{238}U$)i ratio (> 1.5, up to 3); (iii) those that precipitated from cold waters associated with river systems, which are characterized by oxidizing conditions and fed by high-discharge springs recharged by carbonate aquifers; these continental carbonates have intermediate values of U content and a ($^{234}U/^{238}U$)i ratio < 1.2; and (iv) those that precipitated in hydrogeological settings where mixing between groundwaters from different aquifers occurs.

The graphical method of [27,28], suitably implemented with a large amount of data reviewed from the literature, has proved to be effective for investigating the characteristics of parental waters and the related aquifers of travertines and calcareous tufas. This approach might be useful to infer paleo-environmental characteristics of fossil continental carbonate depositing systems.

Author Contributions: Conceptualization, F.G. and M.B.; methodology, F.G. and M.B.; investigation, F.G. and M.B.; writing—original draft preparation, F.G.; writing—review and editing, M.B. All authors have read and agreed to the published version of the manuscript.

Funding: This research received no external funding.

Data Availability Statement: All data are available in the article and Appendix A.

Acknowledgments: We are grateful to Mario Voltaggio for the constructive discussions and long-standing collaboration. We are indebted to all the authors who published uranium data of travertine and calcareous tufa that we have used in this study. We kindly thank the anonymous reviewers for their useful comments and suggestions.

Conflicts of Interest: The authors declare no conflict of interest.

Appendix A

Table A1. Uranium content (in ppb) and $^{234}U/^{238}U$ initial activity ratio from 15 Italian travertine and calcareous tufa deposits reviewed from the literature.

ID	Site	U ppb	$(^{234}U/^{238}U)_i$	Ref.
1	1—Acquasanta T.	4.5937		[31]
2		16.379		[31]
3		13.108		[31]
4		6.09		[31]
5		15.837		[31]
6		8.545		[31]
7		93.1	1.191	[32]
8		288.2	1.065	[32]
9		9.5	1.489	[32]
10		23.2	1.354	[32]
11		9.2	1.105	[32]
12		20.1	1.056	[32]
13		17.7	1.074	[32]
14		10.2	1.304	[32]
15	2—Acque Albule	106	1.02	[33]
16		118	1.5	[33]
17		238	1.11	[33]
18		774	1.12	[33]
19		21	1.05	[33]
20		28	1.02	[33]
21		27	1.02	[33]
22		179	1.02	[33]
23		10	1.1	[33]
24		610	1.06	[33]
25		31.25	1.056	[3]
26		12.25	1.026	[3]
27		26	1.115	[3]
28		34.25	1.281	[3]
29		14.75	1.094	[3]
30		14	1.084	[3]
31		27.25	1.093	[3]
32		24.667	1.054	[3]
33		25.25	1.081	[3]
34		211	0.958	[3]
35		26.5	1.027	[3]
36		27	1.006	[3]
37		458.5	1.107	[3]
38	3—Aniene valley	260	1.06	[34]
39		130	1.028	[34]
40		240	1.1	[34]
41		260	1.059	[34]
42		230	1.037	[34]
43		230	1.086	[34]
44		220	1.098	[34]
45		240	1.021	[34]
46		170	1.082	[34]
47		160	1.12	[34]
48		160	1.066	[34]
49		170	1.078	[34]
50		540	1.042	[34]
51		480	1.042	[34]
52	4—Bagno Vignoni	8	2.879	[35]
53		9.4	1.634	[35]
54		8.5	2.759	[35]
55		7.9	1.91	[35]
56		18.768	2.34	[35]
57		26.421	2.3	[35]
58		9.521	2.553	[35]
59		21.394	2.498	[35]
60		13.049	2.436	[35]
82	7—Etna Mt	1266	0.974	[38]
83		1422	1.133	[38]
84		2286	1.03	[38]
85		978	1.096	[38]
86		1685	1.057	[38]
87		1151	1.008	[38]
88		1073	1.043	[38]
89		1006	1.086	[38]
90		1812	1.025	[38]
91	8—Euganean hills	11.7	2.15	[39]
92		3	2.01	[39]
93	9—Middle Velino Valley	216	1.05	[40]
94		339	1.04	[40]
95		289	1.04	[40]
96		282	1.42	[40]
97		368	1.02	[40]
98		356	1.13	[40]
99	10—Prima Porta	21	1.275	[41]
100		62.3	1.065	[41]
101		39	1.43	[41]
102		15	1.3	[41]
103	11—Rapolano	27.9	1.49	[28]
104		35.1	1.18	[28]
105		17.3	1.36	[28]
106		10.8	1.5	[28]
107		15.9	1.57	[28]
108		24.8	1.6	[28]
109		12.9	1.71	[28]
110		9.7	1.5	[28]
111		7.6	1.108	[42]
112		84	0.91	[42]
113		103	0.983	[42]
114		220	1.08	[42]
115		50	1.96	[42]
116	12—Rocchetta al Volturno	166	1.06	[43]
117		139	1	[43]
118		143	1.12	[43]
119		128	1.16	[43]
120		138	1.13	[43]
121		209	1.02	[43]
122		243	1.1	[43]
123		177	1.08	[43]
124		164	0.99	[43]
125		265	1.07	[43]
126		115	1.14	[43]
127		39	1.17	[43]
128		113	1.1	[43]
129		103	1.17	[43]
130		219	1.06	[43]
131	13—Sarteano	105	1.157	[44]
132		146	1.316	[44]
133		203	1.028	[44]
134		221	1.17	[44]
135		155	1.077	[44]
136		167	1.206	[44]
137	14—Saturnia	129	1.125	[45]
138		208	1.133	[45]
139		18	1.565	[45]
140		13	1.585	[45]
141		10	1.234	[45]

Table A1. *Cont.*

ID	Site	U ppb	$(^{234}U/^{238}U)_i$	Ref.
61	5—Bolsena Lake	990	1.35	[27]
62		1600	1.19	[27]
63		1520	1.18	[27]
64		780	1.03	[27]
65		790	1.03	[27]
66		660	1.07	[27]
67		1010	1.07	[27]
68		6390	1.1	[27]
69		120	1.195	[27]
70		4210	1.021	[27]
71	6—Canino	128	1.349	[36]
72		54	1.311	[36]
73		51	1.426	[36]
74		71	1.297	[36]
75		11	1.566	[36]
76		10	1.34	[36]
77		14	1.33	[36]
78		50	1.09	[36]
79		880	1.02	[27]
80		960	1.06	[27]
81		1410	1.04	[27]
142	15—Semproniano	15.063		[12]
143		29.783		[12]
144		7.287		[12]
145		12.908		[12]
146		7.451		[12]
147		22.212		[12]
148		65.13		[12]
149		30.647		[12]
150		77	1.017	[12]
151		53	1.094	[12]
152		53	1.198	[12]
153		77	1.112	[12]
154		22	1.361	[46]
155		196		[46]
156		53	1.094	[45]
157		53	1.198	[45]
158		77	1.112	[45]
159		31	1.403	[45]
160		12.1	1.449	[45]
161		22	1.361	[45]
162		45	1.196	[45]
163		9.5	1.307	[45]

References

1. Panichi, C.; Tongiorgi, E. Carbon isotopic composition of CO_2 from springs, fumaroles, mofettes and travertines of central and southern Italy: A preliminary prospection method of geothermal area. In Proceedings of the 2nd UN Symposium on the Development and Use of Geothermal Energy, San Francisco, CA, USA, 20–29 May 1975; pp. 815–825.
2. Minissale, A. Origin, transport and discharge of CO_2 in central Italy. *Earth-Sci. Rev.* **2004**, *66*, 89–141. [CrossRef]
3. Faccenna, C.; Soligo, M.; Billi, A.; De Filippis, L.; Funiciello, R.; Rossetti, C.; Tuccinei, P. Late Pleistocene depositional cycles of the lapis Tiburtinus travertine (Tivoli, Central Italy): Possible influence of climate and fault activity. *Glob. Planet. Change* **2008**, *63*, 299–308. [CrossRef]
4. Andrews, J.E. Palaeoclimatic records from stable isotopes in riverine tufas: Synthesis and review. *Earth-Sci. Rev.* **2006**, *75*, 85–104. [CrossRef]
5. Capezzuoli, E.; Gandin, A.; Pedley, H.M. Decoding tufa and travertine (freshwater carbonates) in the sedimentary record: The state of the art. *Sedimentology* **2014**, *61*, 1–21. [CrossRef]
6. Hancock, P.L.; Chalmers, R.M.L.; Altunel, E.; Çakir, Z. Travitonics: Using travertines in active fault studies. *J. Struct. Geol.* **1999**, *21*, 903–916. [CrossRef]
7. Crossey, L.J.; Karlstrom, K.E.; Springer, A.E.; Newell, D.; Hilton, D.R.; Fischer, T. Degassing of mantle-derived CO_2 and He from springs in the southern Colorado Plateau region—Neotectonic connections and implications for groundwater systems. *Geol. Soc. Am. Bull.* **2009**, *121*, 1034–1053. [CrossRef]
8. Priewisch, A.; Crossey, L.J.; Karlstrom, K.E.; Polyak, V.J.; Asmerom, Y.; Nereson, A.; Ricketts, J.W. U-series geochronology of large-volume Quaternary travertine deposits of the southeastern Colorado Plateau: Evaluating episodicity and tectonic and paleohydrologic controls. *Geosphere* **2014**, *10*, 401–423. [CrossRef]
9. Navarro, A.; Font, X.; Viladevall, M. Geochemistry and groundwater contamination in the La Selva geothermal system (Girona, Northeast Spain). *Geothermics* **2011**, *40*, 275–285. [CrossRef]
10. Brogi, A.; Alçiçek, M.C.; Yalçıner, C.Ç.; Capezzuoli, E.; Liotta, D.; Meccheri, M.; Rimondi, V.; Ruggieri, G.; Gandin, A.; Boschi, C.; et al. Hydrothermal fluids circulation and travertine deposition in an active tectonic setting: Insights from the Kamara geothermal area (western Anatolia, Turkey). *Tectonophysics* **2016**, *680*, 211–232. [CrossRef]
11. Frery, E.; Gratier, J.P.; Ellouz-Zimmerman, N.; Deschamps, P.; Blamart, D.; Hamelin, B.; Swennen, R. Geochemical transect through a travertine mount: A detailed record of CO_2-enriched fluid leakage from Late Pleistocene to present-day—Little Grand Wash fault (Utah, USA). *Quat. Int.* **2017**, *437*, 98–106. [CrossRef]
12. Berardi, G.; Vignaroli, G.; Billi, A.; Rossetti, F.; Soligo, M.; Kele, S.; Baykara, M.O.; Bernasconi, S.M.; Castorina, F.; Tecce, F.; et al. Growth of a Pleistocene giant carbonate vein and nearby thermogene travertine deposits at Semproniano, southern Tuscany, Italy: Estimate of CO_2 leakage. *Tectonophysics* **2016**, *690*, 219–239. [CrossRef]
13. Shen, C.-C.; Wu, C.-C.; Cheng, H.; Edwards, R.L.; Hsieh, Y.-T.; Gallet, S.; Chang, C.-C.; Li, T.-Y.; Lam, D.D.; Kano, A.; et al. High-precision and high resolution carbonate 230Th dating by MC-ICP-MS with SEM protocols. *Geochim. Cosmochim. Acta* **2012**, *99*, 71–86. [CrossRef]
14. Bourdon, B.; Turner, S.; Henderson, G.M.; Lundstrom, C.C. Introduction to U-series geochemistry. *Rev. Mineral. Geochem.* **2003**, *52*, 1–21. [CrossRef]
15. Chabaux, F.; Riotte, J.; Dequincey, O. U-Th-Ra fractionation during weathering and river transport. *Rev. Mineral. Geochem.* **2003**, *52*, 533–576. [CrossRef]
16. Langmuir, D. Uranium-solution-mineral equilibria at low temperatures with applications to sedimentary ore deposits. *Geochim. Cosmochim. Acta* **1978**, *42*, 547–569. [CrossRef]

17. Murphy, W.M.; Shock, E.L. Environmental aqueous geochemistry of actinides. In *Uranium: Mineralogy, Geochemistry and the Environment*; Burns, P.C., Finch, R., Eds.; Walter de Gruyter GmbH & Co KG: Berlin, Germany, 1999; Volume 38, pp. 221–254.
18. Pentcost, A. *Travertine*, 1st ed.; Springer: Berlin/Heidelberg, Germany, 2005; p. 446.
19. Sturchio, N.C.; Antonio, M.R.; Soderholm, L.; Sutton, S.R.; Brannon, J.C. Tretravalent uranium in calcite. *Science* **1998**, *281*, 971–973. [CrossRef]
20. Rihs, S.; Condomines, M.; Sigmarsson, O. U, Ra and Ba incorporation during precipitation of hydrothermal carbonates: Implications for 226Ra-Ba dating of impure travertines. *Geochim. Cosmochim. Acta* **2000**, *64*, 661–671. [CrossRef]
21. Reeder, R.J.; Nugent, M.; Tait, D.; Morris, D.E.; Heald, S.M.; Beck, K.M.; Hess, W.P.; Lanzirotti, A. Coprecipitation of Uranium (VI) with Calcite: XAFS, micro-XAS, and luminescence characterization. *Geochim. Cosmochim. Acta* **2001**, *65*, 3491–3503. [CrossRef]
22. Khoury, H.N.; Salameh, E.M.; Clark, I.D. Mineralogy and origin of surficial uranium deposits hosted in travertine and calcrete from central Jordan. *Appl. Geochem.* **2014**, *43*, 49–65. [CrossRef]
23. Ivanovich, M.; Latham, A.G.; Ku, T.-L. Uranium-series disequilibrium applications in geochronology. In *Uranium-Series Disequilibrium Applications to Earth, Marine and Environmental Sciences*; Ivanovich, M., Harmon, R.S., Eds.; Oxford Science Publications; Clarendon Press: Oxford, UK, 1992; pp. 62–94.
24. Fleischer, R.L. Alpha-recoil damage and solution effects in minerals: Uranium isotopic disequilibrium and radon release. *Geochim. Cosmochim. Acta* **1982**, *46*, 2191–2201. [CrossRef]
25. Weyer, S.; Anbar, A.D.; Gerdes, A.; Gordon, G.W.; Algeo, T.J.; Boyle, E.A. Natural fractionation of $^{238}U/^{235}U$. *Geochim. Cosmochim. Acta* **2008**, *72*, 345–359. [CrossRef]
26. Rovan, L.; Lojen, S.; Zuliani, T.; Kanduč, T.; Petrič, M.; Horvat, B.; Rusjan, S.; Štrok, M. Comparison of Uranium Isotopes and Classical Geochemical Tracers in Karst Aquifer of Ljubljanica River catchment (Slovenia). *Water* **2020**, *12*, 2064. [CrossRef]
27. Taddeucci, A.; Voltaggio, M. Th-230 dating of the travertines connected to the Vulsini Mts. volcanism (Northern Latium, Italy): Neotectonics and hydrogeology. *Period. Mineral.* **1987**, *56*, 295–302.
28. Brogi, A.; Capezzuoli, E.; Aquè, R.; Branca, M.; Voltaggio, M. Studying travertines for neotectonics investigations: Middle–Late Pleistocene syn-tectonic travertine deposition at Serre di Rapolano (Northern Apennines, Italy). *Int. J. Earth Sci.* **2010**, *99*, 1383–1398. [CrossRef]
29. Özkul, M.; Kele, S.; Gökgöz, A.; Shen, C.-C.; Jones, B.; Baykara, M.O.; Fórizs, I.; Nemeth, T.; Chang, Y.-W.; Alçiçek, M.C. Comparison of the Quaternary travertine sites in the Denizli Extensional Basin based on their depositional and geochemical data. *Sedim. Geol.* **2013**, *294*, 179–204. [CrossRef]
30. Gandin, A.; Capezzuoli, E. Travertine: Distinctive depositional fabrics of carbonates from thermal spring systems. *Sedimentology* **2014**, *61*, 264–290. [CrossRef]
31. Janssens, N.; Capezzuoli, E.; Claes, H.; Muchez, F.; Yu, T.-L.; Shen, C.-C.; Ellam, R.M.; Swennen, R. Fossil travertine system and palaeofluid provenance, migration and evolution through time: Example from the geothermal area of Acquasanta Terme (Central Italy). *Sediment. Geol.* **2020**, *398*, 105580. [CrossRef]
32. Sembroni, A.; Molin, P.; Soligo, M.; Tuccimei, P.; Anzalone, E.; Billi, A.; Franchini, S.; Ranaldi, M.; Tarchini, L. The uplift of the Adriatic flank of the Apennines since the Middle Pleistocene: New insights from the Tronto River basin and the Acquasanta Terme Travertine (central Italy). *Geomorphology* **2020**, *352*, 106990. [CrossRef]
33. Faccenna, C.; Funiciello, R.; Montone, P.; Parotto, M.; Voltaggio, M. Late Pleistocene strike–slip tectonics in the Acque Albule Basin (Tivoli, Latium). *Mem. Descr. Carta Geol. d'It.* **1994**, *49*, 37–50.
34. Carrara, C.; Branca, M.; Pisegna, E.; Verrubbi, V.; Voltaggio, M. Calcareous tufa deposits of the Aniene valley between Vallepietra and Mandela-Vicovaro (Latium, Central Italy). *Il Quaternario* **2006**, *19*, 19–44.
35. Brogi, A.; Liotta, D.; Capezzuoli, E.; Matera, P.F.; Kele, S.; Soligo, M.; Tuccimei, P.; Ruggieri, G.; Yu, T.-L.; Shen, C.-C.; et al. Travertine deposits constraining transfer zone neotectonics in geothermal areas: An example from the inner Northern Apennines (Bagno Vignoni-Val d'Orcia area, Italy). *Geothermics* **2020**, *85*, 101763. [CrossRef]
36. Carrara, C. I travertini di Canino (Viterbo, Italia Centrale): Elementi di Cronolitostratigrfia, di Geochimica Isotopica e loro significato ambientale e climatico. *Il Quaternario* **1994**, *7*, 73–90.
37. D'Alessandro, W.; Glammanco, S.; Bellomo, S.; Parello, F. Geochemistry and mineralogy of travertine deposits of the SW flank of Mt. Etna (Italy): Relationships with past volcanic and degassing activity. *J. Volcanol. Geotherm. Res.* **2007**, *165*, 64–70. [CrossRef]
38. Romano, R.; Taddeucci, A.; Voltaggio, M. Datazione col metodo del Th-130 di alcuni travertini deposti sul versante sudoccidentale del M.te Etna. *Rend. Soc. Ital. Mineral. Petrol.* **1987**, *42*, 294.
39. Pola, M.; Gandin, A.; Tuccimei, P.; Soligo, M.; Deiana, R.; Fabbri, P.; Zampieri, D. A multidisciplinary approach to understanding carbonate deposition under tectonically controlled hydrothermal circulation: A case study from a recent travertine mound in the Euganean hydrothermal system, northern Italy. *Sedimentology* **2014**, *61*, 172–199. [CrossRef]
40. Soligo, M.; Tuccimei, P.; Barberi, R.; Delitala, M.C.; Miccadei, E.; Taddeucci, A. U/Th dating of freshwater travertine from Middle Velino Valley (Central Italy): Geological and paleoclimatic implications. *Palaeogeogr. Palaeoclimatol. Palaeoecol.* **2002**, *184*, 147–161. [CrossRef]
41. Giustini, F.; Brilli, M.; Di Salvo, C.; Mancini, M.; Voltaggio, M. Multidisciplinary characterization of the buried travertine body of Prima Porta (Central Italy). *Quat. Int.* **2020**, *568*, 65–78. [CrossRef]
42. Carrara, C.; Ciuffarella, L.; Paganin, G. Inquadramento geomorfologico e climatico -ambientale dei travertini di Rapolano Terme (SI). *Il Quaternario* **1998**, *11*, 319–329.

43. Brancaccio, L.; D'Argenio, B.; Ferreri, V.; Stanzione, D.; Taddeucci, D.; Voltaggio, M. I travertini di Rocchetta a Volturno (Molise): Datazioni con ^{230}Th e modello deposizionale. *Mem. Soc. Geol. Ital.* **1988**, *41*, 673–683.
44. Brogi, A.; Capezzuoli, E.; Buracchi, E.; Branca, M. Tectonic Control on Travertine and Calcareous Tufa Deposition in a Low-Temperature Geothermal System (Sarteano, Central Italy). *J. Geol. Soc. London* **2012**, *169*, 461–476. [CrossRef]
45. Vignaroli, G.; Berardi, G.; Billi, A.; Kele, S.; Rossetti, F.; Soligo, M.; Bernasconi, S.M. Tectonics, hydrothermalism, and paleoclimate recorded by Quaternary travertines and their spatio-temporal distribution in the Albegna basin, central Italy: Insights on Tyrrhenian margin neotectonics. *Lithosphere* **2016**, *8*, 335–358. [CrossRef]
46. Billi, A.; Berardi, G.; Gratier, J.P.; Rossetti, F.; Vignaroli, G.; Baykara, M.O.; Bernasconi, S.; Kele, S.; Soligo, M.; De Filippis, L.; et al. First records of syn-diagenetic non-tectonic folding in quaternary thermogene travertines caused by hydrothermal incremental veining. *Tectonophysics* **2017**, *700–701*, 60–79. [CrossRef]
47. Fusari, A.; Carroll, M.R.; Ferraro, S.; Giovannetti, R.; Giudetti, G.; Invernizzi, C.; Mussi, M.; Pennisi, M. Circulation path of thermal waters within the Laga foredeep basin inferred from chemical and isotopic (δ18O, δD, 3H, 87Sr/86Sr) data. *Appl. Geochem.* **2017**, *78*, 23–34. [CrossRef]
48. Carucci, V.; Petitta, M.; Aravena, R. Interaction between shallow and deep aquifers in the Tivoli Plain (Central Italy) enhanced by groundwater extraction: A multiisotope approach and geochemical modeling. *Appl. Geochem.* **2012**, *27*, 266–280. [CrossRef]
49. Chiodini, G.; Frondini, F.; Cardellini, C.; Parello, F.; Peruzzi, L. Rate of diffuse carbon dioxide Earth degassing estimated from carbon balance of regional aquifers: The case of central Apennine, Italy. *J. Geophys. Res.* **2000**, *105*, 8423–8434. [CrossRef]
50. Sappa, G.; Ferranti, F. Pertuso Spring discharge assessment in the Upper Valley of Aniene River (Central Italy). *Int. J. Energy Environ.* **2016**, *10*, 33–40.
51. Cioni, R.; Guidi, M.; Pierotti, L.; Scozzari, A. An automatic monitoring network installed in Tuscany (Italy) for studying possible geochemical precursory phenomena. *Nat. Hazards Earth Syst. Sci.* **2007**, *7*, 405–416. [CrossRef]
52. Magi, F. Isotope geochemistry of rainfall, thermal and non-thermal waters from the Mt. Amiata area (northern-central Italy). *Plinius* **2020**, *46*, 56–63.
53. Mosello, R.; Arisci, S.; Bruni, P. Lake Bolsena (Central Italy): An updating study on its water chemistry. *J. Limnol.* **2004**, *63*, 1–12. [CrossRef]
54. Chiodini, G.; Giaquinto, S.; Frondini, F.; Santucci, A. Hydrogeochemistry and hydrogeology of the Canino hydrothermal system (Italy). *Geothermics* **1991**, *20*, 329–342. [CrossRef]
55. Bellia, C.; Gallardo, A.H.; Yasuhara, M.; Kazahaya, K. Geochemical characterization of groundwater in a volcanic system. *Resources* **2015**, *4*, 358–377. [CrossRef]
56. Gherardi, F.; Panichi, C.; Caliro, S.; Magro, G.; Pennisi, M. Water and gas geochemistry of the Euganean and Berician thermal district (Italy). *Appl. Geochem.* **2000**, *15*, 455–474. [CrossRef]
57. Giustini, F.; Blessing, M.; Brilli, M.; Lombardi, S.; Voltattorni, N.; Widory, D. Determining the origin of carbon dioxide and methane in the gaseous emissions of the San Vittorino plain (Central Italy) by means of stable isotopes and noble gas analysis. *Appl. Geochem.* **2013**, *34*, 90–101. [CrossRef]
58. Petitta, M. Idrogeologia della media valle del fiume Velino e della piana di S. Vittorino (Rieti, Italia centrale) / Hydrogeology of the middle valley of the Velino river and of the S. Vittorino plain (Rieti, Central Italy). *Ital. J. Eng. Geol. Environ.* **2009**, *1*, 157–181.
59. Giustini, F.; Brilli, M.; Mancini, M. Geochemical study of travertines along middle-lower Tiber valley (central Italy): Genesis, palaeo-environmental and tectonic implications. *Int. J. Earth Sci.* **2018**, *107*, 1321–1342. [CrossRef]
60. Brilli, M.; Giustini, F. Geochemical stratigraphy of the Prima Porta travertine deposit (Roma, Italy). *Minerals* **2023**. *this Special Issue*.
61. Minissale, A.; Vaselli, O.; Tassi, F.; Magro, G.; Grechi, G.P. Fluid mixing in carbonate aquifers near Rapolano (central Italy): Chemical and isotopic constraints. *Appl. Geochem.* **2002**, *17*, 1329–1342. [CrossRef]
62. Celico, P. *Idrogeologia dei Massicci Carbonatici, Delle Piane Quaternarie e Delle Aree Vulcaniche Dell'italia Centro-Meridionale (Marche e Lazio Meridionali, Abruzzo, Molise e Campania)*; Cassa per il Mezzogiorno, stampa Grafiche Magliana: Roma, Italy, 1983; p. 236.
63. Cuoco, E.; Colombani, N.; Darrah, T.H.; Mastrocicco, M.; Tedesco, D. Geolithological and anthropogenic controls on the hydrochemistry of the Volturno River (Southern Italy). *Hydrol. Process.* **2016**, *31*, 627–638. [CrossRef]
64. Fancelli, R.; Nuti, S. Studio sulle acque termali e minerali del graben di Siena. *Boll. Soc. Geol. Ital.* **1975**, *94*, 135–155.
65. Panichi, C.; D'Amore, F.; Fancelli, R.; Noto, P.; Nuti, S. Geochemical Survey of the Siena Province. In Proceedings of the Seminar on Geothermal Energy, Bruxelles, Belgium, 6–8 December 1977; Volume 2, pp. 481–503.
66. Bertrami, R.; Cameli, G.M.; Lovari, F.; Rossi, U. Discovery of Latera geothermal field, problems of the exploration and research. Proceedings of Seminar on Utilization of Geothermal Energy for Electric Power Production and Space Heating, Florence, Italy, 14–17 May 1984; pp. 1–18.
67. Barbagli, A.; Brogna, F.N.A.; Callegari, I.; Guastaldi, E.; Liali, G.; Marsico, N.; Rezza, C.; Trotta, M. Approccio multi-isotopico ed idrogeochimico per la caratterizzazione di acque termali: Il caso di Saturnia (GR). *Ital. J. Groundw.* **2013**, *AS07029*, 025–040.
68. Barberi, F.; Innocenti, F.; Ricci, C.A. Il magmatismo dell'Appennino Centro Settentrionale. *Rend. Della Soc. Ital. Di Mineral. Petrol.* **1971**, *27*, 169–210.
69. Peccerillo, A. *Cenozoic Volcanism in the Tyrrhenian Sea region*, 2nd ed.; Springer: Cham, Switzerland, 2017; p. 399.
70. Cocozza, T. Nuovi dati stratigrafici e tettonici sul monte Canino (Viterbo). *Geol. Rom.* **1963**, *2*, 15–40.

71. Viganò, A.; Zampieri, D.; Rossato, S.; Martin, S.; Selli, L.; Prosser, G.; Ivy-Ochs, S.; Campedel, P.; Fedrizzi, F.; Franceschi, M.; et al. Past to present deformation of the central-eastern Southern Alps: From the foreland to the Giudicarie belt. *Geol. Field Trips Maps* **2018**, *10*, 2–78. [CrossRef]
72. Minissale, A.; Kerrick, D.M.; Magro, G.; Murrell, M.T.; Paladini, M.; Rihs, S.; Sturchio, N.C.; Tassi, F.; Vaselli, O. Geochemistry of Quaternary travertines in the region north of Rome (Italy): Structural, hydrologic and paleoclimatic implications. *Earth Planet. Sci. Lett.* **2002**, *203*, 709–728. [CrossRef]
73. Maggi, M.; Cianfarra, P.; Salvini, F.; Coelho de Lima, C. Staircase fractures in microbialites and the role of lamination-related mechanical anisotropy: The example of the Acquasanta Terme travertine deposits (central Italy). *GSA Bull.* **2015**, *127*, 879–896. [CrossRef]
74. Menichetti, M. Assetto strutturale del sistema geotermico di Acquasanta Terme (Ascoli Piceno). *Rend. Online Soc. Geol. Ital. Note Brevi* **2008**, *1*, 118–122.
75. Verdelocco, S.; Turkowsky, P.; Walzer, D. L'incidenza dei fattori geologici e delle variazioni climatiche nell'individuazione di aree ad alto potenziale di radon. *Geol. Tec. Ambient.* **2000**, *1*, 45–52.
76. Aiuppa, A.; Allard, P.; D'Alessandro, W.; Michel, A.; Parello, F.; Treuil, M.; Valenza, M. Mobility and fluxes of major, minor and trace metals during basalt weathering and groundwater transport at Mt. Etna volcano (Sicily). *Geochim. Cosmochim. Acta* **2000**, *64*, 1827–1841. [CrossRef]

minerals

MDPI

Article

Geochemical Stratigraphy of the *Prima Porta* Travertine Deposit (Roma, Italy)

Mauro Brilli * and Francesca Giustini

Istituto di Geologia Ambientale e Geoingegneria (IGAG), Consiglio Nazionale delle Ricerche (CNR), Area della Ricerca di Roma 1, Via Salaria km 29,300, Monterotondo St., 00015 Rome, Italy; francesca.giustini@igag.cnr.it

* Correspondence: mauro.brilli@cnr.it

Abstract: A stratigraphy of a buried travertine deposit was developed using stable isotope geochemistry, trace elements, and radiometric dating. The travertine was identified in a well at *Prima Porta* (north of Rome, Italy), located at the western boundary of the Tiber Valley, a morpho-tectonic depression of extensional origin. It deposited close to a spring that discharged groundwaters from the nearby volcanic aquifer and was associated with the rise of a deep-seated CO_2-rich fluid. The deposition occurred between 53.5 ± 10 ka to 24.2 ± 4.7 ka; its activation was probably coeval with the wettest climatic conditions occurring during Marine Isotope Stage 3, and the end coincided with the cold and arid phase of the last glacial maximum. The chronostratigraphy showed a strong variation in the accumulation rate along the depositional sequence, greater in the lower half and much slower in the upper part, with a sharp decrease in the accumulation rate occurring between 47 and 43 ka. Isotope and chemical stratigraphy described a temporal evolution of events that are correlated to the global climatic variability; palaeoclimatic changes, in fact, influenced the hydrological regime and indirectly the tectonic activity by modulating the emission of deep CO_2, the chemistry of the groundwater, and ultimately the precipitation of the travertine.

Keywords: travertine; stable isotopes; trace elements

Citation: Brilli, M.; Giustini, F. Geochemical Stratigraphy of the *Prima Porta* Travertine Deposit (Roma, Italy). *Minerals* **2023**, *13*, 789. https://doi.org/10.3390/min13060789

Academic Editor: János Haas

Received: 30 April 2023
Revised: 6 June 2023
Accepted: 7 June 2023
Published: 8 June 2023

1. Introduction

Along the north-western edge of the Tiber Valley, a morpho-tectonic depression of extensional origin located in the western sector of central Italy, a subterranean travertine deposit named *Prima Porta* travertine (PP), was recently identified in a core recovery from drilling carried out for irrigation purposes. It was studied and described by [1,2]. It is approximately 3 m thick, lying below 8 m of recent alluvial deposits (Figure 1a,b). This travertine was likely deposited in proximity to a spring mainly fed by groundwater which flowed through a volcanic aquifer and was associated with the rise of a deep-seated CO_2-rich fluid.

Previous studies [1,2] aimed to reconstruct the genetic conditions and processes that led to the formation of the PP travertine deposit using stable isotopes, petrographic features, and U-Th dating. They have shown that PP carbonate geobody is a calcitic thermogenic travertine deposited in a low-to-moderate energy environment, such as gently-dipping, shallow pools of low-angle terraced slopes over its entire depositional history, which covered about 30 ka years between the late-stage of Marine Isotope Stage (MIS) 3 and the beginning of MIS 2.

The favorable conditions that led to the deposition of this travertine fit into the context of extensive tectonics, which involved the formation of the largest morpho-tectonic depressions in central Italy, the Tiber Valley. Along this major regional structure, late Quaternary fossil travertines are widespread, as reported in some of the geological literature [3–5]. The travertine development was mostly controlled by two main factors: climate conditions and tectonic evolution [6,7]. The PP travertine seems to be related mainly to this latter since it formed close to a spring of moderate thermal waters at the western border faults of the graben that housed the paleo-bed of the river Tiber, here oriented NE-SW. It may also have had a climate control because the deposition period (between 54 and 24 ka) encompassed the warm, wet conditions that occurred during MIS 3 and the cold, arid phase of the last glacial maximum, considered climatically adverse to abundant carbonate precipitation when, however, the deposition of travertine ended.

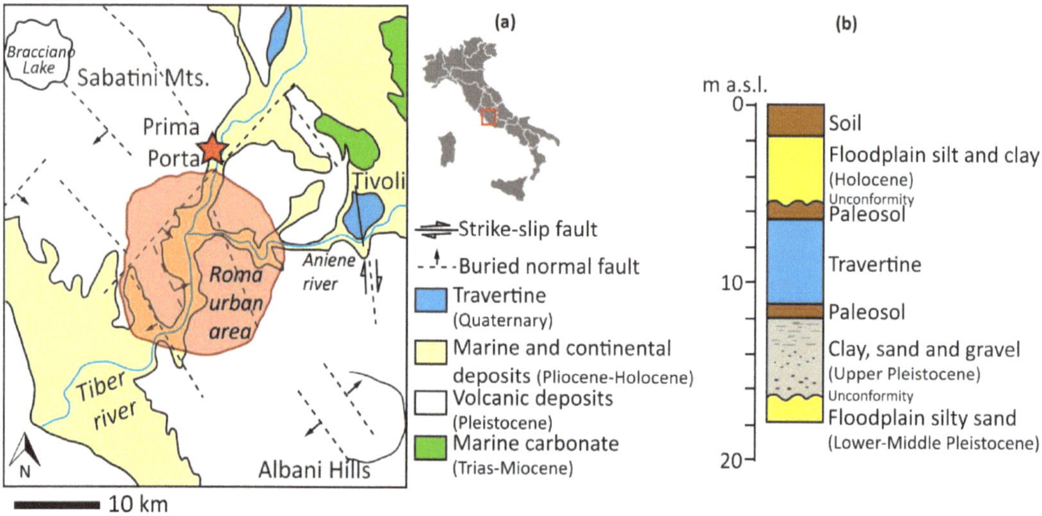

Figure 1. (**a**) Geological sketch of the *Prima Porta* area and (**b**) Schematic section of the core (m a.s.l stands for meters above sea level); the travertine deposit (in blue) is set within the alluvial sediment of the river Tiber between two levels of paleosols that mark periods of aridity before and after its formation.

At first glance, the available stable isotope data in [2] did not show interesting paleoclimate-related changes when plotted versus the core recovery depth, nor events that could be related to the structural evolution of the area. To verify whether the palaeoclimatic signal was hidden in the apparent 30 ka continuous sequence, the stratigraphy of some trace elements was developed in this study and paralleled with the isotopic records. A group of elements, generally considered to substitute to different extents for calcium in the crystal lattice of calcite [8], was analyzed by Inductively Coupled Plasma Atomic Emission Spectrometry (ICP-AES) using the same samples used for measuring isotope ratios in [2].

Extracting palaeoclimatic and palaeoenvironmental information from terrestrial carbonate deposits has been largely applied to Quaternary environmental science research; nevertheless, speleothems ([9] and references therein) or lake sediments ([10] and references therein) were used typically for long and detailed palaeoclimatic records, and [11] and references therein also reported as riverine tufa carbonates can preserve palaeoclimatic information in their geochemical records, emphasizing that they offer an opportunity to study details of relatively short-lived climatic events since calcareous tufas occur in nature with a temporal shorter sequence of a higher resolution due to their rapid accumulation. Stable isotopes, however, were the most used environmental and palaeoclimatic geochemical proxies in the time series of continental carbonate sequences (as they are in the marine stratigraphy), to which chemical stratigraphy has been more rarely applied.

In this paper, we aim for the following targets: First, extracting palaeoclimatic and palaeoenvironmental information from travertine sequences, although the thermogenic nature of these geobodies would lead to arguing that the geochemical features can mainly reflect endogenic processes; Second, evaluating the potential of elemental parameters in travertine stratigraphy and their effectiveness in combination with stable isotopes to infer environmental signals. A multiproxy geochemical approach to sequential carbonate veins deposited by deep, CO_2-rich fluids was successfully used to derive palaeoclimate and fault movement records of the last 135 ka [12].

2. Geology of the Area

The PP travertine is in the lower Tiber Valley to the north of Rome, in the inner sector of the Central Apennine fold-thrust belt, a Meso-Cenozoic carbonate thrust sheet that built up and migrated eastward during the Neogene westward subduction of the Adriatic plate [13]. From the end of the Neogene to the Quaternary, on the Tyrrhenian side of the Apennine belt, extensional back-arc processes have occurred: the reduction of the thickness of the lithosphere, volcanism, extensional basins, and high heat flux [14–19].

Since the Pleistocene (in the last 1.4 Ma), the entire area underwent active uplift [20]. In this period, a juvenile river, the "Paleo-Tiber" [21–24], developed in a palaeo-valley of tectonic origin with a NNW-SSE trend (Paleo-Tiber Graben [25]); the fluvial sedimentation was largely driven by glacio-eustatic sea-level fluctuations [26] and mostly characterized by gravel, sand, and minor amount of mud, deeply entrenched into the Pliocene-Early Pleistocene marine clay and silt [27]. In the middle Pleistocene, about 700 ka, large explosive volcanic activity superimposed fluvial development; it originated from the Albani Hills and Sabatini Mts. and characterized the Roman Area with its peculiar alkaline–potassic eruptive products, remaining intermittently active until recent times [28,29]. The fluvial deposits were covered by thick pyroclastic sequences (ignimbrites, phreatomagmatic ashes, pumice, and ash falls), dating from about 600 to 280 ka [30,31]. This recent tectonic history differs from the previous extensional regime because it is mainly characterized by N-striking right-lateral and NE-striking transtensional-to-normal faults [32,33]. These structures have partially controlled the latest stages of volcanism and related hydrothermal outflows [34]. It is during this phase, characterized by episodic volcano-clastic contributions from the nearby volcanic districts, that the river network changed its original N-S orientation, consistent with the extensional structures, to a transversal direction, wedging its bed between the Albane and Sabatine vulcanites.

The PP travertine is placed at the top of the Upper Pleistocene-Holocene fluvial incised valley, which constitutes the Tiber Depositional Sequence ([35] and references therein), a complex stratigraphic architecture that crosscuts the Paleo-Tiber Graben. The valley fill was mostly controlled by the last cycle of sea level change (from MIS 5d to MIS 1) and is composed of basal channel gravels and overlying channel sand and floodplain mud, respectively ascribed to the late lowstand, still regressive phases (26–14 ka) and to the late transgressive, highstand phases (14–0 ka). Buried and detached depositional terraces (locally preserved below more recent deposits) are attributed to the early lowstand phase (116–26 ka) and record intermittent phases of sedimentation during the general lowering of the base level.

In the study area, the alluvial plain is ~2 km wide and has elevations ranging from 19 to 22 m above sea level; to the west and east, the plain connects to slightly higher sectors (up to ~80 m a.s.l.), corresponding to the volcanic hills. The *Prima Porta* borehole is located on the west side of the plain, close to the hilly western sector belonging to the Sabatini volcanic complex (Figure 1a).

Hydrogeology Features

The regional hydrogeology of the area of the Tiber Valley is controlled by stratigraphic sequences and structural settings. Allochthonous, low-permeability Mesozoic-Cenozoic (Upper Cretaceous- Paleogene) flysch sequences, and post-orogenic clay-rich Miocene-Quaternary sediments, overlying Meso-Cenozoic limestones, represent the regional aquiclude [36]. It underlies the Quaternary volcanic and alluvial sediments.

The area of *Prima Porta* lies at the northern border of the Paleo-Tiber Graben. The graben is filled by an aquifer more than 100 m thick towards the southern edge, consisting of a thin layer of volcanic rocks over layers of sand and gravel. [36]. In the north, around the study area, the aquifer thickness is reduced. The presence of numerous springs, mainly located along the eastern side of the valley where the bedrock is shallower than the right bank, marks the lithological contact between Plio-Pleistocene bedrock and sedimentary or volcanic rocks. The springs are often ipothermal and sometimes associated with gas emissions because fluids flow upward along tectonic discontinuities from deep reservoirs. The chemical composition of the related groundwater is generally dominated by Ca, Mg, and SO_4, which is derived from the dissolution of Mesozoic carbonate rocks and a Triassic anhydride layer at the base of the carbonate succession. Some groundwater is enriched in Na and Cl, probably due to the dissolution of Miocene-Pliocene marine evaporates [3,37].

3. The Previous Studies

As reported by [2], the travertine body was found between 8 and 11 m of an about 18 m core drilled on the alluvial sediments of the Tiber Valley. The stratigraphy of the recovery drill is shown in Figure 1b. Below the travertine, a massive, dark brown/red clayey paleosol layer of one meter is followed by 6 m thick of grey, massive pebbly, silty sand and clay with sparse carbonate encrustations, sediments of the Tiber Depositional Sequence [35] that rest unconformably over yellow-beige, massive, silty sands, attributed to the Paleo-Tiber Unit. Above the travertine body, a paleosol again marks a break in the depositional episode, and the upper part of the core consists of beige, massive silts, and clays, which are associated with a floodplain environment; they likely record the most recent phase of deposition from overbank flooding of the Tiber River. The travertine is white and characterized by horizontal to sub-horizontal, irregular, wavy laminations, with alternating dm-thick planar bedsets and sub-horizontal bedlets; the macro and microscopic observations evidenced a quite repetitive alternation of undulated layers of rafts and shrubs between micrite crusts, which macroscopically are substantiated by the sub-horizontal, irregular, wavy laminations; these structures can be reconducted to a typical low to moderate energy genetic environments, such as terrace pools or flat ponds. Only in 15 cm out of the entire thickness are the depositional surfaces steeply inclined (up to 45° dip), in the interval between 10.15 and 10.30 m depth, indicating that at that position, a slope system, such as the rims and walls of pools in terraced slope dams, has taken place for a certain elapse of time. With such exceptions, it is possible to argue that along the vertical profile of the travertine, the environmental setting did not substantially change during its development, likely remaining subaqueous and devoid of erosional surfaces.

Stable isotopes of carbon and oxygen from 17 samples collected randomly along the travertine sequence were also analyzed by [2], allowing the further definition of the genetic constraints and evidencing the origin of carbon and the environmental temperature (the isotopic data published in [2] are also listed in Table 1). The high $\delta^{13}C$ signature, between +8.8 and +11.2‰, reflects non-soil carbon sources and an origin likely associated with the decarbonation of local carbonate bedrock; the $\delta^{18}O$ range is consistent with a carbonate precipitated from water with an isotopic composition close to that of the present-day local groundwater and temperatures between 8 and 17 °C [38]. The local groundwater, sampled in the well from which the core was drilled, showed a pH between 5.9 to 6.2, conductivity between 3080 and 2752 μS/cm, and temperatures between 18 and 20 °C (data measured at different periods of the year, November 2012 and March 2013—these data were partially published in [1]). The discrepancy between the actual water temperature and the calculated one can be attributed to the distance between the location of the drilling and the now undetermined spring orifice. Therefore, the carbonate-precipitating water may have cooled due to surface flow.

Table 1. Elemental composition, stable isotopes, U/Th ages, and depth of the samples from PP travertine core. (Stable isotope data and U/Th ages are from [2]).

Sample	Depth (m)	$\delta^{13}C$ (‰)	$\delta^{18}O$ (‰)	Fe (ppm)	K (ppm)	Mg (ppm)	Mn (ppm)	Na (ppm)	Sr (ppm)	Age (ka)
TR8 C	7.90	11.84	−3.30	577	819	3710	1291	913	1027	
TR8 E	7.98	11.42	−3.29	477	961	4184	1480	1151	1133	
TR9 C	8.08	10.57	−3.11	752	809	3694	1637	912	850	24.2
TR9 L	8.38	11.65	−3.37	449	1019	3765	1775	1167	1201	
TR9 P	8.55	11.26	−3.25	474	934	3909	1593	1042	894	44.6
TR9 U	8.76	11.24	−3.6	638	830	2947	1395	908	727	
TR9 X	8.84	11.54	−3.41	586	706	2370	1127	672	478	
TR9 Z	8.97	11.06	−3.06	678	667	2463	1551	656	432	
TR10 C	9.10	10.81	−3.15	740	665	2197	1287	656	459	
TR10 E	9.18	10.38	−3.54	491	668	2107	1247	630	445	
TR10 F	9.23	10.83	−3.32	508	656	2248	1179	624	439	50.4
TR10 H	9.32	10.61	−3.16	612	638	1805	1271	593	315	
TR10 O	9.54	11.00	−3.30	590	677	2364	953	667	350	
TR11 G	10.07	11.54	−2.93	541	672	2671	1239	716	336	
TR11 M	10.35	11.52	−2.88	604	700	2992	1176	736	361	53.5
TR11 P	10.46	11.22	−3.08	694	774	2493	1392	757	339	
TR11 R	10.53	11.82	−2.66	790	712	3360	1047	767	412	

The most striking aspect of the isotopes of this travertine is the complete absence of any correlation between the two variables (Figure 2a); this suggests that the environmental conditions remained stable throughout formation unless a thorough diagenetic isotope resetting did involve the entire carbonate body after the completion of deposition. Ref. [2] detected the effects of diagenesis in the secondary deposition of cement only, although not very frequent; some microfractures filled by sparite crystals and re-crystallization forming darker micrite surrounded by cloudy micrite or sparite crystals were observed. These cements were randomly analyzed for stable isotopes; they resulted similarly or identically to the primary calcium carbonate as if it may be constituted by early diagenetic fabrics connected to syn-sedimentary fluids. Post-depositional precipitation of sparry calcite within the porosity was revealed in a singular case by the total carbon isotope reset due to organic carbon incorporation. The largely maintained depositional pore network and the absence of primary precipitation of aragonite suggest that post-depositional dissolution and reprecipitation events did not occur on a large scale; finally, considering that (i) the quantity of cement that lines the voids and pores is quite limited and, (ii) the isotope composition of the cement is similar to that of the bulk rock and micrite samples in most cases, cementation probably did not cause a significant diagenetic resetting of the isotope signature. Instead, re-crystallization is generally considered a process relating to early diagenesis [39] or a syn-depositional process [40]. In these cases, it is reasonable to assume that circulating fluids had the same characteristics (e.g., temperature and $\delta^{18}O$) as the water from which carbonate precipitation occurred and, therefore, the observed re-crystallization had no significant impact on the isotope ratios.

U/Th dating of the travertine of four approximately equally spaced levels yielded the following ages: 53.5 ± 10 ka, 50.4 ± 8.7 ka, 44.6 ± 6.4 ka, and 24.2 ± 4.7 ka, which were respectively placed at 10.35, 9.23, 8.55, 8.08 m from the top of the core (data already published in [2] and reported in Table 1). The deposition history of the travertine deposit encompassed a period of great climatic variability (intercepting both the MIS 3 and the MIS 2) within a general cooling trend which had its peak at the last glacial maximum.

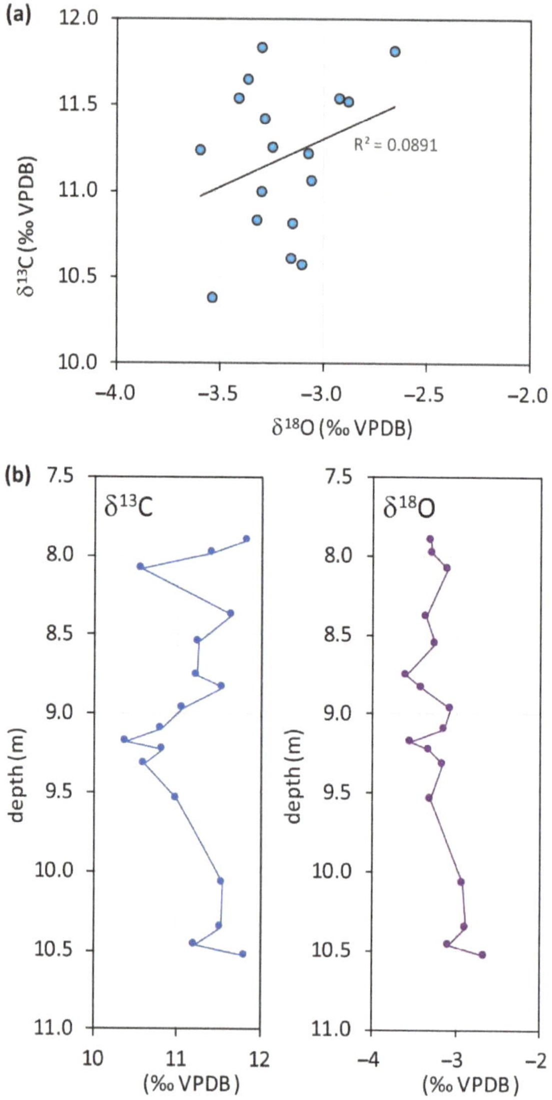

Figure 2. (**a**) Scatterplot of stable oxygen and carbon isotope compositions from *Prima Porta* travertine samples, data are from [2]; (**b**) $\delta^{13}C$ and $\delta^{18}O$ profiles of the travertines of *Prima Porta* versus depth of recovery.

4. Sampling and Analytical Method

Sampling was described by [2]. On the 17 samples in which stable isotope ratios were determined, the concentrations of Sr, Mn, Mg, Fe, Na, and K were analyzed by ICP-AES with a Varian Vista RL CCD Simultaneous ICP-AES spectrometer with an axial torch (Department of Earth Sciences, Sapienza University, Rome, Italy).

A 3% solution of hydrochloric acid (HCl) was added at room temperature to about 200 mg of each powdered sample and maintained under stirring. The acid was introduced in successive small additions while monitoring the pH. Upon completion of the acid–carbonate reaction, after about 1 h, the solution was filtered (0.2 μ). Following the principles described by [41], this approach aimed to minimize the dissolution of cations from silicate impurities in the bulk carbonate sample, whereas oxide compounds were at least partially dissolved. All solvents and reagents were of the highest purity grade commercially available. Deionized water (resistivity 18 MΩ cm^{-1}) obtained from a Milli-Q purification system was used to prepare all standard and sample solutions that were eventually analyzed by the ICP instrument. Emission lines for ICP-AES analysis were chosen according to previous interference studies. Internal standards were added to compensate for any effects from acid or instrument drift. Analytical precision was calculated as the percent relative standard deviation of a set of ten repetitions of the same sample of travertine calcite in any single session of analysis. Results indicated precision values always less than 5% for Sr, Fe, K, and Na elements, of which the concentrations were between 2 and 5 mg/L in the sample digestion acidic solutions; always less than 3% for Mg and Mn which were between 5 and 20 mg/L. Accuracy was evaluated on standard solutions and was less than 10% for all elements. The measurement conditions and emission lines are listed in Tables A1 and A2 of the Appendix A.

5. Results

As already determined in [1,2], the PP travertine is totally calcite, with no relics of aragonite crystal structures (radiated fibers or needle aggregates) upon observation of thin sections. The mineralogy of the $CaCO_3$ phase is the prime factor determining the concentration of a trace element in its lattice [8].

Table 1 shows the results of the chemical analysis together with the results of isotopic analysis and U/Th dating.

Covariation plots of travertine chemistry allow us to establish when the presence of cations is related to a common process that, for Sr and Mg, is dominantly the substitution for Ca in the calcite crystal lattice. Figure 3 shows that these two elements are strongly, positively correlated. Na and K are strongly positively correlated with each other and, in turn, correlated with Mg and Sr. All these elements seem to mainly be substitutes for Ca. Since their behavior is essentially in agreement and since it is reasonable to assume that their partition coefficients between water and solid phase respond differently to temperature, deposition rate, and salinity, it is possible to conclude that their abundance is mainly related to the respective concentrations in the travertine-precipitating waters. Consequently, the elemental variations as a function of time reflect the different rates of silicate mineral dissolution from the volcanic aquifer.

Whereas iron is definitely uncorrelated with other cations, Mn shows intermediate behavior. It must be concluded that iron may represent a detrital fraction present in the travertine or a residual component, made of oxides and hydroxides and produced during travertine deposition or post-syn-sedimentary travertine dissolution; manganese is a mix between the calcite crystal lattice incorporation and other provenances.

Chemical data, when stratigraphically plotted versus depth (Figure 4), show common features, as obvious for the highly correlated elements (Mg, Sr, Na, K). These trends can be summarized as low concentration levels in the lower section, between 10.5 and 9.3 m, and an apparent gradual increase towards the top. For Mn, this trend is barely visible, and it is absent for iron.

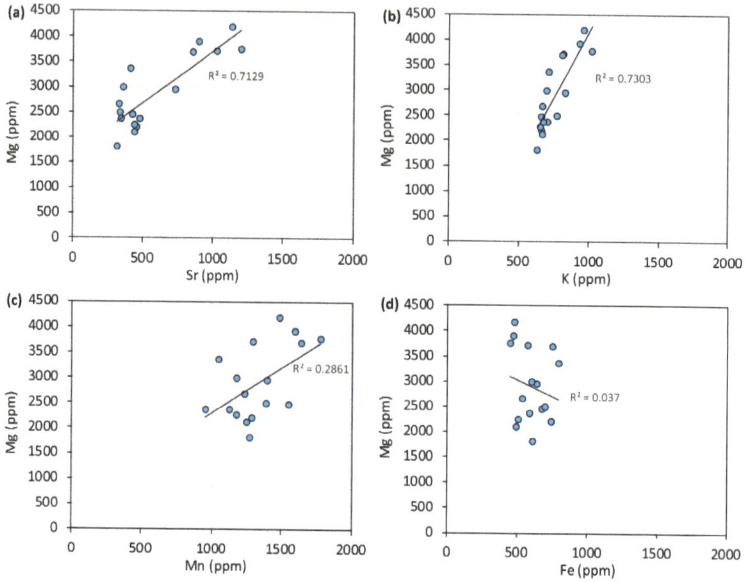

Figure 3. Covariation plots of PP travertine chemistry. Sr and K are strongly correlated with Mg and, in turn, are strongly correlated. Na is not presented in these scatterplots as it behaves similarly to K.

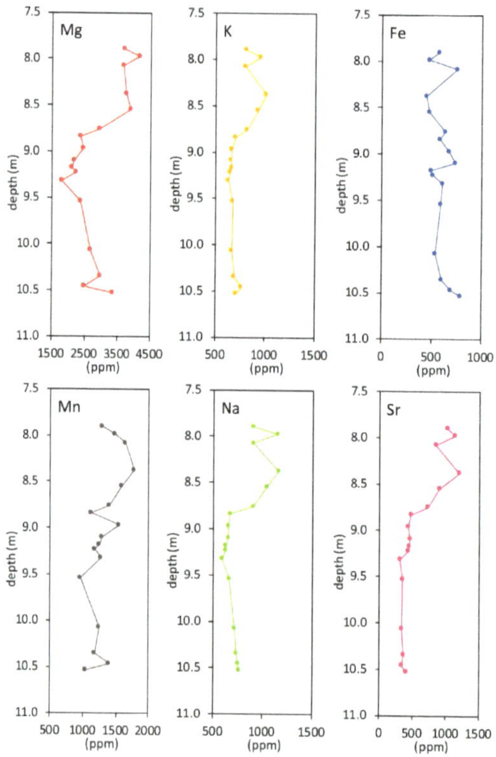

Figure 4. Chemical data plotted versus depth.

6. Discussion

Minor and trace elements Sr, Mn, Mg, Fe, Na, and K were analyzed as they can potentially substitute, to varying degrees, for Ca in the $CaCO_3$ (calcite) crystal lattice. Sr and Mg were selected because they are abundant and the most common substitutes for Ca. The partition of Mg between calcite and water has been extensively studied in marine solutions, and it depends on several factors; nevertheless, in freshwater environments with low ionic strength and low Mg/Ca ratios (<1), it is possible to argue that its content in the calcite precipitates is determined only by the Mg/Ca ratio and the temperature of the parental water [42,43]. The variations in Mg content observed in the PP travertine may be partly due to temperature changes. Still, its strong correlation with Sr, whose behavior as a function of temperature is generally irrelevant [44,45], suggests that the main factor for forcing the Mg and Sr variations is their content fluctuation in the water. Na and K incorporation is thought to occur predominantly by substitution for Ca, as the correlation with Sr and Mg would indicate; this substitution occurs with a charge imbalance that had to be compensated by further chemical substitution. Their abundant presence in the travertine (on average 800 and 750 ppm) indicates that travertine parental groundwater largely interacted with the aquifer of the nearby Sabatini volcanic complex, one of the largest centers of the alkali potassium Roman volcanic province. The water chemistry reported by [1] measured at the PP well also shows a high concentration of Na and, especially, K (146 and 203 mg/L, respectively). Finally, iron and manganese, which in principle can substitute for Ca in the calcite lattice, as Fe (II) and Mn (II), do not seem to have this provenance, as they are uncorrelated with the other elements. Their presence in the aqueous solutions that precipitate the PP travertine could be relatively abundant as the field parameters measured in local groundwater in different periods show slightly reducing features and a pH around 6; after the emergence, the presumable progressive transition to oxidative features of these waters may have led to the oxidation of iron and its precipitation as insoluble oxides and hydroxides (when it is not already precipitated as insoluble $FeCO_3$); whereas Mn (II), which is stable at this oxidation state in natural surface environments, is likely precipitated as $MnCO_3$ even if a certain possibility of incorporation into the calcite crystal lattice cannot be totally excluded. Furthermore, iron may also have concentrated in residual products after post- or syn-sedimentary processes of travertine compaction and dissolution. The same fate may have involved Mn, given the difference of several orders of magnitude in the solubility product between calcite and rhodochrosite mineral phases [46]. Since the acid attack used for the analysis can dissolve both the oxide and the carbonate phases, the results likely reflect the cationic substitutes of the calcite and the sedimentary residual fractions. The results of the iron and manganese contents may have been affected by the latter components.

6.1. PP Travertine Chrono-Stratigraphy

Using the four U/Th dates, an age model was drawn and shown in Figure 5a. The points, representing ages versus depths, have been linearly interpolated, and consequently, a constant accumulation rate is assumed; it is displayed, expressed in centimeters per thousand years, as a three-stepped line in Figure 5b. This representation has the evident drawback of introducing artificial discontinuities in the sedimentation rate but the advantage of making evident the remarkable difference between the bottom and the top of the travertine sequence. Simple calculations allow us to estimate that after the first 10 ka, the average accumulation rate varied from 35 to 12 cm/ka, whereas in the last 20 ka, it dropped by one order of magnitude, on average, to 2.4 cm/ka. It is evident that most of the travertine was deposited on terraced sediments between 55 and 45 ka; from 45 ka onwards, the precipitation was still active until at least 24 ka and potentially until 16 ka, but at a considerably slower rate; the upper 50 cm have the longest duration, exceeding 20 ka. The period over which the PP travertine formed corresponds to MIS 3 (57–29 ka) and part of MIS 2 (29–14 ka) [47]. Climatically, this period was characterized by considerable variability; millennial-scale Dansgaard–Oeschger (DO) cycles (abrupt temperature rise followed by gradual cooling) [48,49], episodically interrupted by cold phases (Heinrich events), superimposed on a cooling trend which culminated in the last glacial maximum (LGM) (26.5–19 ka) [50], when the global ice volume reached its maximum expansion and the eustatic sea level was at its lowest.

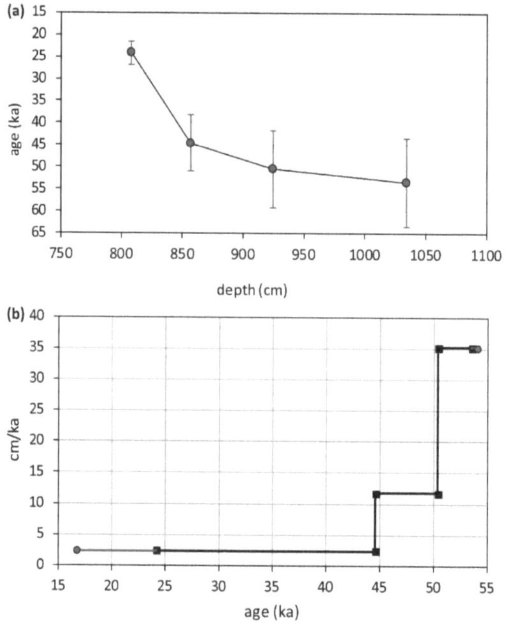

Figure 5. (**a**) Age model developed from the U/Th ages and the depths of the samples which were measured for U/Th dating technique, age model dots have been linearly interpolated; (**b**) Accumulation rate of PP travertine derived from the age model (blue prolongations of the line represent the linear extrapolation of accumulation rate outside the measured age interval).

The age model allowed the conversion of the chemical and isotopic data from the depth to the time domain (Figure 6). The elemental records versus time are clearly similar (Figure 6), except for iron (not reported in the figure). Concentrations in ppm show lower values between the 55 and 45 ka interval, after which a sudden positive shift is evident. The duration of the shift is about 4 ka. Hereafter the curves maintain a constant trend up to 16 ka, which marks the end of travertine deposition according to the linear extrapolation of the accumulation rate. The oxygen isotopes behave concordantly with the Ca-substitute elemental profiles: they show a decline in the 55–47 ka interval and a steady state after 43 ka; the variations recorded by the isotopic curve are very limited and are always less than one delta unit.

Referring to the detailed Greenland ice core data record (NGRIP) of global-scale climate evolution during the period of formation of the PP travertine sequence, it is possible to correlate the lower end of the sequence with the sharp climatic amelioration that took place at 54 ka coinciding with Dansgaard–Oeschger (DO)-14. This climatic event could mark the onset of PP deposition; the subsequent cooling trend of the DO cycle could be found in the concave profile of some PP records, such as those of Mg and isotopes. The upward recovery of the curves and the termination of this interval coincide perfectly with the abrupt temperature increase related to DO-12 at 46.5 ka. In this perspective, the oxygen isotope record can be interpreted to reflect the oscillation of oxygen isotopes in the precipitation waters that fed the local aquifer, and the variability in the calcite elemental concentrations must therefore reflect the variation of the ratio of trace elements to calcium (X/Ca ratio) in the groundwater. From the 47 ka onwards, the NGRIP record shows successive DO-cycles superimposed to a general trend leading to the LGM; the last 26 ka recorded in the travertine data have a much minor time resolution due to the dramatic fall in the accumulation rate, inferior to the minimum to detect the next climatic oscillations. The general trend does not allow for even the identification of the LGM. It shows a slight decrease in elemental concentrations of Ca-substitutes, especially evident for Sr, and a slight negative peak for $\delta^{13}C$.

The palynological signature for the last 300 ka from the Valle di Castiglione site [51], only about 25 km southeast of the PP travertine location, marked major local vegetation events that can be correlated with global climate change. The curve in Figure 6 represents one aspect of the processing of palynological data: the ratio (in percentage) between arboreal and non-arboreal pollens, which provides a record of humidity evolution. The positive peaks can be interpreted as corresponding to relatively humid times (more arboreal pollens), whereas depressions represent relatively dry times (less arboreal pollens). In the development period of the PP travertine, the pollen record shows a decrease in humidity from a peak placed before the onset of PP deposition (not shown in the figure); within the interval between 54 and 47 ka, a rise of humidity and its successive decline is apparent (Figure 6). After the period when the maximum development of PP travertine ends, the pollen recorded its driest conditions, which, aside from episodic weak recoveries, remained stable until the LGM.

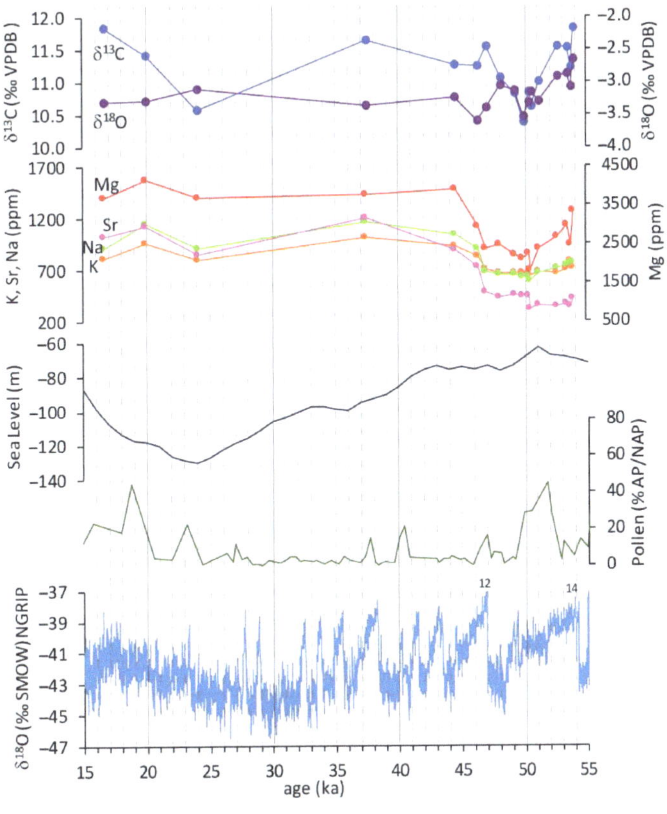

Figure 6. Comparison of different palaeoclimatic records between 15 and 55 ka. From the top: isotopic records of PP travertine; alkali and alkaline earth metal records expressed in ppm of PP travertine; sea level change from [52]; arboreal to non-arboreal pollens ratio expressed in percentage from [51]; NGRIP $\delta^{18}O$ record (numbers represent the DO events) from [48].

6.2. Climate Evolution in PP Travertine Records

The climatic evolution described by the Greenland ice core data record [48], Valle di Castiglione palynological stratigraphy [51], and sea level record from [52], displayed in the lower half of Figure 6, has influenced the chemistry of the PP travertine-precipitating groundwater. The variations in all elemental and isotopic records can be traced to variations in climatic changes on a global scale, as the correlation with the NGRIP record, especially between 54 and 43 ka, would demonstrate. The cooling event after DO-14, which started at 54 ka, was recorded in the oxygen isotope record as a decrease in $\delta^{18}O$ of precipitation waters; it was, however, a period of relatively high humidity. The $\delta^{13}C$ record has a trend concordant with the $\delta^{18}O$ record showing an inflection that is expected to reflect either a decrease in spring CO_2 outgassing related to atmospheric cooling or a decrease in the deep CO_2 component of the gas emission, the latter leading, in turn, to an increased contribution of more negative soil CO_2. After 50 ka, both isotopic records showed a modest recovery as if to restore the initial conditions. However, the superimposed trend of climatic deterioration that led to the glacial phase created cold/arid conditions and a progressive decrease in sea level, which resulted in a reduction in the availability of groundwaters. The alkali and alkaline earth metal records showed little to no change at the first cooling as a mostly humid interglacial climate persisted. It evidently caused only a decrease in Mg/Ca ratios in groundwater or a possible temperature-dependent decrease in partition coefficient. At about 48 ka, when the oxygen record stabilizes at its lowest level (around −3.3‰), all elemental concentrations show a sudden shift to higher values. Minerals containing Ca have greater resistance to weathering due to water-rock interaction, so this shift reflects a greater dissolution of metals from volcanic aquifer rocks relative to Ca. A decrease in water aggressiveness to the aquifer rocks can be inferred. The increase in X/Ca ratios in groundwater can be interpreted as a result of the increasing pH due to a reduction of the deep CO_2 input at the shallower crustal levels. This event may have followed a change in the hydraulic conductivity related to a fracture restriction driven by locally reduced pore pressure resulting from decreased groundwater recharge. The reduction of CO_2 surface emission reduced the input of carbon from deep sources, and the lowering of the regional water table greatly influenced water availability; both events dramatically reduced the precipitation rate of the PP travertine.

7. Conclusions

A recent study on the product of drilling carried out on the right bank of the Tiber River Valley, immediately northeast of Rome (*Prima Porta*), revealed the presence of a 3 m thick travertine body lying below 8 m of alluvial deposits. It was studied for isotope and chemical stratigraphy and U/Th dating. This travertine was deposited near a spring fed by groundwaters from the volcanic aquifer of the Sabatini mountains. It was associated with the rise of thermal, CO_2-dominated fluids from the deep crust through direct faulting related to the graben hosting the Tiber Valley. U/Th dating has shown that this travertine body was formed in the period 55–16 ka, and that about 46 ka ago, the accumulation rate of the travertine sequence recorded a sharp decline. Age data from *Prima Porta*, although having a high level of uncertainty, suggest that travertine deposition started approximately coeval with the wettest/high-temperature conditions of the DO-14 event, whereas the end of deposition would coincide with the cold and arid phase of the LGM. Isotope records of the travertine layers described a palaeoclimate coherent to this evolution, even if the signal cannot be interpreted from 43 ka onwards because of the reduced sampling resolution. Chemical records seem to reflect the element variations (at least for those which substitute for Ca in the travertine calcites) in the travertine-precipitating groundwaters: after an approximately constant level in the 55–47 ka interval, a remarkable shift towards higher content occurred lasting about 4 ka; this shift was coeval with the accumulation rate reduction and with the definitive establishment of arid and cold climatic conditions. After this event, the elemental records remain steady until the top of the sequence. This event has a remarkable palaeoclimate connotation as shown by its correlation with the following events: (i) the abrupt temperature increase that marks the DO-12 at 46.5 ka and its subsequent descent, (ii) the end of the warm, humid peak registered by pollen record of Valle di Castiglione, (iii) the onset of a definite sea level decrease (and a climate trend that led from MIS3 to MIS2 and to the LGM). Its clearest evidence in the chemical records and accumulation rate rather than in isotopes is because this event is related to effects that have had a strong impact on the hydrological system. The relatively rapid settling down of the glacial and arid climate triggered a series of events resulting from the lowering of the regional aquifer water table. The decrease in water availability caused a reduction in the hydrostatic pressure and fracture openings in the crust; this may have contributed to changing the hydraulic conductivity of the faults with a consequent reduction of CO_2 input into the groundwater, which became less aggressive towards the rocks of the aquifer; the reduced dissolving capacity of the groundwater changed its chemistry which was recorded in the carbonates deposited upon its emergence.

The establishment of glacial and arid climatic conditions, therefore, had simultaneous effects on the groundwater level and the fault movements, with consequences on the depositional–erosive process of the PP travertine.

This paper demonstrates the efficacy of a multimethod approach, using trace elements in combination with isotopes and dating techniques, to address the difficulties of untangling the palaeoclimatic signals present in travertine deposits. These are continental sedimentary geo-bodies often scarcely considered for palaeoclimatic investigation compared to speleothems (or calcareous tufas) because they are endogenous and, therefore, less influenced by surface environments. Nevertheless, the information they provide can be even more organic if we want to deduce the evolution of the climate together with its interaction with the earth and the environmental dynamics.

Author Contributions: Conceptualization, M.B. and F.G.; methodology, M.B. and F.G.; investigation, M.B. and F.G.; writing—original draft preparation, M.B.; writing—review and editing, M.B. and F.G. All authors have read and agreed to the published version of the manuscript.

Funding: This research was funded by the project of Italian CNR: DTA.AD002.707–Lab. Isotopi Stabili (Coordinator M. Brilli).

Data Availability Statement: All data are available in the article.

Acknowledgments: We sincerely thank Azienda Agricola Orti del Canottiere, Giuseppe Capelli, and Livio Manzetti for providing a travertine core and permitting access to the study area. We also thank the technician of the ICP-AES lab Teodoro Coppola of the Department of Earth Science of the Sapienza University of Rome, for the assistance in the analysis. We are indebted to Asmaa Fayek of Helwan University, Cairo, Egypt, for her continuous help in the laboratory work.

Conflicts of Interest: The authors declare no conflict of interest.

Appendix A

Table A1. Main operating conditions for ICP-AES analysis.

RF power	1.3 kW
Gas	Ar 99.999%
Plasma Ar flux	16.5 L min^{-1}
Auxiliary Ar flux	1.5 L min^{-1}
Nebulizer Ar flux	0.70 L min^{-1}
Sample aspiration rate	1 mL min^{-1}
reading	30 s
number of replicates	3

Table A2. Analytical characteristics of the ICP-AES method.

	λ [nm]	LOD * [mg kg^{-1}]
Na	589.592	2
K	766.490	4
Ca	396.847	0.06
Mg	279.553	0.04
Fe	259.940	0.8
Mn	257.610	0.08
Sr	407.771	0.05

* LOD Limit of detection (3s detection).

References

1. Giustini, F.; Brilli, M.; Mancini, M. Geochemical study of travertines along middle-lower Tiber valley (central Italy): Genesis, palaeo-environmental and tectonic implications. *Int. J. Earth Sci.* **2018**, *107*, 1321–1342. [CrossRef]
2. Giustini, F.; Brilli, M.; Di Salvo, C.; Mancini, M.; Voltaggio, M. Multidisciplinary characterization of the buried travertine body of Prima Porta (Central Italy). *Quat. Int.* **2020**, *568*, 65–78. [CrossRef]
3. Minissale, A.; Kerrick, D.M.; Magro, G.; Murrell, M.T.; Paladini, M.; Rihs, S.; Sturchio, N.C.; Tassi, F.; Vaselli, O. Geochemistry of Quaternary travertines in the region north of Rome (Italy): Structural, hydrologic and paleoclimatic implications. *Earth Planet. Sci. Lett.* **2002**, *203*, 709–728. [CrossRef]
4. Mancini, M.; Girotti, O.; Cavinato, G.P. Il Pliocene e il Quaternario della Media Valle del Tevere (Appennino Centrale). *Geol. Rom.* **2004**, *37*, 175–236.
5. Manfra, L.; Masi, U.; Turi, B. La composizione isotopica dei travertini del Lazio. *Geol. Rom.* **1976**, *15*, 127–174.
6. Faccenna, C.; Soligo, M.; Billi, A.; De Filippis, L.; Funiciello, R.; Rossetti, C.; Tuccinei, P. Late Pleistocene depositional cycles of the lapis Tiburtinus travertine (Tivoli, Central Italy): Possible influence of climate and fault activity. *Glob. Planet. Chang.* **2008**, *63*, 299–308. [CrossRef]
7. De Filippis, L.; Faccenna, C.; Billi, A.; Anzalone, E.; Brilli, M.; Özkul, M.; Soligo, M.; Tuccimei, P.; Villa, I.M. Growth of fissure ridge travertines from geothermal springs of Denizli basin, western Turkey. *Geol. Soc. Am. Bull.* **2012**, *124*, 1629–1645. [CrossRef]
8. Veizer, J. Chapter 8. Trace elements and isotopes in sedimentary carbonate. In *Carbonates: Mineralogy and Chemistry*; Reeder, R., Ed.; De Gruyter: Berlin, Germany; Boston, MA, USA, 1983; pp. 265–300.
9. McDermott, F. Palaeo-climate reconstruction from stable isotope variations in speleothems: A review. *Quat. Sci. Rev.* **2004**, *23*, 901–918. [CrossRef]
10. Leng, M.J.; Marshall, J.D. Palaeoclimate interpretation of stable isotope data from lake sediment archives. *Quat. Sci. Rev.* **2004**, *23*, 811–831. [CrossRef]
11. Andrews, J.E. Palaeoclimatic records from stable isotopes in riverine tufas: Synthesis and review. *Earth Sci. Rev.* **2006**, *75*, 85–104. [CrossRef]
12. Kampman, N.; Burnside, N.M.; Shipton, Z.K.; Chapman, H.J.; Nicholl, J.A.; Ellam, R.M.; Bickle, M.J. Pulses of carbon dioxide emissions from intracrustal faults following climatic warming. *Nat. Geosci.* **2012**, *5*, 352–358. [CrossRef]
13. Boccaletti, M.; Ciaranfi, N.; Cosentino, D.; Deiana, G.; Gelati, R.; Lentini, F.; Massari, F.; Moratti, G.; Pescatore, T.; Ricci Lucchi, F.; et al. Palinspastic Restoration and Paleogeographic Reconstruction of the Peri-Tyrrhenian Area during the Neogene. *Palaeogeogr. Palaeoclimatol. Palaeoecol.* **1990**, *77*, 41–50. [CrossRef]
14. Funiciello, R.; Locardi, E.; Parotto, M. Lineamenti geologici dell'area sabatina orientale. *Boll. Soc. Geol. Ital.* **1976**, *83*, 831–849.
15. Barchi, M.; Minelli, G.; Pialli, G. The CROP 03 profile: A synthesis of results on deep structures of the Northern Apennines. *Mem. Soc. Geol. Ital.* **1998**, *52*, 383–400.
16. Jolivet, L.; Faccenna, C.; Goffé, B.; Mattei, M.; Rossetti, F.; Brunet, C.; Storti, F.; Funiciello, R.; Cadet, J.P.; D'Agostino, N.; et al. Midcrustal shear zones in postorogenic extension: Example from the northern Tyrrhenian Sea. *J. Geophys. Res.* **1998**, *103*, 12123–12160. [CrossRef]
17. Chiodini, G.; Cardellini, C.; Amato, A.; Boschi, E.; Caliro, S.; Frondini, F.; Ventura, G. Carbon dioxide Earth degassing and seismogenesis in central and southern Italy. *Geophys. Res. Lett.* **2004**, *31*, L07615. [CrossRef]
18. Acocella, V.; Funiciello, R. Transverse Systems along the Extensional Tyrrhenian Margin of Central Italy and Their Influence on Volcanism. *Tectonics* **2006**, *25*, TC2003. [CrossRef]
19. Billi, A.; Tiberti, M.M.; Cavinato, G.P.; Cosentino, D.; Di Luzio, E.; Keller, J.V.A.; Kluth, C.; Orlando, L.; Parotto, M.; Praturlon, A.; et al. First results from the CROP-11 deep seismic profile, central Apennines, Italy: Evidence of mid-crustal folding. *J. Geol. Soc. Lond.* **2006**, *163*, 583–586. [CrossRef]
20. Mancini, M.; D'Anastasio, E.; Barbieri, M.; De Martini, P.M. Geomorphological, paleontological and 87Sr/86Sr isotope analyses of Early Pleistocene paleoshorelines to define the uplift of Central Apennines (Italy). *Quat. Res.* **2007**, *67*, 487–501. [CrossRef]

21. Funiciello, R.; Giuliani, R.; Marra, F.; Salvi, S. The influence of volcanism and tectonics on Plio-Quaternary regional landforms in the Southeastern Sabatinian area (Central Italy). *Mem. Descr. Carta Geol. d'It.* **1994**, *49*, 323–332.
22. Marra, F.; Rosa, C.; De Rita, D.; Funiciello, R. Stratigraphic and tectonic features of the Middle Pleistocene sedimentary and volcanic deposits in the area of Rome (Italy). *Quat. Int.* **1998**, *47*, 51–63. [CrossRef]
23. Giordano, G.; Esposito, A.; De Rita, D.; Fabbri, M.; Mazzini, I.; Trigari, A.; Rosa, C.; Funiciello, R. The sedimentation along the Roman coast between Middle and Upper Pleistocene: The interplay of eustatism, tectonics and volcanism. New data and review. *Il Quat.* **2003**, *16*, 121–129.
24. Girotti, O.; Mancini, M. Plio-Pleistocene stratigraphy and relations between marine and non-marine successions in the Middle Valley of the Tiber River. *Il Quat.* **2003**, *16*, 89–106.
25. Marra, F.; Florindo, F. The subsurface geology of Rome: Sedimentary processes, sea-level changes and astronomical forcing. *Earth Sci. Rev.* **2014**, *136*, 1–20. [CrossRef]
26. Milli, S. Depositional setting and high-frequency sequence stratigraphy of the Middle-Upper Pleistocene to Holocene deposits of the Roman Basin. *Geol. Romana* **1997**, *33*, 99–136.
27. Funiciello, R.; Giuliani, R.; Marra, F.; Salvi, S. Superfici strutturali plio-quaternarie al margine sud-orientale del Distretto Vulcanico Sabatino. *Studi Geol. Camerti. Spec. CROP* **1991**, *11*, 301–304.
28. De Rita, D.; Funiciello, R.; Parotto, M. *Geological Map of the Colli Albani Volcanic Complex, 1:50,000 Scale*; SELCA: Florence, Italy, 1988.
29. De Rita, D.; Faccenna, C.; Funiciello, R.; Rosa, C. Structural and geological evolution of the Colli Albani volcanic district. In *The Volcano of the Alban Hills*; Trigila, R., Ed.; Tipografia SGS: Rome, Italy, 1995; pp. 33–71.
30. Barberi, F.; Buonasorte, G.; Cioni, R.; Fiordelisi, A.; Foresi, L.; Iaccarino, S.; Laurenzi, M.A.; Sbrana, A.; Vernia, A.; Villa, I.M. Plio–Pleistocene geological evolution of the geothermal area of Tuscany and Latium. *Mem. Descr. Carta Geol. It.* **1994**, *49*, 77–135.
31. Karner, D.B.; Marra, F.; Renne, P.R. The history of the Monti Sabatini and Alban Hills volcanoes: Groundwork for assessing volcani-tecnonic hazards for Rome. *J. Volcanol. Geotherm. Res.* **2001**, *107*, 185–219. [CrossRef]
32. Alfonsi, L.; Funiciello, R.; Mattei, M.; Girotti, O.; Maiorani, A.; Preite Martinez, M.; Trudu, C.; Turi, B. Structural and geochemical features of the Sabina strike-slip fault (Central Apennines). *Boll. Soc. Geol. Ital.* **1991**, *110*, 217–230.
33. Faccenna, C. Structural and hydrogeological features of Pleistocene shear zones in the area of Rome (Central Italy). *Ann. Geofis.* **1994**, *37*, 121–133. [CrossRef]
34. Faccenna, C.; Funiciello, R.; Mattei, M. Late Pleistocene N-S shear zones along the Latium Tyrrhenian margin: Structural characters and volcanological implications. *Boll. Geofis. Teor. Appl.* **1994**, *36*, 507–522.
35. Milli, S.; Mancini, M.; Moscatelli, M.; Stigliano, F.; Marini, M.; Cavinato, G.P. From river to shelf, anatomy of a high-frequency depositional sequence: The late Pleistocene to Holocene Tiber depositional sequence. *Sedimentology* **2016**, *63*, 1886–1928. [CrossRef]
36. Di Salvo, C.; Di Luzio, E.; Mancini, M.; Moscatelli, M.; Capelli, G.; Cavinato, G.P.; Mazza, R. GIS-based hydrostratigraphic modeling of the city of Rome (Italy): Analysis of the geometric relationships between a buried aquifer in the Tiber Valley and the confining hydrostratigraphic complexes. *Hydrogeol. J.* **2012**, *20*, 1549–1567. [CrossRef]
37. Minissale, A. Origin, transport and discharge of CO_2 in central Italy. *Earth Sci. Rev.* **2004**, *66*, 89–141. [CrossRef]
38. Kele, S.; Breitenbach, S.F.M.; Capezzuoli, E.; Meckler, A.N.; Ziegler, M.; Millan, I.M.; Kluge, T.; Deák, J.; Hanselmann, K.; John, C.M.; et al. Temperature dependence of oxygen- and clumped isotope fractionation in carbonates: A study of travertines and tufas in the 6–95 °C temperature range. *Geochim. Cosmochim. Acta* **2015**, *168*, 172–192. [CrossRef]
39. Chafetz, H.S.; Folk, R.L. Travertines: Depositional morphology and the bacterially constructed constituents. *J. Sediment. Petrol.* **1984**, *54*, 289–316.
40. Andrews, J.E.; Brasier, A.T. Seasonal records of climatic change in annually laminated tufas: Short review and future prospects. *J. Quat. Sci.* **2005**, *20*, 411–421. [CrossRef]
41. Claes, H.; Huysmans, M.; Soete, J.; Dirix, K.; Vassilieva, E.; Marques Erthal, M.; Vandewijngaerde, W.; Hamaekers, H.; Aratman, C.; Özkul, M.; et al. Elemental geochemistry to complement stable isotope data of fossil travertine: Importance of digestion method and statistics. *Sediment. Geol.* **2019**, *386*, 118–131. [CrossRef]
42. Huang, Y.-M.; Fairchild, I.J. Partitioning of Sr and Mg into calcite under karst-analogue experimental conditions. *Geochim. Cosmochim. Acta* **2001**, *65*, 47–62. [CrossRef]
43. Ihlenfeld, C.; Norman, M.D.; Gagan, M.K.; Drysdale, R.N.; Maas, R.; Webb, J. Climatic significance of seasonal trace element and stable isotope variations in a modern freshwater tufa. *Geochim. Cosmochim. Acta* **2003**, *67*, 2341–2357. [CrossRef]
44. Gabitov, R.I.; Sadekov, A.; Leinweber, A. Crystal growth rate effect on Mg/Ca and Sr/Ca partitioning between calcite and fluid: An in situ approach. *Chem. Geol.* **2014**, *367*, 70–82. [CrossRef]
45. Day, C.D.; Henderson, G.M. Controls on trace-element partitioning in cave-analogue calcite. *Geochim. Cosmochim. Acta* **2014**, *120*, 612–627. [CrossRef]
46. Jensen, D.L.; Boddum, J.K.; Tjell, J.C.; Christensen, T.H. The solubility of rhodochrosite ($MnCO_3$) and siderite ($FeCO_3$) in anaerobic aquatic environments. *Appl. Geochem.* **2002**, *17*, 503–511. [CrossRef]
47. Lisiecki, L.E.; Raymo, M.E. A Pliocene-Pleistocene stack of 57 globally distributed benthic δ18O records. *Paleoceanography* **2005**, *20*, PA1003. [CrossRef]
48. Johnsen, S.J.; Clausen, H.B.; Dansgaard, W.; Fuhrer, K.; Gundestrup, N.; Hammer, C.U.; Iversen, P.; Jouzel, J.; Stauffer, B.; Steffensen, J.P. Irregular glacial interstadials recorded in a new Greenland ice core. *Nature* **1992**, *359*, 311–313. [CrossRef]

49. Dansgaard, W.; Johnsen, S.; Clausen, H.B.; Dahl-Jensen, D.; Gundestrup, N.S.; Hammer, C.U.; Hvidberg, C.S.; Steffensen, J.P.; Sveinbjörnsdottir, A.E.; Jouzel, J.; et al. Evidence for general instability of past climate from a 250-kyr ice-core record. *Nature* **1993**, *364*, 218–220. [CrossRef]
50. Clark, P.U.; Dyke, A.S.; Shakun, J.D.; Carlson, A.E.; Clark, J.; Wohlfarth, B.; Mitrovica, J.X.; Hostetler, S.W.; McCabe, A.M. The last glacial maximum. *Science* **2009**, *325*, 710–713. [CrossRef] [PubMed]
51. Tzedakis, P.C.; Andrieu, V.; de Beaulieu, J.-L.; Birks, H.J.B.; Crowhurst, S.; Follieri, M.; Hooghiemstra, H.; Magri, D.; Reille, M.; Sadori, L.; et al. Establishing a terrestrial chronological framework as a basis for biostratigraphical comparisons. *Quat. Sci. Rev.* **2001**, *20*, 1583–1592. [CrossRef]
52. Spratt, M.; Lisiecki, L.E. A Late Pleistocene sea level stack. *Clim. Past* **2016**, *12*, 1079–1092. [CrossRef]

minerals

MDPI

Article

Age and Depositional Temperature of Quaternary Travertine Spring Mounds from Slovakia

Daniella S. C. Vieira [1,2], Daniel Pivko [3], László Rinyu [4], László Palcsu [4], Gabriella I. Kiss [4], Hsun-Ming Hu [5,6,7], Chuan-Chou Shen [6,7] and Sándor Kele [2,8,*]

1 Institute of Geography and Earth Sciences, Faculty of Science, Eötvös Loránd University, 1117 Budapest, Hungary; daniellacvieirah@gmail.com
2 Research Centre for Astronomy and Earth Sciences, Institute for Geological and Geochemical Research, Eötvös Loránd Research Network, Budaörsi út 45, 1112 Budapest, Hungary
3 Department of Geology and Paleontology, Faculty of Natural Sciences, Comenius University, Ilkovičova 6, 842 15 Bratislava, Slovakia; daniel.pivko@uniba.sk
4 Isotope Climatology and Environmental Research Centre (ICER), Institute for Nuclear Research, Eötvös Loránd Research Network, Bem tér 18/c, 4026 Debrecen, Hungary; rinyu.laszlo@atomki.hu (L.R.); palcsu.laszlo@atomki.hu (L.P.); kiss.gabriella@atomki.hu (G.I.K.)
5 Radiogenic Isotope Facility, School of Earth and Environmental Sciences, The University of Queensland, Brisbane, QLD 4072, Australia; hsunming.hu@gmail.com
6 High-Precision Mass Spectrometry and Environment Change Laboratory (HISPEC), Department of Geosciences, National Taiwan University, Taipei 10617, Taiwan; river@ntu.edu.tw
7 Research Center for Future Earth, National Taiwan University, Taipei 10617, Taiwan
8 CSFK, MTA Centre of Excellence, Konkoly Thege Miklós út 15-17, 1121 Budapest, Hungary
* Correspondence: kele.sandor@csfk.org

Abstract: Travertine spring mounds are common in Slovakia; however, their age and depositional temperature are still poorly known. Our study is the first multimethodological investigation involving stable carbon, oxygen, and clumped isotope (Δ_{47}) analyses and U-Th age determination of travertine mounds from different locations in Slovakia (Santovka, Dudince, Čerin, Bešeňová, Liptovský Ján, Liptovské Sliače, Vyšné Ružbachy, Gánovce, and Sivá Brada) to provide information about their age, origin, precipitation conditions, and temperature. The positive $\delta^{13}C$ values imply that the parent water was charged with heavy CO_2 of deep origin. The $\delta^{18}O$ values of spring waters range between −11.4‰ and −8.9‰, whereas the $\delta^{2}H$ values vary from −80.5‰ to −58.3, indicating a meteoric origin for spring waters. Clumped isotope compositions (Δ_{47}) correspond to a deposition temperature between 4 °C and 32 °C. The U-Th age data of the studied travertines vary from 1.2 (Liptovské Sliače) to 301 ka (Dudince). Our results can serve as a basis for further detailed geochronological and geochemical studies to reconstruct the paleoclimate and paleoenvironment during travertine deposition periods in Slovakia since the mid-Pleistocene transition.

Keywords: travertine spring mounds; Slovakia; stable and clumped isotopes; U-Th dating

Citation: Vieira, D.S.C.; Pivko, D.; Rinyu, L.; Palcsu, L.; Kiss, G.I.; Hu, H.-M.; Shen, C.-C.; Kele, S. Age and Depositional Temperature of Quaternary Travertine Spring Mounds from Slovakia. *Minerals* **2023**, *13*, 794. https://doi.org/10.3390/min13060794

Academic Editors: Francesca Giustini and Mauro Brilli

Received: 5 April 2023
Revised: 1 June 2023
Accepted: 8 June 2023
Published: 10 June 2023

1. Introduction

The interest in travertine deposits as a paleoenvironmental and paleoclimatic tool has been well documented around the world. Recently, studies on travertines in Slovakia have compiled information on the forms, environments, facies, and ages of travertine sites [1,2], with a primary focus on the northern region of the country [3–5]. However, detailed information on the age, internal structure, and depositional environment is scarce in other regions of the country, where prior research has been limited to geothermal potential, hydrogeochemical, paleobotanical, and paleontological studies [6–10].

Situated in a tectonically complex area of the Western Carpathians and the northern edge of the Pannonian Basin, Slovakia has a favourable geological and tectonic structure for the occurrence of mineral and thermal waters in its territory, as well as several natural travertine that form springs [1,2,11]. Some of the Slovakian springs present unique features, such as mounds and cones (e.g., Santovka, Dudince, Bojnice, Vyšné Ružbachy), and exhibit a rather low temperature of precipitation (e.g., Mičinské travertíny −8 °C) [12].

Spring mounds are spectacular forms of travertine deposits characterized by a circular topographic rise that develops around a spring vent. From the surface of a travertine mound, the water flows downward in depressed routes, precipitating calcium carbonate, calcite, and aragonite [13,14]. Travertine morphology is controlled by various factors, such as basement relief, potentiometric surface, CO_2 pressure, water chemistry, and temperature [4,14].

The potential preservation of physical, chemical, and microbial signatures, due to rapid precipitation, makes the travertine deposits valuable archives of environmental, geothermal, and geological processes that occur during periods of spring activity, in addition to allowing the identification of the temperature of carbonate-forming fluids, as well as the timing of precipitation [15–20]. In the present paper, we provide new stable isotope and radiometric age for active and inactive travertine mounds in Slovakia. The research focuses on the stable isotope geochemical study of these mounds, including the estimation of the deposition temperature and oxygen isotope composition of the travertine precipitating paleofluids using a carbonate-clumped isotope thermometry method.

2. Geological Settings

Slovakia (Figure 1A) is part of the Western Carpathians and was formed as a result of multiple tectonic processes involving the Variscan and Alpine tectogenesis [21,22]. The internal portion of the Western Carpathians is built by a nappe stack, which is represented by thick-skinned tectonic units (Gemericum, Veporicum, Tatricum) covered by a thin-skinned nappe system predominantly composed of Mesozoic variable carbonate rocks (Silicicum, Hronicum, Tatricum, Fatricum). Mineral springs are abundant in the country and are primarily associated with the Mesozoic carbonates and evaporites, as well as the Tertiary marine sediments. In several areas, travertine have been formed along deep faults, where some, such as the Variscan and Alpine faults, also provide escape routes for CO_2 [23].

The southern part of the territory is characterized by vast lowlands, extensions of the Neogene Pannonian Basin [24]. The travertine mounds located in this area (1—Santovka, 2—Dudince; Figure 1B) are situated on the Neogene volcanites that cover the nappe stacking formations of the Internal Western Carpathians. The basement of the volcanites is the Hronicum nappe near Santovka and the Veporicum tectonic units near Dudince. The Triassic part of these units consists mainly of limestone, which is the source of the dissolved carbonate content of the travertine-depositing mineral waters [8].

In the central region of Slovakia, the Čerín spring mounds (3, Figure 1B) rest directly on the Pliocene fluvial and the Holocene deluvial deposits in the Sliač-Hron region, which is part of the Neogene volcanites. The volcanic rocks lie on the deformed Mesozoic nappe system predominantly composed of Triassic limestone and dolomite of the Hronicum and Silicicum units [22]. Travertine springs at Liptovské Sliače (4), Bešeňová (5), and Liptovský Ján (6) (Figure 1B) were formed at different times and locations along tectonic faults in the Liptov Basin [25]. The basin is an intramountainous depression in the Internal Western Carpathians and is filled with Palaeogene sediments. The Mesozoic nappes, specifically the Triassic limestone and dolomite of the Hronicum unit and the Cretaceous limestone and marlstone of the Fatricum unit, form the basement of the Paleogene deposits and serve as sources for the dissolved carbonate content of mineral and thermal waters that feed the Liptov travertine springs [26].

The spring mounds located in northeastern Slovakia (7—Gánovce, 8—Vyšné Ružbachy, and 9—Sivá Brada; Figure 1B) are directly situated on the Holocene deluvial and fluvial deposits. The springs are recharged from limestone, dolomite, and evaporites of the Mesozoic nappe system. At Sivá Brada, the recharge source is derived from the Hronicum, Tatricum, and Veporicum units, while at Vyšné Ružbachy and Gánovce, it is acquired from the Fatricum and Hronicum units [27].

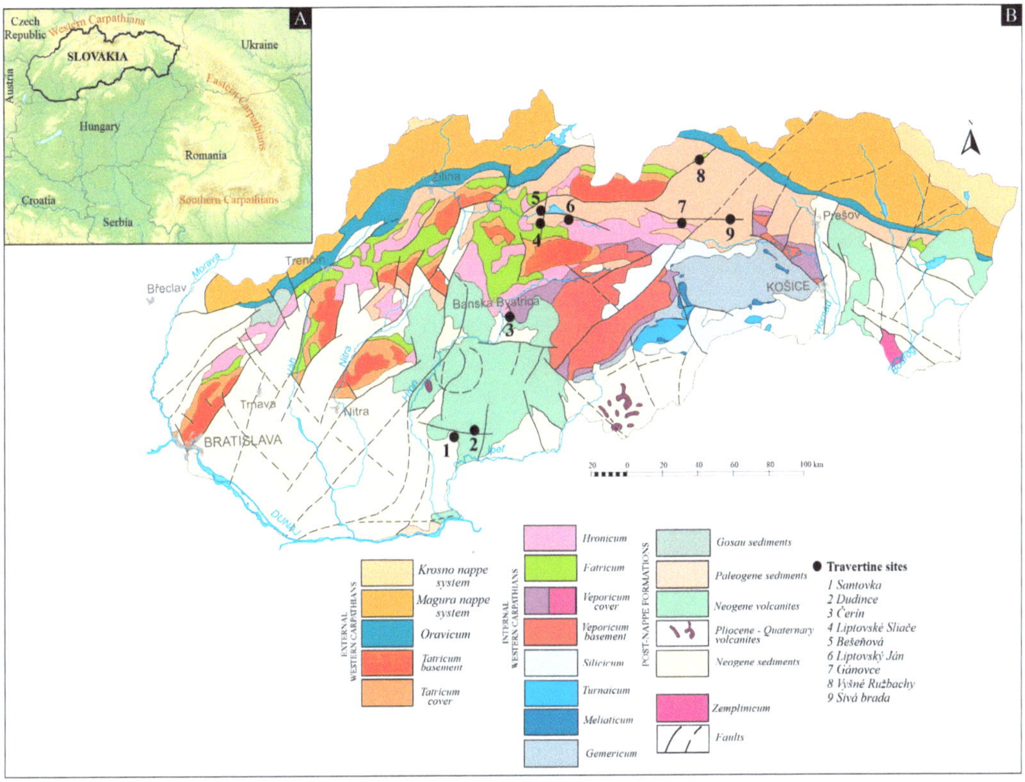

Figure 1. (**A**) Geographical setting of Slovakia and (**B**) tectonic sketch of Slovakia, modified from Hók et al. [26].

3. Materials and Methods

3.1. Field Studies and Sampling

The field research of this study was focused on active and inactive travertine spring mounds that occur at Santovka, Dudince, Čerin, Bešeňová, Liptovský Ján, Liptovské Sliače, Vyšné Ružbachy, Gánovce, and Sivá Brada (Figure 2). The sampling covered sections from the top to the base of each spring mound, starting with the vent and moving along the channels towards the distal parts downstream. In the case of active deposits, spring waters were also sampled at the vent and collected in 100 mL plastic bottles for stable isotope analyses. The basic physicochemical parameters of the water (pH, temperature) were measured in situ with a multi-parameter portable meter (WTW ProfiLine pH/Cond 3320).

Figure 2. Sampling sites in Slovakia. (**A**) Spring mound in Santovka Village; (**B**) Hostečný mound in Dudince spa; (**C**) Čerín spring mound (Mičinské travertíny); (**D**) Mound slopes with cascades in Bešeňová; (**E**) Kaďa natural crater in Liptovský Ján; (**F**) Vyšný Sliač crater; (**G**) Gánovce travertine spring; (**H**) Vyšné Ružbachy mound lake travertine; (**I**) Sivá Brada mound with recent orifices.

3.2. $\delta^{18}O$, $\delta^{13}C$, and $\delta^{2}H$ Analysis

The stable carbon and oxygen isotopic compositions ($\delta^{13}C$, $\delta^{18}O$) of the travertine samples and the oxygen and hydrogen isotopic compositions ($\delta^{18}O_w$, $\delta^{2}H$) of the water samples were performed using a Thermo Scientific Delta plus XP mass spectrometer (Thermo Scientific, Waltham, MA, USA) equipped with an automated GasBench II in the stable isotope laboratory of the Institute for Geological and Geochemical Research (IGGR), Research Centre for Astronomy and Earth Sciences (Budapest, Hungary). A total of samples were selected, powdered, and homogenized using an agate mortar and pestle, and stable carbon and oxygen isotope measurements were carried out on the bulk carbonate samples using the continuous flow technique with the H_3PO_4 digestion method [28]. The $\delta^{18}O$ values of the water samples ($\delta^{18}O_w$) were measured using the CO_2–water equilibration method [29]. The hydrogen isotope compositions were determined using the Pt-assisted H_2–H_2O equilibration [30].

Isotopic compositions of all the carbonate samples were measured in duplicate, and the water samples were measured in triplicate. The mean values are reported in the standard δ-notation in parts per thousand (‰) relative to Vienna PeeDee Belemnite (V-PDB; $\delta^{13}C$ and $\delta^{18}O$) and Vienna Standard Mean Ocean Water (V-SMOW; $\delta^{18}O_w$, $\delta^{2}H$). The reproducibility is better than ±0.1‰ for the $\delta^{13}C$ and $\delta^{18}O$ values of carbonates and ±0.1‰ and ±2‰ for $\delta^{18}O_w$ and $\delta^{2}H$, respectively. The accuracy was routinely verified by replicating the measurements of the laboratory standards calibrated to NBS19 and LSVEC.

3.3. Clumped Isotope Analysis

Clumped isotope analyses of the carbonates were carried out at the Isotope Climatology and Environmental Research Centre (ICER), Institute for Nuclear Research (ATOMKI), Debrecen. The analysis of the carbonate samples was performed on a Thermo Scientific 253 Plus 10 kV Isotope Ratio Mass Spectrometer after phosphoric acid digestion at 70 °C using a Thermo Scientific Kiel IV automatic carbonate device.

The negative background, which is caused by secondary electrons, was corrected by the application of the pressure-sensitive baseline correction [31] on all the raw beam signals. Approximately 100–120 μg aliquots of each carbonate sample measurement were replicated 11–12 times and measured alongside the carbonate standard samples. ETH1, ETH2, and ETH3 were used as normalization standards, and IAEA-C2 was used as a monitoring sample to determine the long-term reproducibility of the instrument. Simultaneously with the clumped isotope analysis, the conventional carbonate stable isotope composition was also determined for the same samples.

The data evaluation was carried out with Easotope software (release 20190125, concept by Cédric John, programmed by Devon Bowen) [32] using a CO_2-clumped ETH PBL replicate analysis method with the revised IUPAC parameters for ^{17}O correction [33–40]. The Δ_{47} results are given in the I-CDES90 scale [41], and the apparent temperatures in °C were calculated based on the Δ_{47}-temperature calibration from Anderson et al. [42] with one standard error.

3.4. U-Th Dating

The travertine samples were dated with the U-Th technique at the High–Precision Mass Spectrometry and Environmental Change Laboratory (HISPEC), Department of Geosciences, (National Taiwan University, Taipei, Taiwan) and at the Isotope Climatology and Environmental Research Centre (ICER), Institute for Nuclear Research (ATOMKI), Debrecen, Hungary. Both laboratories followed similar analytical procedures, where the selected samples were dissolved in a concentrated HNO_3 solution, and a triple spike (^{229}Th-^{233}U-^{236}U) was added. Later, 0.5 mL of $HClO_4$ was added to the sample solution to decompose the organic matter. Uranium and thorium final separation was achieved by a chemical method [43] and the protocol techniques for multi-collector inductively coupled plasma mass spectrometry [44] were used for determining the U-Th isotopic compositions and content.

Uncertainties in the U-Th isotopic data (corrections for blanks, multiplier dark noise, abundance sensitivity, contents of the four nuclides in a spike solution) were calculated at the 2σ level [43], and age corrections were calculated using an estimated atomic $^{230}Th/^{232}Th$ ratio of 4 (±2) × 10^{-6}. The half-lives of the U-Th nuclides used for age determination are reported by [45].

4. Results

4.1. U-Th Geochronological Data

The uranium and thorium isotopic compositions and the U-Th ages are summarized in Table 1 and Figure 3. The U-Th age data of this study cover a wide span between 301,072 ± 18,317 and 1156 ± 518 yr BP.

Table 1. U-Th dating results for travertine mounds in the study area.

Locality	^{238}U (ppb)	^{232}Th (ppt)	$\delta^{234}U$ [a]	$[^{230}Th/^{238}U]$ Activity [c]	$^{230}Th/^{232}Th$ (ppm)	Age (yr Ago) Uncorrected	Age (yr Ago) Corrected [c,d]	Age (yr BP) *	$\delta^{234}U_{initial}$ Corrected [b]
Santovka	71	3814 ± 8	1286 ± 5	0.1534 ± 0.0018	47 ± 1	7530 ± 91	6912 ± 323	6841 ± 323	1311 ± 5
	396	32,618 ± 3	471 ± 3	0.0856 ± 0.0005	19 ± 0	6515 ± 39	4880 ± 1157	4808 ± 1157	478 ± 4
	63	3190 ± 7	1027 ± 21	0.1400 ± 0.0020	46 ± 1	7759 ± 141	7104 ± 356	7032 ± 356	1048 ± 21
Dudince	139	52,503 ± 14	2518 ± 8	0.8431 ± 0.0017	37 ± 0	28,917 ± 95	25,925 ± 2124	25,853 ± 2124	2710 ± 18
	97	1335 ± 24	1213 ± 3	2.4304 ± 0.0403	2900 ± 71	301,239 ± 18,330	301,144 ± 18,317	301,072 ± 18,317	2838 ± 161
Lip. Sliače	370	15,187 ± 1	638 ± 4	0.0292 ± 0.0005	15 ± 0	1958 ± 35	1228 ± 518	1156 ± 518	640 ± 4
Lip. Ján	239	3686 ± 1	493 ± 3	0.1463 ± 0.0006	155 ± 1	11,188 ± 58	10,891 ± 218	10,819 ± 218	508 ± 4
Vyšné Ružbachy	789	20,995 ± 1	79 ± 2	0.9287 ± 0.0017	557 ± 1	204,146 ± 1871	203,465 ± 1919	203,393 ± 1919	140 ± 4
	806	835 ± 0	96 ± 2	0.1069 ± 0.0005	864 ± 13	11,174 ± 63	11,147 ± 66	11,075 ± 66	100 ± 3
	1320	130 ± 0	90 ± 2	0.1122 ± 0.0005	1806 ± 780	11,825 ± 65	11,827 ± 65	11,755 ± 65	93 ± 2
	978	8026 ± 2	106 ± 3	0.1006 ± 0.0007	193 ± 2	10,385 ± 80	10,169 ± 172	10,097 ± 172	109 ± 3
	1173	650 ± 0	89 ± 2	0.1060 ± 0.0005	1153 ± 26	11,149 ± 63	11,134 ± 64	11,062 ± 64	92 ± 2
Gánovce	163	1007 ± 0	448 ± 3	0.0979 ± 0.0005	240 ± 1	7612 ± 46	7489 ± 98	7417 ± 98	457 ± 3
	261	1207 ± 0	459 ± 3	0.0558 ± 0.0004	186 ± 2	4243 ± 34	4151 ± 73	4079 ± 73	465 ± 3

Analytical errors are 2σ of the mean. [a] $[^{238}U] = [^{235}U] \times 137{,}818$ (±0.65‰); $\delta^{234}U = ([^{234}U/^{238}U]_{activity} - 1) \times 1000$. [b] $\delta^{234}U_{initial}$ corrected was calculated based on ^{230}Th age (T), i.e., $\delta^{234}U_{initial} = \delta^{234}U_{measured} \times e^{l234xT}$ and T is corrected age. [c] $[^{230}Th/^{238}U]activity = 1 - e^{-l230T} + (\delta^{234}U_{measured}/1000)[\lambda_{230}/(\lambda_{230} - \lambda_{234})](1 - e^{-(\lambda230-\lambda234)T})$, where T is the age. Decay constants are 9.1705 × 10^{-6} yr^{-1} for ^{230}Th, 2.8221 × 10^{-6} yr^{-1} for ^{234}U, and 1.55125 × 10^{-10} yr^{-1} for ^{238}U. [d] Age corrections, relative to chemistry date on 4 October 2021, were calculated using an estimated atomic $^{230}Th/^{232}Th$ ratio of 4(±2) ppm. Those are the values for a material at secular equilibrium with a crustal $^{232}Th/^{238}U$ value of 3.8. The errors are arbitrarily assumed to be 50%. * BP stands for "Before Present", where the "Present" is defined as the year 1950 A.D.

The oldest travertine (301,072 ± 18,317 yr BP), the Tatar Spring (Tatársky prameň) at Dudince, is characterized by a semi-spherical spring mound with a diameter of 860 m and a height of about 8 m, and a crater on top (Figure 3B). At Dudince, a younger age of 25,853 ± 2124 yr BP was also obtained for the largest travertine terrace with a height of approximately 5 m. The terrain was formed by several mounds joined together, and this age refers to the top of the crater of the original spring, preserved on the highest flat spring mound (Figure 3C).

Figure 3. Sampled points for U-Th dating and obtained ages (±2σ) at the studied sites. Ages are reported in years before present (yr BP). (**A**) Spring mound in Santovka Village; (**B**) Tatarsky spring mound at Dudince; (**C**) Rímske Kúpele mound at Dudince; (**D**) Vyšný Sliač spring mound crater; (**E**) "White house" inactive travertine crater in Vyšné Ružbachy; (**F**) Vyšné Ružbachy crater-like orifice and small spring mound downstream; (**G**) Gánovce travertine mound; (**H**) Inactive travertine mound in Liptovský Ján.

At Vyšné Ružbachy, two generations of travertine deposits are recognized. The oldest sample at Vyšné Ružbachy (Figure 3F) is an old crater with a depth of 3.5 m, which is currently filled with water at a temperature of about 22 °C. The age determination of this sample yields an age of 203,393 ± 1919 yr BP. The youngest travertine generation is represented by a small travertine mound 3 m from the crater (Figure 3F) with an age of 11,075 ± 66 yr BP and the "White House" inactive crater (Figure 3E), with ages of 11,755 ± 65, 11,062 ± 64 and 10,097 ± 172 yr BP, obtained from the base, middle, and top, respectively. An inactive mound at Liptovský Ján (Figure 3H) shows a similar age of 10,819 ± 218 yr BP.

At Santovka, a large mound 5 m high and 40 m in diameter, and a small crater at the top (Figure 3A) were also sampled from the base to the top, yielding ages of 7032 ± 356, 6841 ± 323, and 4808 ± 1157 yr BP, respectively. Similarly, younger ages were obtained from the youngest part of the Gánovce mound (7417 ± 98 and 4079 ± 73 yr BP, Figure 3G). In Liptovské Sliače, an inactive broad orifice and two small craters of different sizes were found, but only the small one (Figure 3D) was successfully dated, providing the youngest age of 1156 ± 518 yr BP.

4.2. Isotopic Composition of Carbonates ($\delta^{18}O$, $\delta^{13}C$, and Δ_{47})

The carbon and oxygen isotope compositions of all the travertine samples are reported in Table S1 (see the supplementary materials) and displayed in Figure 4. The results show positive $\delta^{13}C$ values for all the sites, ranging from 4.9 to 12.3 (‰, V-PDB). The $\delta^{18}O$ values (V-PDB) are negative and range between −12.2 and −5.8 (‰, V-PDB).

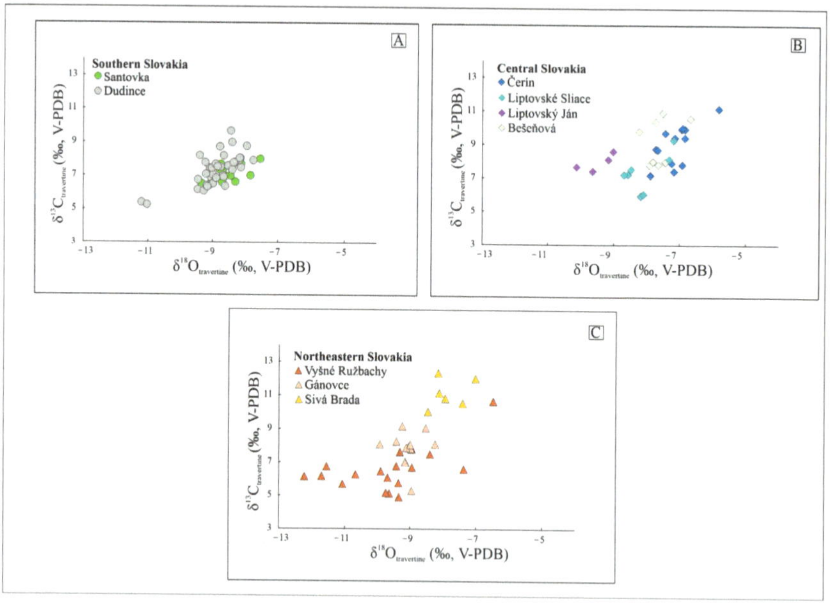

Figure 4. Stable isotope composition of the Slovakian travertine spring mounds (See Table S1) of (**A**) Southern Slovakia, (**B**) Central Slovakia and (**C**) Northeastern Slovakia.

The clumped isotope composition (Δ_{47}) of the inactive travertine mounds was measured in the samples collected from the orifices (presenting close to equilibrium conditions, as suggested by Kele et al. [46]) and from the base of each selected mound to reconstruct the palaeotemperature of the depositing fluids. The $\Delta_{47(I\text{-}CEES90)}$ values range from 0.571 to 0.668‰, whereas the calculated deposition temperatures varied between 10 and 32 °C,

4 and 18 °C, and 17 and 21 °C for southern, central, and northeast Slovakia, respectively (Table 2 and Table S3).

Table 2. Clumped isotope results and calculated $\delta^{18}O_W$.

Locality	ID	Lithofacies	$\delta^{13}C$ (V-PDB) ‰	$\delta^{18}O$ (V-PDB) ‰	$\Delta_{47\ (I\text{-}CDES90)}$ (‰)	$\Delta_{47\ (I\text{-}CDES90)}$ std Error (‰)	$T\Delta_{47}$ (°C)	±	$\delta^{18}O_W$	$\delta^{13}C_{CO2}$	T °C
1	STV5	Vent	6.4	−9.0	0.571	0.0111	32	4	−8.1	−2.9	11.2–26 *
	STV38	Distal Slope	7.2	−8.6	0.638	0.0089	10	3	−12.7	−1.9	
2	DUD1	Vent	7.7	−8.2	0.608	0.0148	20	5	−10.0	−1.3	10.7–30 *
	DUD2	Vent	7.8	−8.0	0.604	0.0089	21	3	−9.6	−1.1	
	DUD6	Proximal Slope	7.9	−8.2	0.601	0.0148	22	5	−9.5	−1.0	
	DUD15	Distal Slope	8.0	−8.1	0.614	0.0070	18	2	−10.3	−0.9	
	DUD18	Vent	8.0	−8.4	0.588	0.0120	26	4	−8.8	−0.9	
	DUD23	Vent	7.6	−8.9	0.631	0.0103	12	3	−12.5	−1.4	
	DUD29	Parasite Vent	8.3	−8.8	0.582	0.0100	28	4	−8.7	−0.6	
3	C7	Vent	7.5	−7.3	0.660	0.0104	4	3	−13.0	−1.5	7.9–13.3
4	S14	Vent	6.5	−7.8	0.655	0.0140	6	4	−13.0	−2.8	17.4
	S19	Distal Slope	7.5	−8.6	0.613	0.0136	18	4	−10.8	−1.5	
5	L24	Distal Slope	8.4	−9.1	0.650	0.0124	7	3	−13.9	−0.5	18.7–24.5
6	B32	Proximal Slope	7.2	−7.3	0.652	0.0070	6	2	−12.4	−1.8	14–15.6
	B34	Distal Slope	7.8	−8.0	0.632	0.0063	12	2	−11.6	−1.2	
7	N12	Distal Slope	6.5	−10.1	0.604	0.0099	21	3	−11.6	−2.7	20.7–22.2
	N13	Proximal Slope	6.2	−12.1	0.609	0.0115	19	4	−14.0	−3.0	
8	N27	Distal Slope	8.0	−8.9	0.615	0.0104	17	3	−11.4	−0.9	22.4–23.8
	N29	Distal Slope	9.2	−9.3	0.618	0.0108	17	3	−11.7	0.5	
9	N32	Vent	10.1	−8.5	0.608	0.0162	20	5	−10.2	1.7	5.5–13.6

1. Santovka, 2. Dudince, 3. Čerín, 4. Liptovské Sliače, 5. Liptovský Ján, 6. Bešeňová, 7. Vyšné Ružbachy, 8. Gánovce, 9. Sivá brada. Δ_{47}-based temperature ($T\Delta_{47}$) was calculated using the equation presented by Anderson et al. [42]. $\delta^{18}O_W$ was calculated using the equation of Kele et al. [46]. $\delta^{13}C_{CO2}$ was calculated using the empirical equation of Panichi and Tongiorgi [47]. T°C is temperatures measured nowadays. * Data taken from Bačová et al. [48].

4.3. Isotopic Composition of Water ($\delta^{18}O_w$, δ^2H)

The stable hydrogen and oxygen isotope composition of the water sampled at active travertine spring mounds, together with the measured physicochemical parameters are reported in Table S2. The δ^2H values of the water are quite high, ranging between −80.5‰ and −58.3‰ (V-SMOW), whereas the $\delta^{18}O_W$ values vary from −11.4‰ to −8.9‰ (V-SMOW). When plotted in the $\delta^2H/\delta^{18}O_W$ diagram (Figure 5), they fall along the trend defined by local (LMWL) [49] and global (GMWL) [50] meteoric water lines.

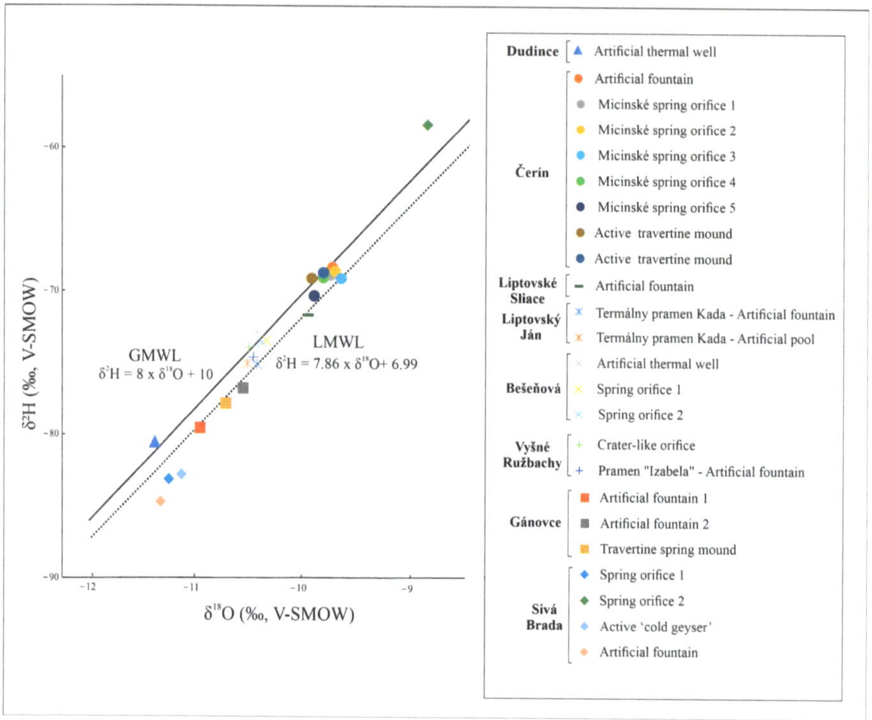

Figure 5. Isotopic compositions of water samples of travertine-forming springs in Slovakia, indicating the global (GMWL) and local meteoric line (LMWL) for comparison.

5. Discussion

5.1. U-Th Ages

Several travertine occurrences can be found in Slovakia [1], which are often formed on fault intersections and are related to extensional tectonics. Detailed data about their age are very limited and thus far, the time of their deposition has mainly been assumed based on paleobotanical, malacological, and paleontological finds and geomorphological data [26,51–60]. The U-Th data were only published by Kovanda et al. [61] for the Skalka travertine mound, and Gradziński et al. [3] for sites located at Vyšné Ružbachy, Lúčky, and Bešeňová. Pivko and Vojtko [1] summarized the published age data of Slovakian travertine.

The geochronological data of travertine can provide information about the paleoclimate of continental environments since travertine deposition is commonly linked with warm and wet climate conditions during the Quaternary period [11,62–64]. Changes in climatic conditions, including cold (glacial) and warm (interglacial) periods, can impact the development of mineral and thermal waters. To investigate the possible relationship between travertine deposition and the paleoclimate in the study area, Figure 6 presents the $\delta^{18}O$ isotope stack record [65] and indicates the ages of the studied travertine. These data show that Slovakian travertine precipitated in both glacial and interglacial periods, with a tendency towards warmer climate conditions. The samples were dated to the MIS 1, 2, 7, and 9 periods. It is important to note that the MIS 9 sample has a relatively high error margin, which suggests that it could potentially extend partially into MIS 8.

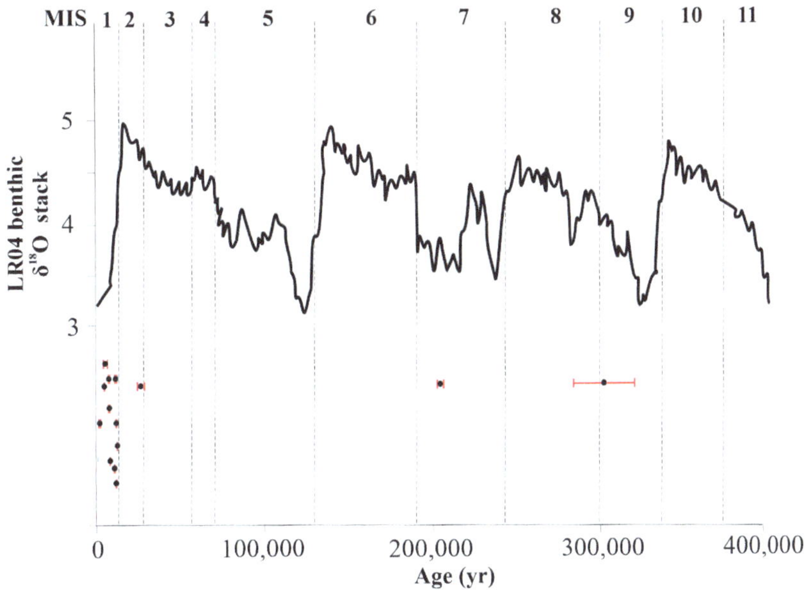

Figure 6. U-Th ages of Slovakian travertine compared with LR04 benthic isotope stack record (Lisiecki and Raymo [65]). Dated samples (black circle) display error bars in red. Global glacial periods (white bars) and interglacial periods (shaded bars) are shown by marine isotope stages (MIS) defined by Lisiecki and Raymo [65].

Although our geochronological data are in agreement with previous studies in Slovakia, suggesting deposition during warm and humid periods of the Holocene [66], the correlation with climate phases seems uncertain, as shown by travertine ages that fit MIS 2 (glacial) and MIS 8/9 (glacial/interglacial) periods.

Dudince (301,072 ± 18,317 yr BP) and Vyšné Ružbachy (203,393 ± 1919 yr BP) appear to be correlated with interglacial stages (Figure 6) in the Pleistocene. However, the Dudince site also shows a U-Th age of 25,853 ± 2124, suggesting another deposition in MIS 2 during the last glacial maximum (LGM) (26.5–19 ka) [67]. The correspondence between the travertine ages and glacial/interglacial times suggests that travertine deposition was not completely absent during glacial periods, but likely reduced [68].

Travertine deposition during dry glacial periods can indicate tectonic activity, rather than climate influence, since faults control the movement of CO_2-rich water toward the surface [69,70]. Deeply derived CO_2 is expected to be present during both glacial and interglacial periods, and fluid recharge related to global and regional climate would provide the necessary water for the observed travertine deposition in dry periods. In Slovakia, the distribution of mineral water is closely related to well-known deep faults and geological structures in the country [23], which facilitate long-distance communication between carbon dioxide sources and mineral water collectors.

The Holocene is well documented by our age data measured from the travertine deposits of Santovka, Liptovský Ján, Liptovské Sliače, Vyšné Ružbachy, and Gánovce with a time interval between 1156 ± 518 and 11,755 ± 65 yr BP, reflecting a time of a warm and wet climate [64,66]. During this period, the recharge of a groundwater aquifer could have been increased and travertine deposition occurred during glacial terminations and continued into interglacial times.

5.2. Travertine Classification and Isotope Signature

The $\delta^{13}C$ values are used to geochemically classify the travertine deposits into two main groups [71]. Thermogene travertine show $\delta^{13}C$ values between −3‰ and +10‰, while meteogene travertine are characterized by lower $\delta^{13}C$ values between −12‰ and +2‰ [14]. The $\delta^{13}C$ data of the studied Slovakian travertine (+4.9 to +12.3‰, V-PDB) show a typical thermogene isotope value similar to the ones found in previous studies for travertine in Slovakia [3,5], Italy [72–74], and Turkey [75,76].

The positive $\delta^{13}C$ and negative $\delta^{18}O$ values obtained for all the samples analyzed suggest a combination of hydrothermal fluids and meteoric water with a contribution of heavy CO_2 formed by the thermal decomposition of carbonate rocks at depth [75]. The $\delta^{13}C$ values observed in the active travertine deposits at Čerin, Bešeňová, Vyšné Ružbachy, and Sivá brada are more positive than +10‰ (Table S1). The shift in the $\delta^{13}C$ values up to +11.7‰ V-PDB are thought to be a result of more rapid CO_2-degassing, which is associated with fast flowing water on the steeper parts of the downslope [75,76]. In the case of the vent samples of active springs, the shift can be attributed to the presence of bacterial mats, since these organisms can locally remove isotopically light CO_2 [75].

The $\delta^{13}C$ data of travertine can be used to determine the source of carbon dioxide once the $\delta^{13}C$ value of the travertine is controlled by the carbon species dissolved in the groundwater. Based on that, the $\delta^{13}C$ of the CO_2 was calculated using the equation $\delta^{13}C_{(CO2)} = 1.2(\delta^{13}C_{(Trav)}) - 10.5$ proposed by Panichi and Tongiorgi [47], to calculate the original $^{13}C/^{12}C$ ratios of CO_2 from the $\delta^{13}C$ values measured in the fossil travertine. These calculated values correspond to the $\delta^{13}C$ of the CO_2 released from the water during travertine deposition [17,77]. The calculated $\delta^{13}C_{CO2}$ of our samples (Table 2) range from −3 to 1.7‰.

Generally, the $\delta^{13}C$ values of CO_2 originating from magmatic sources show values from −7 to −5‰ [78] and measurements of the $^3He/^4He$ ratios confirmed the presence of mantle-derived gas in the Carpathian Basin [79–81]. However, the $\delta^{13}C$ values calculated from the measured travertine values are more positive (−3 to 1.7‰) than the $\delta^{13}C$ value of CO_2 coming from a pure igneous source, which suggests a mixed origin.

Cornides and Kecskés [79] examined CO_2 discharges in Slovakia, obtaining $\delta^{13}C$ values in the range of −6 to −3‰, indicating the presence of CO_2 of a mantle origin. Despite this, the final stable carbon isotope composition of travertine can be influenced by the carbon dissolved from the primary carbonate rock. Since in the Carpathian Basin the carbonates in the bedrock show $\delta^{13}C$ values in a range of 0 to 3‰ [79], the positive values calculated for the Slovakian travertine (Table S3) could derive from isotopically heavy CO_2 liberated during thermometamorphic decarbonation of carbonate basement rocks, such as Triassic limestone from Silicicum, Hronicum, Tatricum, and Fatricum units.

5.3. Paleotemperatures

The temperature of the fluid was determined using clumped isotope thermometry and shows variations across different sites. The comparison with the temperature of the modern fluids measured in situ at active springs (5.5 to 26.9 °C) is shown in Table 2.

In southern Slovakia, the Santovka clumped isotope paleotemperatures were found to be 32 ± 4 °C at the vent and 10 ± 3 °C for a sample collected from a distant slope of the mound. Bačová et al. [48] obtained a temperature range of 11.2 to 26 °C measured in situ, which is similar to our calculated paleotemperatures (Table 2). At the Dudince site, three different vent samples showed paleotemperatures of 21 ± 3, 12 ± 3, and 28 ± 4 °C. The temperatures of 12 ± 3 and 28 ± 4 °C were calculated from samples collected from the same mound, but the higher temperature was obtained from a sample collected from a younger parasite vent located on the slope of the mound. The current temperature measurements at the site range between 10.7 °C and 30 °C [48], indicating no significant difference from the paleotemperatures. The colder temperature (12 ± 3 °C) could indicate a mixture with cold precipitation or the formation of a pool at the top of the vent.

In central Slovakia, the Čerín clumped isotope paleotemperature was found to be 4 ± 3 °C, which is consistent with the current temperatures within the range of error. At Vyšný Sliač, a sample collected from a broad orifice of approximately 25 m in diameter showed a paleotemperature of 6 ± 4 °C, while a sample collected from the distal part of a different mound showed a temperature of 18 ± 4 °C, similar to the temperature of 17.4 °C measured at an artificial fountain on the slope of this mound. The paleotemperature of 6 ± 4 °C was collected on the edge of the 25 m diameter orifice. Considering that water temperature tends to decrease as the distance from the spring orifice increases, this sample may be located at a considerable distance from the actual spring orifice. Here, kinetic effects, such as fractionation resulting from water evaporation and CO_2 degassing, may have influenced the isotope data. These effects could have led to an enrichment of ^{18}O in the calcite, lowering the calculated temperature.

The same case is applied to the Bešeňová and Liptovský Ján data. At Bešeňová, paleotemperatures were calculated from samples collected at the top (6 ± 2 °C) and base (12 ± 2 °C) of the so-called "Rock tower" travertine, which is approximately 9 m high, and consists of a layered deposit. Close to this site, the present spring water temperature ranges between 14 °C and 15.6 °C. At Liptovský Ján, the clumped isotope paleotemperature calculated for a sample collected at the distal slope of an inactive mound was 7 ± 3 °C, while the current water temperature measurements range from 18.7 °C to 24.5 °C. Since the exact location of the vent could not be identified, it is not possible to make a reliable comparison with the measured temperatures.

In northeastern Slovakia, paleotemperatures were determined at Vyšné Ružbachy, with results of 19 ± 4 °C and 21 ± 3 °C. These values are comparable to recent temperature measurements at the site, which range from 20.7 °C to 22.2 °C. At Gánovce, the paleotemperature found was 17 ± 3 °C, which is similar to the current temperature range of 22.4 °C to 23.8 °C, when considering the error (± 3). Sivá Brada exhibited paleotemperatures of 20 ± 5 °C for vent samples, which are significantly higher than the current temperatures of 5.5–13.6 °C. These differences in temperature could be attributed to several factors, such as changes in the hydrothermal system, mixing with cooler surface waters, or variations in the thermal input. Further detailed studies of each site would be needed to determine the exact cause of the temperature discrepancy.

5.4. Isotopic Signature of Paleofluids

The isotopic composition of water provides useful information about its origin, mixing between different sources, and water–rock interaction processes [82]. The $\delta^{18}O_W$ values were plotted against the δ^2H values for the study area (Figure 5) and have been compared to the global meteoric water line (GMWL) [50] and the local meteoric water line (LMWL), with the latter defined by the equation $\delta^2H = 7.86 \times \delta^{18}O + 6.99$ [49]. The LMWL is considered more accurate than the GMWL given the climatic/topographic conditions of the study area.

The meteoric origin of spring water is strongly suggested by the linear distribution of δ^2H and $\delta^{18}O_W$ along the GMWL and LMWL lines (as shown in Figure 5). The values of the mineral waters fall along the lines, with the lower part reflecting colder climate conditions during the infiltration processes [83]. One water sample from the Sivá Brada site ($\delta^{18}O_W$: −8.9, δ^2H: −58.3) displays the highest stable isotope composition and lies slightly above the GMWL. This could indicate a warmer climate during infiltration and mixing with young, fresh water.

We calculated the $\delta^{18}O_W$ of the travertine precipitating fluid (Table 2) from the measured carbonate $\delta^{18}O$ values and the $T_{\Delta 47}$ values using the equation of Kele et al. [46]. This empirical equation is expressed as $1000\ln\alpha_{\text{(calcite-water)}} = (20 \pm 2)\ 1000/T - (36 \pm 7)$, where α (calcite-water) = $(\alpha^{18}O_{\text{calcite}} + 1000)/(\alpha^{18}O_{\text{water}} + 1000)$ and T is the temperature of the mineralizing $CaCO_3$-rich fluids ($T_{\Delta 47}$) expressed in K.

These data indicate that the aquifer water forming the travertine deposits could have had $\delta^{18}O_W$ values between −14 and −8.1‰. The water of the modern springs has an isotope composition range of −80.5‰ to −58.3‰ (V-SMOW) and −11.4‰ to −8.9‰ (V-SMOW) for δ^2H and $\delta^{18}O_w$, respectively. This δ^{18}O variability between past and recent values could be interpreted as an influence of present-day precipitation (−10.4‰ to −8.7‰) on the water in this region [84].

6. Conclusions

This study investigated travertine mound springs of different regions of Slovakia through water and carbonates isotope geochemistry, clumped isotope, and U-Th geochronology. Integration of these new data permitted us to outline the general features of paleofluid circulation in this region and the deposition age of these travertine.

Based on their positive δ^{13}C values, the travertine samples are of thermogene origin. The deposition temperatures estimated from clumped isotope analysis (Δ_{47}) range from 4 ± 3 to 32 ± 4 °C, characteristic of cold to warm springs. The clumped isotope paleotemperature data obtained for Slovakian travertine have provided valuable insights into the past temperature trends of paleospring systems in the region. The similarities observed between paleo and current temperatures suggest that the paleospring systems have not change significatively over the time, but have remained relatively stable. In the case of Sivá Brada, the discrepancy in temperatures could be attributed to several factors (changes in the hydrothermal system, mixing with cooler surface waters, or variations in the thermal input), and more research is needed to better understand these variations.

Slovakian travertine mounds may have formed due to the deposition of meteoric-derived fluids, which were able to rise upwards along basement-penetrating faults. The travertine depositional age provided by the U-Th method varies in a wide range from 1.2 to 301 ka and demonstrates that it has been deposited mostly in warm and wet periods, but also in cold and dry periods.

Our results can contribute to the knowledge of the formation of the mineral water and travertine in Slovakia and could serve as a base for further detailed geochronological and geochemical studies to reconstruct the paleoclimate and paleoenvironment during their deposition.

Supplementary Materials: The following supporting information can be downloaded at: https://www.mdpi.com/article/10.3390/min13060794/s1, Table S1: Stable isotope composition of the Slovakia travertine spring mounds. Table S2: Temperature, pH and stable isotope composition of the water samples. δ^{18}O and δ^2H values are expressed in the δ notation (‰) against V-SMOW. * Data taken from Franko et al. [84]. Table S3: Complete clumped isotope results and calculated $\delta^{18}O_W$. Table S4: Travertine locations with GPS coordinates and brief description of each site.

Author Contributions: D.S.C.V. wrote the manuscript and prepared the figures and tables. D.S.C.V., S.K. and D.P. performed field work. S.K. supervised the research. L.R. performed clumped isotope analysis of carbonate. D.S.C.V., L.P. and G.I.K. performed U-series analysis at ATOMKI. H.-M.H. and C.-C.S. performed U-series analysis at HISPEC. All authors have read and agreed to the published version of the manuscript.

Funding: This research was funded by the European Union and the State of Hungary and co-financed by the European Regional Development Fund in the project of GINOP-2.3.2-15-2016-00009 'ICER'. This research was also supported by Tempus Public Foundation, Stipendium Hungaricum Scholarship and the Doctoral School of Earth Sciences at Eötvös Loránd University (ELTE).

Data Availability Statement: All data are available in the article.

Acknowledgments: We kindly thank the anonymous reviewers for their valuable comments, which have significantly improved the quality of the manuscript.

Conflicts of Interest: The authors declare no conflict of interest.

References

1. Pivko, D.; Vojtko, R. A review of travertines and tufas in Slovakia: Geomorphology, environments, tectonic pattern, and age distribution. *Acta Geol. Slovaca* **2021**, *13*, 49–78.
2. Pivko, D. A review of Slovak travertine and tufa facies and their environment. *Acta Geol. Slovaca* **2021**, *13*, 129–166.
3. Gradziński, M.; Dulinski, M.; Hercman, H.; Stworzewicz, E.; Holúbek, P.; Rajnoga, P.; Wroblewski, W.; Kováčová, M. Facies and age of travertines from Spiš and Liptov regions (Slovakia)—Preliminary results. *Slov. Kras (Acta Carsologica Slovaca)* **2008**, *46*, 31–40.
4. Gradziński, M.; Wroblewski, W.; Bella, P. Cenozoic freshwater carbonates of the Central Carpathians (Slovakia): Facies, environments, hydrological control and depositional history. In *Guidebook Field Trips Accompanying, Proceedings of the 31st IAS Meeting of Sedimentology, Kraków, Poland, 22–25 June 2015*; Polish Geological Society: Kraków, Poland, 2015; pp. 217–245.
5. Gradziński, M.; Wróblewski, W.; Duliński, M.; Hercman, H. Earthquake-affected development of a travertine ridge. *Sedimentology* **2014**, *61*, 238–263. [CrossRef]
6. Kriš, J.; Marton, J.; Skultétyová, I. Mineral and geothermal waters of Slovakia. *GeoJournal* **1995**, *35*, 431–442. [CrossRef]
7. Hyánková, K.; Melioris, L. The unusual chemism of mineral waters at Dudince. *Geol. Carpathica* **1993**, *44*, 123–131.
8. Melioris, L. Mineral and thermal waters of the Ipelská Pahorkatina hillyland. *Environ. Geol.* **2000**, *39*, 448–462. [CrossRef]
9. Šolcová, A.; Petr, L.; Hájková, P.; Petřík, J.; Tóth, P.; Rohovec, J.; Bátora, J.; Horsák, M. Early and middle Holocene ecosystem changes at the Western Carpathian/Pannonian border driven by climate and Neolithic impact. *Boreas* **2018**, *47*, 897–909. [CrossRef]
10. Dabkowski, J.; Frodlová, J.; Hájek, M.; Hájková, P.; Petr, L.; Fiorillo, D.; Dudová, L.; Horsák, M. A complete Holocene climate and environment record for the Western Carpathians (Slovakia) derived from a tufa deposit. *Holocene* **2019**, *29*, 493–504. [CrossRef]
11. Pentecost, A. Geochemistry of carbon dioxide in six travertine-depositing waters of Italy. *J. Hydrol.* **1995**, *167*, 263–278. [CrossRef]
12. Scheuer, G.; Schweitzer, F. A Kárpát-Medence környéki édesvizi mészköelöfordulások összehasonlitésa a hazai adottságokkal, I. Szlovákia. *Földtani Közlöny* **1981**, *111*, 453–471.
13. Chafetz, H.S.; Folk, R.L. Travertines; depositional morphology and the bacterially constructed constituents. *J. Sediment. Res.* **1984**, *54*, 289–316. [CrossRef]
14. Pentecost, A. *Travertine*; Springer Science & Business Media: Berlin/Heidelberg, Germany, 2005.
15. Fouke, B.W.; Farmer, J.D.; Des Marais, D.J.; Pratt, L.; Sturchio, N.C.; Burns, P.C.; Discipulo, M.K. Depositional Facies and Aqueous-Solid Geochemistry of Travertine-Depositing Hot Springs (Angel Terrace, Mammoth Hot Springs, Yellowstone National Park, U.S.A.). *J. Sediment. Res.* **2000**, *70*, 565–585. [CrossRef] [PubMed]
16. Andrews, J.E.; Riding, R.; Dennis, P.F. The stable isotope record of environmental and climatic signals in modern terrestrial microbial carbonates from Europe. *Palaeogeogr. Palaeoclimatol. Palaeoecol.* **1997**, *129*, 171–189. [CrossRef]
17. Minissale, A.; Kerrick, D.M.; Magro, G.; Murrell, M.T.; Paladini, M.; Rihs, S.; Sturchio, N.C.; Tassi, F.; Vaselli, O. Geochemistry of Quaternary travertines in the region north of Rome (Italy): Structural, hydrologic and paleoclimatic implications. *Earth Planet. Sci. Lett.* **2002**, *203*, 709–728. [CrossRef]
18. Liu, Z.; Zhang, M.; Li, Q.; You, S. Hydrochemical and isotope characteristics of spring water and travertine in the Baishuitai area (SW China) and their meaning for paleoenvironmental reconstruction. *Environ. Geol.* **2003**, *44*, 698–704. [CrossRef]
19. Kano, A.; Okumura, T.; Shiraishi, F.; Takashima, C. *Geomicrobiological Properties and Processes of Travertine: With a Focus on Japanese Sites*, 1st ed.; Springer: Singapore, 2019; p. 181.
20. Brogi, A.; Capezzuoli, E.; Moretti, M.; Olvera-García, E.; Matera, P.F.; Garduno-Monroy, V.-H.; Mancini, A. Earthquake-triggered soft-sediment deformation structures (seismites) in travertine deposits. *Tectonophysics* **2018**, *745*, 349–365. [CrossRef]
21. Bezák, V.; Biely, A.; Elečko, M.; Konečný, V.; Mello, J.; Polák, M.; Potfaj, M. A new synthesis of the geological structure of Slovakia—The general geological map at 1:200,000 scale. *Geol. Q.* **2011**, *55*, 1–8.
22. Hók, J.; Pelech, O.; Teťák, F.; Németh, Z.; Nagy, A. Outline of the geology of Slovakia (W. Carpathians). *Miner. Slovaca* **2019**, *51*, 31–60.
23. Franko, O.; Franko, J. Thermal waters of the Hornonitrianska kotlina depression and their utilization. *Environ. Geol.* **2000**, *39*, 501–515. [CrossRef]
24. Bodiš, D.; Remšík, A.; Černák, R.; Marcin, D.; Ženišová, Z.; Fl'aková, R. Geothermal and hydrogeological conditions, geochemical properties and uses of geothermal waters of the Slovakia. In *Geothermal Water Management, 1 ed.*; Bundschuh, J., Tomaszewska, B., Eds.; CRC Press: Boca Raton, FL, USA, 2018; pp. 41–62.
25. Chrobak-Žuffová, A.; Papčo, P. Liptov region and its geotourism wealth. In Proceedings of the GEOTOUR & IRSE 2014, Miskolc, Hungary, 16–18 October 2014; pp. 98–108.
26. Hók, J.; Kováč, M.; Rakús, M.; Kováč, P.; Nagy, A.; Kováčová-Slamková, M.; Sitár, V.; Šujan, M. Geologic and tectonic evolution of the Turiec depression in the Neogene. *Slovak Geol. Mag.* **1998**, *4*, 165–176.
27. Michalko, J.; Fendek, M. Environmental isotopes in groundwaters of Levočskê Kotlina Basin. In Proceedings of the XVII Congress of Carpathian-Balkan Geological Association, Bratislava, Slovakia, 1–4 September 2002.
28. Spötl, C.; Vennemann, T.W. Continuous-flow isotope ratio mass spectrometric analysis of carbonate minerals: Letter to the Editor. *Rapid Commun. Mass Spectrom.* **2003**, *17*, 1004–1006. [CrossRef]

29. Epstein, S.; Mayeda, T. Variation of O^{18} content of waters from natural sources. *Geochim. Cosmochim. Acta* **1953**, *4*, 213–224. [CrossRef]
30. Coplen, T.B.; Wildman, J.D.; Chen, J. Improvements in the gaseous hydrogen-water equilibration technique for hydrogen isotope-ratio analysis. *Anal. Chem.* **1991**, *63*, 910–912. [CrossRef]
31. Bernasconi, S.M.; Hu, B.; Wacker, U.; Fiebig, J.; Breitenbach, S.F.M.; Rutz, T. Background effects on Faraday collectors in gas-source mass spectrometry and implications for clumped isotope measurements: Background effects on Faraday collectors in mass spectrometry. *Rapid Commun. Mass Spectrom.* **2013**, *27*, 603–612. [CrossRef] [PubMed]
32. John, C.M.; Bowen, D. Community software for challenging isotope analysis: First applications of 'Easotope' to clumped isotopes: Community software for challenging isotope analysis. *Rapid Commun. Mass Spectrom.* **2016**, *30*, 2285–2300. [CrossRef] [PubMed]
33. Baertschi, P. Absolute^{18}O content of standard mean ocean water. *Earth Planet. Sci. Lett.* **1976**, *31*, 341–344. [CrossRef]
34. *IAEA-TECDOC—825;* Standards and Intercomparison Materials Distributed by the International Atomic Energy Agency for Stable Isotope Measurements. International Atomic Energy Agency: Vienna, Austria, 1995; pp. 13–29.
35. Meijer, H.A.J.; Li, W.J. The Use of Electrolysis for Accurate δ^{17}O and δ^{18}O Isotope Measurements in Water. *Isot. Environ. Health Stud.* **1998**, *34*, 349–369. [CrossRef]
36. Assonov, S.S.; Brenninkmeijer, C.A.M. A redetermination of absolute values for17RVPDB-CO2 and17RVSMOW. *Rapid Commun. Mass Spectrom.* **2003**, *17*, 1017–1029. [CrossRef] [PubMed]
37. Brand, W.A.; Assonov, S.S.; Coplen, T.B. Correction for the ^{17}O interference in $\delta(^{13}C)$ measurements when analyzing CO_2 with stable isotope mass spectrometry (IUPAC Technical Report). *Pure Appl. Chem.* **2010**, *82*, 1719–1733. [CrossRef]
38. Bernasconi, S.M.; Müller, I.A.; Bergmann, K.D.; Breitenbach, S.F.M.; Fernandez, A.; Hodell, D.A.; Jaggi, M.; Meckler, A.N.; Millan, I.; Ziegler, M. Reducing Uncertainties in Carbonate Clumped Isotope Analysis Through Consistent Carbonate-Based Standardization. *Geochem. Geophys. Geosystems* **2018**, *19*, 2895–2914. [CrossRef] [PubMed]
39. Daëron, M.; Blamart, D.; Peral, M.; Affek, H.P. Absolute isotopic abundance ratios and the accuracy of Δ47 measurements. *Chem. Geol.* **2016**, *442*, 83–96. [CrossRef]
40. Schauer, A.J.; Kelson, J.; Saenger, C.; Huntington, K.W. Choice of ^{17}O correction affects clumped isotope (Δ_{47}) values of CO_2 measured with mass spectrometry: ^{17}O correction affects CO_2 clumped isotopes. *Rapid Commun. Mass Spectrom.* **2016**, *30*, 2607–2616. [CrossRef] [PubMed]
41. Bernasconi, S.; Daëron, M.; Bergmann, K.D.; Bonifacie, M.; Meckler, A.N. InterCarb: A community effort to improve inter-laboratory standardization of the carbonate clumped isotope thermometer using carbonate standards. *Geochemistry* **2020**, *22*, e2020GC009588.
42. Anderson, N.T.; Kelson, J.R.; Kele, S.; Daëron, M.; Bonifacie, M.; Horita, J.; Mackey, T.J.; John, C.M.; Kluge, T.; Petschnig, P.; et al. A Unified Clumped Isotope Thermometer Calibration (0.5–1100 °C) Using Carbonate-Based Standardization. *Geophys. Res. Lett.* **2021**, *48*, e2020GL092069. [CrossRef]
43. Shen, C.-C.; Lawrence Edwards, R.; Cheng, H.; Dorale, J.A.; Thomas, R.B.; Bradley Moran, S.; Weinstein, S.E.; Edmonds, H.N. Uranium and thorium isotopic and concentration measurements by magnetic sector inductively coupled plasma mass spectrometry. *Chem. Geol.* **2002**, *185*, 165–178. [CrossRef]
44. Shen, C.-C.; Wu, C.-C.; Cheng, H.; Lawrence Edwards, R.; Hsieh, Y.-T.; Gallet, S.; Chang, C.-C.; Li, T.-Y.; Lam, D.D.; Kano, A.; et al. High-precision and high-resolution carbonate 230Th dating by MC-ICP-MS with SEM protocols. *Geochim. Cosmochim. Acta* **2012**, *99*, 71–86. [CrossRef]
45. Cheng, H.; Lawrence Edwards, R.; Shen, C.-C.; Polyak, V.J.; Asmerom, Y.; Woodhead, J.; Hellstrom, J.; Wang, Y.; Kong, X.; Spötl, C.; et al. Improvements in 230Th dating, 230Th and 234U half-life values, and U–Th isotopic measurements by multi-collector inductively coupled plasma mass spectrometry. *Earth Planet. Sci. Lett.* **2013**, *371–372*, 82–91. [CrossRef]
46. Kele, S.; Breitenbach, S.F.M.; Capezzuoli, E.; Meckler, A.N.; Ziegler, M.; Millan, I.M.; Kluge, T.; Deák, J.; Hanselmann, K.; John, C.M.; et al. Temperature dependence of oxygen- and clumped isotope fractionation in carbonates: A study of travertines and tufas in the 6–95 °C temperature range. *Geochim. Cosmochim. Acta* **2015**, *168*, 172–192. [CrossRef]
47. Panichi, C.; Tongiorgi, E. Carbon isotopic composition of CO_2 from springs, fumaroles, mofettes and travertines of central and southern Italy: A preliminary prospection method of geothermal areas. In Proceedings of the 2nd United Nations Symposium on the Development and Use of Geothermal Resources, San Francisco, CA, USA, 20 May 1975.
48. Bačová, N.; Nemeth, Z.; Repčiak, M. Mineral Waters of the Dudince Spa. *Slovak Geol. Mag.* **2016**, *16*, 125–147.
49. Holko, L.; Dóša, M.; Michalko, J.; Šanda, M. Isotopes of oxygen-18 and deuterium in precipitation in Slovakia/Izotopy kyslíka-18 A deutéria v zrážkach na Slovensku. *J. Hydrol. Hydromech.* **2012**, *60*, 265–276. [CrossRef]
50. Craig, H. Standard for reporting concentrations of deuterium and oxygen-18 in natural waters. *Science* **1961**, *133*, 1833–1834. [CrossRef]
51. Petrbok, J. Mekkýši travertinù slovenského Krasu, Gánovcù s okolím, Spiše a Ružbachù. *Rozpr. II. Tøídy Èeské Akad.* **1937**, *46*, 1–16.
52. Ložek, V.; Prošek, F. Krasové zjevy v travertinech a jejich statigrafický význam. *Československý Kras* **1957**, *10*, 145–158.
53. Del Tredici, P. Ginkgos and multituberculates: Evolutionary interactions in the Tertiary. *Biosystems* **1989**, *22*, 327–339. [CrossRef] [PubMed]

54. Ložek, V. Nové interglaciální malakofauny ze Slovenska. (Neue interglaziale Molluskenfaunen in der Slowakei). *Anthropozoikum* **1958**, *7*, 37–45.
55. Holec, P. Teeth casts of Mastodont species Mammut borsoni (Hays, 1834) from the Dreveník travertine near Spišské Podhradie. *Miner. Slovaca* **1992**, *24*, 467–469.
56. Fordinál, K.; Nagy, A. Hlavina member—Marginal Upper Pannonian sediments of the Rišňovce depression. *Miner. Slovaca* **1997**, *29*, 401–406.
57. Kaminská, L. Príspevok k poznaniu micoquienu na Slovensku. In *Ecce Homo, in Memoriam Jan Fridrich*; Krigl Jan: Praha, Czech Republic, 2010; pp. 90–94. ISBN 9788086912455.
58. Tóth, C.; Krempaská, Z. Pliocene Proboscidea remains from travertine Dreveník site (near Spišské Podhradie, Slovakia). In Proceedings of the 6th Meeting of The European Association of Vertebrate Palaeontologists, Spišská Nová Ves, Slovakia, 30 June–5 July 2008; p. 116.
59. Nemergut, A. Sídlisková Geografia Považia a Ponitria v Staršej Dobe Kamennej. Ph.D. Thesis, Masarykova Univerzita v Brně, Brno, Czech Republic, 2011.
60. Wróblewski, W.; Gradziński, M.; Hercman, H. Suggestions on the allochthonous origin of terra rossa from Drevenik Hill (Spiš, Slovakia). *Slov. Kras* **2010**, *48*, 153–161.
61. Kovanda, J.; Smolíkovà, L.; Ford, D.; Kaminská, L.; Ložek, V.; Horácek, I. The Skalka travertine mound at Hôrce-Ondrej near Poprad (Slovakia). *Sborník Geol. Ved–Antropozoikum* **1995**, *22*, 113–140.
62. Frank, N.; Braum, M.; Hambach, U.; Mangini, A.; Wagner, G. Warm Period Growth of Travertine during the Last Interglaciation in Southern Germany. *Quat. Res.* **2000**, *54*, 38–48. [CrossRef]
63. Rihs, S.; Condomines, M.; Poidevin, J.-L. Long-term behaviour of continental hydrothermal systems. *Geochim. Cosmochim. Acta* **2000**, *64*, 3189–3199. [CrossRef]
64. Soligo, M.; Tuccimei, P.; Barberi, R.; Delitala, M.C.; Miccadei, E.; Taddeucci, A. U/Th dating of freshwater travertine from Middle Velino Valley (Central Italy): Paleoclimatic and geological implications. *Palaeogeogr. Palaeoclimatol. Palaeoecol.* **2002**, *184*, 147–161. [CrossRef]
65. Lisiecki, L.E.; Raymo, M.E. A Pliocene-Pleistocene stack of 57 globally distributed benthic $\delta^{18}O$ records. *Paleoceanography* **2005**, *20*, 9–11. [CrossRef]
66. Gradziński, M.; Hercman, H.; Jaśkiewicz, M.; Szczurek, S. Holocene tufa in the Slovak Karst: Facies, sedimentary environments and depositional history. *Geol. Q.* **2013**, *57*, 769–788. [CrossRef]
67. Clark, P.U.; Dyke, A.S.; Shakun, J.D.; Carlson, A.E.; Clark, J.; Wohlfarth, B.; Mitrovica, J.X.; Hostetler, S.W.; McCabe, A.M. The last glacial maximum. *Science* **2009**, *325*, 710–714. [CrossRef]
68. Bertini, A.; Minissale, A.; Ricci, M. Use of Quaternary travertines of central-southern Italy as archives of paleoclimate, paleohydrology and neotectonics. *Alp. Mediterr. Quat.* **2008**, *21*, 99–112.
69. Ricketts, J.W.; Ma, L.; Wagler, A.E.; Garcia, V.H. Global travertine deposition modulated by oscillations in climate. *J. Quat. Sci.* **2019**, *34*, 558–568. [CrossRef]
70. Uysal, I.T.; Feng, Y.-X.; Zhao, J.-X.; Isik, V.; Nuriel, P.; Golding, S.D. Hydrothermal CO_2 degassing in seismically active zones during the late Quaternary. *Chem. Geol.* **2009**, *265*, 442–454. [CrossRef]
71. Pentecost, A.; Viles, H. A review and reassessment of travertine classification. *Géogr. Phys. Quat.* **1994**, *48*, 305–314. [CrossRef]
72. Gandin, A.; Capezzuoli, E. Travertine versus calcareous tufa: Distinctive petrologic features and stable isotopes signatures. *Alp. Mediterr. Quat.* **2008**, *21*, 125–136.
73. Berardi, G.; Vignaroli, G.; Billi, A.; Rossetti, F.; Soligo, M.; Kele, S.; Baykara, M.O.; Bernasconi, S.M.; Castorina, F.; Tecce, F.; et al. Growth of a Pleistocene giant carbonate vein and nearby thermogene travertine deposits at Semproniano, southern Tuscany, Italy: Estimate of CO_2 leakage. *Tectonophysics* **2016**, *690*, 219–239. [CrossRef]
74. Della Porta, G.; Hoppert, M.; Hallmann, C.; Schneider, D.; Reitner, J. The influence of microbial mats on travertine precipitation in active hydrothermal systems (Central Italy). *Depos. Rec.* **2022**, *8*, 165–209. [CrossRef]
75. Kele, S.; Özkul, M.; Fórizs, I.; Gökgöz, A.; Baykara, M.O.; Alçiçek, M.C.; Németh, T. Stable isotope geochemical study of Pamukkale travertines: New evidences of low-temperature non-equilibrium calcite-water fractionation. *Sediment. Geol.* **2011**, *238*, 191–212. [CrossRef]
76. Özkul, M.; Kele, S.; Gökgöz, A.; Shen, C.-C.; Jones, B.; Baykara, M.O.; Fórizs, I.; Németh, T.; Chang, Y.-W.; Alçiçek, M.C. Comparison of the Quaternary travertine sites in the Denizli extensional basin based on their depositional and geochemical data. *Sediment. Geol.* **2013**, *294*, 179–204. [CrossRef]
77. Kele, S.; Vaselli, O.; Szabó, C.; Minissale, A. Stable isotope geochemistry of Pleistocene travertine from Budakalász (Buda Mts, Hungary). *Acta Geol. Hung.* **2003**, *46*, 161–175. [CrossRef]
78. Hoefs, J. *Stable Isotope Geochemistry*; Springer: Berlin/Heidelberg, Germany, 1997.
79. Cornides, I.; Kecskés, Á. Deep-seated carbon dioxide in Slovakia: The problem of its origin. *Geol. Carpathica* **1982**, *33*, 183–190.
80. Cornides, I.; Kecskes, A. Deep-seated carbon dioxide in Slovakia: Additional comments on the problem of its origin. *Geol. Zb.* **1987**, *38*, 429–435.
81. Cornides, I.; Takaoka, N.; Nagao, K.; Matsuo, S. Contribution of mantle-derived gases to subsurface gases in a tectonically quiescent area, the Carpathian Basin, Hungary revealed by noble gas measurements. *Geochem. J.* **1986**, *20*, 119–125. [CrossRef]

82. Clark, I.D.; Fritz, P. *Environmental Isotopes in Hydrogeology*; CRC Press/Lewis Publishers: Boca Raton, FL, USA, 1997.
83. Trček, B.; Albrecht, L. Overview of isotopic investigations of groundwaters in a fractured aquifer system near Rogaška Slatina, Slovenia. *Geologija* **2017**, *60*, 49–60. [CrossRef]
84. Franko, O.; Šivo, A.; Richtáriková, M.; Povinec, P.P. Radiocarbon Ages of Mineral and Thermal Waters of Slovakia. *Acta Phys. Univ. Comen.* **2008**, *49*, 111–124.

MDPI AG
Grosspeteranlage 5
4052 Basel
Switzerland
Tel.: +41 61 683 77 34
www.mdpi.com

Minerals Editorial Office
E-mail: minerals@mdpi.com
www.mdpi.com/journal/minerals

www.ingramcontent.com/pod-product-compliance
Lightning Source LLC
Chambersburg PA
CBHW041414210926
51850CB00013B/3269